THE ART AND SCIENCE
OF HDR IMAGING

Wiley-IS&T Series in Imaging Science and Technology

Series Editor:
Michael A. Kriss

Consultant Editor:
Lindsay W. MacDonald

Reproduction of Colour (6th Edition)
R. W. G. Hunt

Colour Appearance Models (2nd Edition)
Mark D. Fairchild

Colorimetry: Fundamentals and Applications
Noboru Ohta and Alan R. Robertson

Color Constancy
Marc Ebner

Color Gamut Mapping
Ján Morovič

Panoramic Imaging: Sensor-Line Cameras and Laser Range-Finders
Fay Huang, Reinhard Klette and Karsten Scheibe

Digital Color Management (2nd Edition)
Edward J. Giorgianni and Thomas E. Madden

The JPEG 2000 Suite
Peter Schelkens, Athanassios Skodras and Touradj Ebrahimi (Eds.)

Color Management: Understanding and Using ICC Profiles
Phil Green (Ed.)

Fourier Methods in Imaging
Roger L. Easton, Jr.

Measuring Colour (4th Edition)
R.W.G. Hunt and M.R. Pointer

The Art and Science of HDR Imaging
John McCann and Alessandro Rizzi

Published in Association with the Society for Imaging Science
and Technology

THE ART AND SCIENCE OF HDR IMAGING

John J. McCann
McCann Imaging, USA

Alessandro Rizzi
Università degli Studi di Milano, Italy

A John Wiley & Sons, Ltd., Publication

Cover photographs: John McCann made these images using R.Sobol's HP945 Retinex image processing: l/r Tasmanian Meadow, Hilton Head, Stockholm, Seville, and Canyon de Chelle, NM.

Library of Congress Cataloging-in-Publication Data
McCann, John J.
   The art and science of HDR imaging / John J. McCann, Alessandro Rizzi.
      p. cm.
   Includes bibliographical references and index.
   ISBN 978-0-470-66622-7 (cloth)
   1. High dynamic range imaging. I. Rizzi, Alessandro, 1965- II. Title.
   TR594.M33 2012
   771'.44-dc23
                                                    2011020577

A catalogue record for this book is available from the British Library.

Print ISBN: 9780470666227
ePDF ISBN: 9781119951476
oBook ISBN: 9781119951483
ePub ISBN: 9781119952121
Mobi ISBN: 9781119952138

Set in 9 on 11 pt TimesNewRoman by Toppan Best-set Premedia Limited
Printed and bound in Singapore by Fabulous Printers Pte Ltd

To Mary,
scientist, colleague, wife, mother,
great cook, and fellow adventurer.

John at Yosemite, 1981

Contents

About the Authors

John McCann received a degree in Biology from Harvard College in 1964. He worked in, and managed, the Vision Research Laboratory at Polaroid from 1961 to 1996. He has studied human color vision, digital image processing, large format instant photography, and the reproduction of fine art. His publications and patents have studied Retinex theory, color constancy, color from rod/cone interactions at low light levels, appearance with scattered light, and HDR imaging. He is a Fellow of the Society of Imaging Science &Technology (IS&T) and the Optical Society of America (OSA). He is a past President of IS&T and the Artists Foundation, Boston. He is currently consulting and continuing his research on color vision. He is the IS&T/OSA 2002 Edwin H. Land Medalist, and IS&T 2005 Honorary Member.

Alessandro Rizzi obtained his PhD in Information Engineering, Università di Brescia in 1999, and is an Associate Professor in the Department of Information Science and Communication at the University of Milano. He teaches fundamentals of digital imaging, multimedia video and human-computer interaction. Since 1990, he has studied digital imaging and human vision. His research focuses on issues regarding vision when combined with digital imaging. He has worked on computational models of color appearance for standard and high-dynamic-range images when applied to image enhancement, movie and picture restoration, medical imaging, and computer vision. He is one of the founders of the Italian Color Group, and member of several program committees of conferences related to color and digital imaging. He serves as Co-Chair of the Color Conference in IST&T/SPIE Electronic Imaging in which he introduced "The Dark Side of Color".

Preface

Vision and visual reproduction are a giant puzzle. By comparison, it is much more complicated than a jigsaw puzzle that lies on a flat table. We know a lot about the flat puzzle piece before we start. We know the surface that lays on the table, and just need to figure out its location and orientation.

Vision is more like a wooden three-dimensional puzzle in which the craftsman passes the jigsaw through a volume of wood at many angles and adds many curves. All the pieces of such solid block puzzles have complex irregular shapes. To solve these puzzles we need to look at each piece in all possible ways. We need to inspect each surface to recognize its shape, and then search for its matching surface. We rotate each piece in our hands to see it from all possible directions.

Vision has far more than three dimensions. However, the principle is the same. We need to look at the properties of vision from all possible directions to see if we can find pieces that fit together.

All HDR imaging is the result of having variable illuminations falling on objects. This variable illumination is the result of the 3-D world that has multiple illuminants, shadows, specular and diffuse reflections from other objects. The scene components in a 3-D world play a big part in controlling the light coming from them. This is quite different from flat 2-D surfaces in uniform illumination. Just as with jigsaw puzzles, 3-D scenes are more interesting and complex than spots of light, and 2-D flat scenes.

This text integrates three topics that are thought of as independent disciplines: artists' renditions of scenes and their illumination, color and lightness constancy, and digital HDR imaging. Each is a rich field of study; each is a piece of the puzzle. The shapes of these topics fit together. Their contours match because measurements of appearances in each give the same results. This book is about those measurements of appearances, and models of the results.

This text differs from others in the Wiley-IS&T Series in Imaging Science & Technology in that it concentrates on measuring color appearances in complex scenes. It integrates and extends the work of two photographers who were close friends: Edwin Land and Ansel Adams. One began as an optical physicist and inventor, and then built a great company around instant photography. The other began as a pianist, and turned to photography to express his incredible artistic talent. Both Adams and Land were IS&T Fellows; both were awarded the United States Presidential Medal of Freedom.

Adams taught his Zone System to many thousands of photographers. His books have many scientific plots of film-response functions measured with spots of light. The Zone System taught photographers HDR image capture. That is the record of the scene; it is the analog of a musical score. Making the print, the analog of performing the score, departed from the physics of spots of light and turned to implement the skills of painters in controlling the spatial information content. By controlling the light from the negative in each spatial region of the print, Adams taught how to perform HDR compression. He not only fit the very large dynamic range of scenes in the U.S. southwest into a lower dynamic range print, he synthesized the print's appearance using spatial image processing. Adams did not accurately reproduce the HDR world; he captured it, transformed it, and rendered his artistic visualization of it in the print.

At the same time that Adams started to write about the Zone System, his friend Edwin Land started to create instant photography. Ten years later, after receiving hundreds of photographic patents, Land became fascinated by what he saw by accident. The addition of light in complex images generated different color appearances from those observed in spots of light. Spatial content, not retinal receptor response controls color. He was quick to realize the fundamental difference between pixel-based chemical properties of film, and the spatial-based neural properties of vision.

This text builds on the teaching of Adams and Land. Adams made his visualized print by spatial manipulation, Land set out to understand human vision, to make cameras that mimicked vision.

The appearance of complex scenes varies with scene content. That means that HDR scene reproductions must incorporate scene dependent algorithms. This text describes the work on spatial image processing for scene rendition in the spirit of Adams and Land.

This text expands the approach of measuring complex scenes to find the actual information captured from HDR scenes. It turns out that dynamic range depends on the particular content of each scene. All pixels, whose radiance is below 10 % of the scene maximum, have variable camera digital values that are dependent on the radiances of all the other pixels in the scene. Camera veiling glare, and not scene radiance, limits the captured dynamic range.

The lesson from painters is that accurate reproduction is unnecessary. Photographic studies showed it undesirable, and psychophysics showed it irrelevant.

By concentrating on the spatial properties of artistic rendering, digital HDR, vision and computer models of vision, we can see how these parts of the puzzle fit together so beautifully.

JMC, AR

Series Preface

The Art and Science of HDR Imaging by John McCann and Alessandro Rizzi is the twelfth offering in the Wiley-IS&T Series on Imaging Science and Technology. High Dynamic Range Imaging refers to the ability to accurately capture the complete range of scene content, from the deepest shadows to the brightest highlights, and then reproduce them in a display or hardcopy in such a way that the observer recalls seeing the same image as when he or she took the picture. As the authors have pointed out in their preface, this is not a new challenge. Artists have made their names by creating pleasing and realistic reproductions by using color, texture, contrast and their acquired knowledge of simultaneous contrast and visual edge effects. They were far more limited than today's graphic artists who can use advanced scanners, high dynamic range film or digital cameras and sophisticated software to transform the captured image into a "life-like" electronic display or reflection print (either by halftone or continuous print technology). The HDR challenge lies in several stages of the image chain, including the imaging optics, the dynamic range of the photosensitive imaging material and the corresponding dynamic range of the final display or print along with the viewing conditions.

The imaging optics need to be as "glare free" as possible since any local or long-range glare will fill in the shadows, hiding subtle detail. At the same time the optics must have an effectively small optical point spread function so that high intensity specular reflections, like catch lights in an eye, are not smeared out, thus limiting the "snap" of image highlights and overall print. The imaging media must be able to cover a wide exposure range. Photographic researchers demonstrate that nature provides a luminance ratio that can exceed 1000:1. This means that a film (negative or reversal) or an electronic sensor (CCD or CMOS device) must have a range of ten stops ($2^{10} = 1024$). Some film systems can obtain this dynamic range by using a combination of "slow", "medium" and "fast" layers that use varying silver halide grain sizes and geometries, and sensitizing dyes and doping agents like gold and other elements. Electronic image sensors can obtain high dynamic range by having large photoelectron storage areas, which are closely associated with the size of the active area of the individual pixel elements. Using today's CCD or CMOS technology, a 9 or 10 micron square pixel is required to achieve the required 10-stops of dynamic range. By far the biggest problem in obtaining a high dynamic range image is the display or reflection print. Some of the modern LCD or DLP (using micro mirrors) displays claim to have the required 1000:1 ratio required to reproduce a realistic image, but reflection prints have dynamic ranges below 100:1, far short of that required to hold both the highlights and shadows without the aid of vision based scene luminance compression. New printing techniques have extended the dynamic range for reflection prints; a mordant is placed on a metal sheet (polished aluminum) and an inkjet dye image is transferred to the mordant under heat and pressure. These prints have luminosity not found in other practical printing technologies.

The human visual system, through the control of the iris and adaptation, can cover a range of well over 10,000:1. This means an observer can scan a scene and record in his or her memory the full dynamic range of the scene even if all of the scene is not "seen" in its entirety at one time. Consider talking to friends inside your home and then glancing out the window to see a deer in your backyard. Your eye adjusts to the higher illumination level and you "see" both scenes equally well, even if a film or digital camera cannot record both the inside and outside scene. Using fill flash can solve this problem by setting the camera exposure for the outside and letting the flash bring the interior to the same level. Even if the camera can capture the full dynamic range, one cannot "fit" it onto the reflection print. In the photographic or digital darkroom one can compress the tone scale of the image selectively by area (using segmentation and tone scale adjustment) and form a composite image that "looks" like you remember it when you scanned it with your adaptive visual system.

All the above, and many more issues associated with High Dynamic Range Imaging are covered by John McCann and Alessandro Rizzi in *The Art and Science of HDR Imaging.* The authors provide both a historical and technical review of the methods used to "mimic human vision" to render the appearance of the original scene on media that cannot reproduce the actual scene luminance. The text will provide the reader with the current image processing technology used to "automatically" compress the captured tone scale into image data that can be used by the print or display system in question and how color and the human visual system interact in complex images. Any researcher in the broad field of imaging science and technology will find this a useful and enlightening text.

On a personal note, I have known John McCann for over 40 years. During this period of time I have become a great "fan" of his work on the perception of color images and his constant curiosity with the human visual system and how it works. He continues to provide fascinating visual examples of how humans "see things" and with his wide assortment of colleagues builds models that attempt to explain what is seen in his visual experiments. He has been the perfect mentor for young researchers like Alessandro Rizzi was as they have delved into the mysteries of High Dynamic Range images and the visual models that help elucidate the visual mechanism that control how we "see" them.

Michael A. Kriss
*Formerly of the Eastman
Kodak Company and the
University of Rochester*

Acknowledgements

This work was possible because of the inspiration and support from Edwin Land, Ansel Adams, Daniele Marini, and Mary McCann.

This work was in collaboration with, Nigel Daw, Bill Roberson, Jeanne and Steve Benton, Sholly Kagan, Len Ferrari, Vaito Eloranta, Suzanne McKee, Tom Taylor, Bob Savoy, Jay Scarpetti, John Hall, Jon Frankle, Hugo Liepmann, Alan Stiehl, Jim Burkhardt, Karen Houston, Bill Wray, Ken Kiesel, Sasha Petrov, Jay Thornton, Norbert Herzer, Marzia Pezzetti, Ivar Farup, Carinna Parraman, Vassilios Vonikakis, John Meyer, Bob Sobol, Irwin Sobel, Paul Hubel, Gabriel Marcu, Brian Funt, Florian Ciurea, Carlo Gatta, Massimo Fierro, Edoardo Provenzi, Luca De Carli, Cristian Bonanomi, Davide Gadia, Maurizio Rossi, Majed Chambah, Silvia Zuffi, Marcelo Bertalmio, Vicent Caselles, Tiziano Agostini, Alessandra Galmonte, and Reiner Eschbach.

Section A

History of HDR Imaging

1

HDR Imaging

1.1 Topics

The chapter lists many disciplines that help us understand High-Dynamic-Range (HDR) imaging. It describes the work of artists, scientists, and engineers that have developed techniques to capture HDR scenes and render them in a wide variety of media. This chapter points out the difference between traditional physics, in which light is always measured at a pixel, and human vision, in which spatial image content is important. In vision it is the relationship between pixels, rather than their individual values, that controls appearance. This turns out to be a central theme for successful HDR imaging and will be discussed throughout the text.

1.2 Introduction

High Dynamic Range (HDR) imaging is a very active area of research today. It uses the advances of digital image processing as its foundation. It is used in camera design, software image processing, digital graphic arts and making movies. Everywhere we turn we see that HDR imaging is replacing conventional photography.

If we want to understand HDR imaging we need an interdisciplinary approach that incorporates the many ways of making images. Further, we need to understand exactly what we mean by conventional photography. This text starts with the history of painting and presents an integrated view of the arts, image science, and technology leading to today's HDR. It ends with a discussion of possibilities for the future.

One of the most important factors in understanding imaging is the careful definition of the goal of the image. This is called *rendering intent*. We might, at first, think that images of scenes have the same goal – copy the scene. As we will see in the text, there are a great many different rendering intents. We could have as our goal:

Reproduce exactly the light from the scene

Match the appearance of the scene

Calculate the surface reflectance of objects in the scene

The Art and Science of HDR Imaging, First Edition. John J. McCann, Alessandro Rizzi.
© 2012 John Wiley & Sons, Ltd. Published 2012 by John Wiley & Sons, Ltd.

Calculate the appearance of the scene and print the appearance

Abstract the important features of the scene

Introduce a personal style to the image of a scene

Use the scene for inspiration

Make a pretty picture

It is important to distinguish whether one's goal is to apply skill and craftsmanship to make a personal rendition of a particular scene, or to use science and technology to incorporate a general solution of HDR imaging for all ranges and types of photographic scenes.

The 19th century saw the rapid growth of a cottage industry that required the photographer to make his own plates, take the picture, process the negative, expose the print, develop it, and make it insensitive to light. In the 20th century, the Kodak slogan "You take the picture, we do the rest", changed photography into a major industry. (Chapter 5) It took a great amount of superb scientific research, engineering, and manufacturing development to expand the set of photographers from a devoted few in the 1850s to today's nearly global participation.

There is a parallel in HDR imaging today. In the past few years, we have seen considerable interest in using multiple exposures to capture wide dynamic ranges of scene information, 16 bit quantizations (RAW images), and software tools to recombine the different exposures in a new rendition. These techniques work well for most individual scenes where the photographer uses these tools to make the unique rendition of a particular image. There is a second very important objective, namely to apply art, science, and technology to finding the general solution for using HDR imaging principles for all scenes. That is the goal of this text.

Digital cameras capture scenes using millions of light sensor segments to make *picture elements* (*pixels*). We may like to think that the considerable technological advances in 21st century digital imaging make it possible now to accurately reproduce any scene. On further study, we find that imaging remains the best compromise for the rendering intent we have chosen. Despite all the remarkable accomplishments, we cannot capture and reproduce the light in the world exactly. (Section B) With still further study, we find that accurate reproduction is usually not desirable. This is why we need an interdisciplinary study of image making to find the best compromises in rendering the scene.

After all, rendering the high dynamic range of scenes has been a problem since the beginning of scene reproduction. Paintings of scenes date back 160 centuries to colored wall art in the Lascaux caves. Printing has early examples of stone characters for making clay tablets dating back 50 centuries; color wood block plates for printing uniform colors – 14 centuries; perspective in painting – seven centuries; chiaroscuro rendition of apparent illumination – five centuries; and LeBlon's hand-made color-separation printing plates – three centuries – 160 years before James Clerk Maxwell invented the first mechanical color photography.

1.3   Replicas and Reproductions

Human vision responds to a narrow band of wavelengths in the electromagnetic spectrum using rod-shaped and three types of cone-shaped cells in the retina. These cells generate three channels of color information that are identified with red, green, and blue light. Figure 1.1 is a three-dimensional cube that represents the *color space* of an image in computer memory. It shows R, G, B axes that combine to make white, red, yellow, green, cyan, blue, magenta, and black.

To understand the underlying problems in scene reproduction, we need to differentiate *replicas* and *reproductions*. A replica is a copy of a painting, by the original's artist, using the same media, or color materials. Replicas use the same media, and have the same physical properties for controlling light.

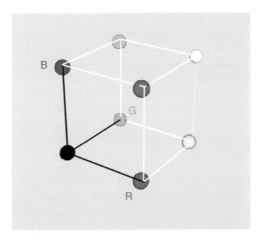

Figure 1.1   A three-dimensional color space with R, G, B axes.

Replicas share the identical color space with the original. Reproductions are copies of original art in different media, such as a computer-screen copy of a painting. The challenge in color reproduction is to capture the information contained in the entire original 3-D color space and to make a copy in a different size and shape reproduction space.

The problem is very similar to moving everything in your house to a new one. The original is one's current house. It is defined by the tools that physics uses to measure light, such as wavelength, and photon count. The *reproduction* house has different dimensions for the length (amount of red), width (amount of green) and height (amount of blue). Reproductions move everything in the old house into the new house, keeping all contents in corresponding rooms, even though the dimensions of the entire house, and each room, are different. Good reproductions are never exact physical copies of the original, because that is not possible. Good reproductions capture the appearance and relationships of objects in the scene. The original and the reproduction have different 3-D color spaces with different sizes and local shapes, because they use different colorants. If the reproduction reproduces the pixels that are in gamut accurately, then the rest of the out-of-gamut pixels make unwanted problems, such as highly visible artifacts. Good reproductions render all the information in the interior of the original's color space, often using different colorimetric colors at corresponding pixels. Reproductions cannot make colorimetric copies of all the pixels in the original because the media are different. The problem with colorimetric reproduction is that it must select the pixels that cannot be reproduced accurately, and do something with them.

1.4   A Choice of Metaphors for HDR Reproduction

Those of us interested in scene reproduction have a choice of two very different paradigms for the process. One paradigm starts with Euclidean geometry that is an essential part of physics and of 19th century psychophysics. The other paradigm starts in the Renaissance and has been amplified by 20th century neurophysiology and image processing.

1.4.1   Pixel-based Reproduction

The first paradigm is the *pixel-based*, or the film photography model. Traditional physics often starts with the Euclidean point. Geometry teaches us about position defined by dimensionless points. Electronic imaging defines picture elements with small, but well defined dimensions, as pixels. Film photography is a triumph of chemistry and materials science using light sensitive crystals that become light-stable images with chemical development. Over the expanse of the sheet of film all pixels have the same response to light, hence the name, *pixel-based* imaging.

Josef Maria Elder's (1905) *History of Photography* describes early Greek and Roman reports of the effects of light on matter, such as cinnabar, indigo blue and rare purple dyes made by exposing snail mucus to light. There were many early silver halide systems prior to 1800 that were the work of notable chemists and physicists. Elder describes the sequence of events leading to Schultze's experiments with silver salts in 1732. In 1737 Hellot, in Paris, made an invisible ink with silver nitrate that turned black with exposure to light (Mees, 1920).

Thomas Wedgwood and Sir Humphrey Davy (1802) did research on silver halide processes at the Royal Institution, London. At that time Thomas Young was Professor of Natural History there and was writing a book. In Young's famous Bakerian Lecture (1802), he made the suggestion:

> "Now, as it is almost impossible to conceive each sensitive point of the retina to contain an infinite number of particles, each capable of vibrating in perfect unison with every possible undulation, it becomes necessary to suppose the number limited, for instance, to the three principal colours, red, yellow, and blue, . . . : and each sensitive filament of the nerve may consist of three portions, one for each principal colour."

This statement is regarded as the foundation of pixel-based *Colorimetry*. More than 50 years later James Clerk Maxwell performed color matching experiments that are the basis of all CIE color standards. He measured how much long-, middle-, and short-wave light is needed to match any wavelength. Even today, colorimetry calculations limit their input information from the scene to measurements of single pixels.

Analyzing images a pixel at a time could come from Euclidean geometry, the chemical nature of silver halide photography, or the influence of Thomas Young's tricolor hypotheses. It may be a combination of all three. Was Thomas Young influenced by the study of photography by his colleagues at the Royal Institution, Woodward and Davy?

1.4.2 Spatial Reproduction

An alternative paradigm is building images out of *spatial comparisons*. Around 1500, Leonardo da Vinci wrote:

> "Of colours of equal whiteness that will seem most dazzling which is on the darkest background, and black will seem most intense when it is against a background of greater whiteness.
>   Red also will seem most vivid when against a yellow background, and so in like manner with all the colours when set against those which present the sharpest contrasts." [CA, 184 V.C] (da Vinci, 1939)

Leonardo da Vinci began the study of how appearance is influenced by the content of the image. Experiments by physicists, Otto von Guericke and Count Rumford (1875); writers, von Goethe (1810); chemists, Chevreul (1837); psychologists, Hering (1872); and designers, Albers (1963) amassed a great variety of evidence that appearance was not pixel-based. The rest of the image affected appearance. Perhaps the most compelling argument is that all of the studies of primate neural mechanisms show a series of complex spatial interactions (Hubel & Wiesel, 2005; McCann, 2005). The studies of 20th century psychophysics of Campbell, Gibson, Land, Gilchrist, Adelson, and many others all show that the spatial content of the image controls appearance.

Throughout this text we will compare and contrast the *pixel-based* and *spatial comparisons* paradigms of human vision and scene reproduction.

1.5   Reproduction of Scene Dynamic Range

Some reproductions involve small transformations of the interior of the original's color space, such as a photographic print of an oil painting in uniform illumination. For reflective materials, such as oil paints, and photographic paper, the range of light between white and black is roughly 32:1. Both ranges are limited by the front-surface reflections, rather than the absorption of light behind the surface. Real-life scenes, both indoor and outdoor, are almost always in non-uniform illumination. Illumination introduces a major challenge to reproduction by having an extremely large original color space. On a clear day shadows cast by the sun are 32 times darker than direct sunlight. The 32:1 range of reflectances in a 32:1 range of illumination creates a 1024:1 range of light. Real-life scene reproduction is analogous to moving a castle into a cottage.

Imaging techniques can record scene information over a High Dynamic Range (HDR) of light. The range of captured information is much more than the 32:1 range possible between white and black in a reflective print. In photography, Ansel Adam's Zone System (Chapter 6) provides the logical framework for capturing the wide range of light in natural scenes and rendering them in a smaller dynamic-range print. Adams described a three step process: measuring scene range, adjusting image capture to record the entire scene range, and locally manipulating the print exposure to render the high-range scene into the low-range print. Adams visualized the final image before exposing the negative. He assigned appearances from white to black to image segments. Once the negative recorded all the information, Adams controlled the local print contrast for each part of the image (manually dodging and burning) to render all the desired information in the high dynamic range scene. Not only can these techniques preserve detail in high- and low-exposures, they can be used to assign a desired tone value to any scene element. Adams described the local contrast control in detail for many of his most famous images. He used chemical and exposure manipulations to spatially control appearances between white and black. Today, we use Photoshop® in digital imaging.

Painters have used spatial techniques since the Renaissance to render HDR scenes, and photographers have done so for 160 years. Land and McCann's Retinex algorithm used the initial stage of Adam's wide-range-information capture for its first stage. Instead of using aesthetic rendering, it adopted the goal that image processing should mimic human vision. The Retinex process writes calculated visual sensations onto prints, rather than writing a record of light from the scene. To this aim, Retinex substitutes the original light values at each pixel with ratios of scene information. This approach preserves the content in the original found in the interior of the color space.

Electronic imaging made it possible, and practical, to manipulate images spatially. Automatic spatial processing is not possible in silver-halide photography because film responds locally. Silver grains count the photons. The same quanta catch produces the same film density. Hence, Adams had to manipulate his images by hand. Digital image processing made it possible for each pixel to influence each other pixel (Section F). Details in the shadows are necessary to render objects in shade to humans. The accuracy of their light reproduction is unimportant: the spatial detail of objects in shadows is essential. Spatial-comparison image processing has been shown to generate successful rendering of HDR scenes. Such processes make use of the improved differentiation of the scene information. By preserving the original scene's edges, observers can see details in the shadows that are lost in conventional photography.

The image reproduction industry has developed high-chroma colorants, ingenious tone-scale and spatial-image-processing techniques, so that they can make excellent reproductions. The secret is that they do not reproduce the original's stimulus. What they do is more like reconstructing the furniture for each room so that it has the same spatial relationship with other objects in the new size of room. Good reproductions retain the original's color spatial relationships in the interior of the color space.

1.6 HDR Disciplines

Many disciplines have contributed to our understanding of HDR imaging. Imaging began by painting. The 17th century interest in the physics of light developed the tools that allow us to measure light. Starting around 1800 the science and technology of photography developed photochemical means of recording scenes. Video, digital and computer graphic imaging followed in the 20th century. All of these disciplines are listed in Table 1.1 in blue horizontal rows. The green bars, below, list the different disciplines in understanding human vision. Psychophysics got its beginning in the second half of the 19th century. Its goal was to apply the advances in physical science directly to understanding human sensory mechanisms. It developed the science of colorimetry that measures when two patches of light appear the same.

In the 20th century most of the research in vision emphasized the effects of spatial contents of images. Much of the information from neurophysiology in the 20th century reports on the spatial interaction of neurons. The final section of Table 1.1 lists the disciplines of spatial reproduction in yellow.

1.6.1 Interactions of Light and Matter

Table1.1 lists the disciplines in the first column that are grouped as art (gray); photography (blue); vision (green); and spatial reproduction (yellow).

The stimulus falling on sensors in cameras and vision depends on the interactions of light (*illumination*) and matter (*reflectances*) of objects. For all disciplines, except painting, the image making process begins with counting photons. The physics of photons provides a single description of light: the energy of a photon expressed as the product of Plank's constant and frequency (hv). The sensor's response depends on its spectral sensitivity and the number of photons falling on the sensor. The response, or *quanta catch*, is the measurement principle for all disciplines. Objects in scenes reflect, transmit, and emit variable quantities of light.

1.6.2 Light Sensors

While the physics of light provides us with a singular description across disciplines, the next two columns in Table 1.1 (Sensors, Color Sensors) have different characteristics. Physics uses the narrowest possible spectral band of wavelength for its measurements. Cameras, that depend on short exposures, use broad bands of wavelengths to capture more light. As well, their spectral windows have minimal overlap to improve color separation. Human sensors have extremely broad spectral response with great overlap, making very poor color separations.

1.6.3 Image Processing

We see a major departure in the image processing column. The tradition in the physics of light is that light must be measured as individual pixels. Light is measured by counting the photons captured on each small picture element *(pixel quanta catch)*. All pixels in a photographic film have the same response to the same quant catch.

Human vision is different. Two image segments with identical quanta catch may, or may not, appear the same. Appearance is controlled by spatial interactions. The study of spatial imaging is shown in the

Table 1.1 The disciplines and characteristics that help us understand HDR imaging.

Discipline	Light	Matter	Sensor	Color Sensor	Image Processing	Rendition	Chapter
Art	perspective chiaroscuro impressionism	reflection transmission emission	artist's choice	artist's choice	spatial vision	painting etching mezzotint	4
Physics	hv	reflection transmission emission	grease spot photoelectric CCD/CMOS	monochromator filters [very narrow-band]	pixel-based	radiance luminance	2
Photography	hv	reflection transmission emission	silver halide(AgX) orthochromatic panchromatic	sensitizing dye on AgX surface [broad band]	pixel-based	print transparency	5–6
Video & Digital	hv	reflection transmission emission	silicon + filter	filters [broad band]	pixel-based	print display	7
Computer Graphics	hv	reflection transmission emission	artist's choice	artist's choice	pixel-based	print display	8
Psychophysics	hv	reflection transmission emission	luminosity: scotopic/photopic	color match [overlapping bands]	mixed pixel/spatial	metric value	2
Neurophysiology	hv	reflection transmission emission	luminosity: scotopic/photopic	cone sensitivities [extreme overlap]	spatial processing	electric response	7
Spatial Vision	hv	reflection transmission emission	luminosity: scotopic/photopic	cone sensitivities [extreme overlap]	spatial processing	predict vision	14–30
Spatial Reproduction	hv	reflection transmission emission	luminance	filters [broad band]	spatial processing	best reproduction	32–35

stippled boxes at the bottom of the column. (Sections D, E, and F) Electronic imaging removes the pixel-based limitations found in film. Although the sensors count photons, the subsequent electronics can introduce spatial processing, that makes it possible to mimic vision.

1.6.4   Image Rendition

In each discipline there are a variety of rendition techniques that integrate all the properties of capture and processing to make the renditions that can be paintings, prints and displays with a host of different rendering intents.

1.7   Outline of the Text

This text, *The Art and Science of HDR Imaging,* is a survey of all of High Dynamic Range imaging. It involves all the disciplines listed in Table 1.1. For almost all of them, the text will describe a high-level, general review of these disciplines. The last column on the right of Table 1.1 identifies the chapters that discuss the discipline and its role in HDR imaging.

1.7.1   Section A – History of HDR Imaging

The text begins with a discussion of the definitions and tools we need to understand HDR. It describes HDR in natural scenes, painting, and silver-halide photographs. Ansel Adams described the Zone System to capture and reproduce the natural HDR scene in LDR prints. Electronic HDR image processing began to make spatial comparisons in the late 1960s to render HDR scenes and computer graphics came of age in the 1990s with sufficient processing power to manipulate and synthesize illumination. All these fields have made important contributions to understanding HDR.

1.7.2   Section B – Measured Dynamic Ranges

After the historical review, the text includes a number of detailed experiments that drill down to a much deeper level of scientific measurements. These measurements identify the limits that cameras can capture accurately in HDR imaging, as well as the limits of stimuli that humans can detect in images. These in-depth experiments show that the physical limits of cameras and those of human vision correlate with the range of light falling on their sensors. Veiling glare in their optical systems limits the dynamic range of light. These physical limits of high-dynamic-range imaging are highly scene dependent.

1.7.3   Section C – Separating Glare and Contrast

Building on the measurements of what cameras can capture, and what humans can see, we explore the scene-dependent properties of HDR imaging. The first scene-dependent property is the scatter of light in the ocular media. It reduces the contrast of edges depending upon the content of the scene. The second scene-dependent property is spatial comparison, or neural image processing. It increases the apparent contrast of edges depending upon scene content. Although the mechanisms have entirely different physiological causes, and they have very different spatial properties, they tend to cancel each other. The lower the physical contrast of the scene on the retina, due to scatter, the higher the apparent contrast to the observer. These two counteracting spatial mechanisms, *glare* and *neural contrast* play important roles in human vision.

1.7.4   Section D – Scene Content Controls Appearances

Ever since Hipparchus observed the appearance, or stellar magnitude, of stars in the 2nd century BC, many people, in many disciplines, have been studying the relationship of scene appearances with the light from the scene. Vision begins with counting the photons falling on the retina, as photographic films and solid-state detectors do. However, vision is different. Vision compares the quanta catch at one receptor with others at different spatial locations.

This section summarizes many different experiments measuring spatial interactions. This information is helpful in evaluating our thinking about HDR image processing algorithms. If we assume that the best way to render HDR scenes is to mimic human vision, then we need to have a better understanding of how vision works. This section describes the small changes in visual appearance of the maxima with large changes in light. Further, it also describes the large change of appearance with small changes in light from areas darker than the maxima.

1.7.5   Section E – Color HDR

This section reviews the many Color Mondrian experiments using complex scenes made of arbitrary, unrecognizable shapes. These experiments test the validity of pixel-based and spatial comparison color constancy algorithms. It reviews experiments that measure color appearances with particular attention to departures from perfect constancy. As well, it describes the conditions that shut off color constancy in complex images. Finally, it reviews recent experiments 3-D Mondrians using low-dynamic-range and high-dynamic-range illumination. Spatial content of the display plays a major role in color HDR and color constancy.

1.7.6   Section F – HDR Image Processing

The final section of the text describes and compares a large variety of HDR algorithms. One way to categorize these algorithms is to just look at the image information used in the calculation. We will describe in Section F four classes of algorithms based on the scene pixels used in image processing.

The first uses a single pixel in the input image to modify its output pixel. The tone scale approach, inherited from film photography, requires that every scene have a unique adjustment. Even then, the results are less than optimal.

Spatial image processing algorithms, that were impossible with film, are now practical with digital image processing. The output of each pixel can be influenced by the other input pixels from the scene.

The second class of algorithm uses some of the pixels in the input image to modify a pixel. (A pixel is modified by its surrounding pixels).

The third uses all of the input pixels in the image to modify a pixel. An example is using all the scene's image to calculate the spatial frequency representation (Fourier and Cosine transforms); followed by a spatial frequency filter; and retransform of the result to the image domain.

The fourth uses all of the pixels, but uses this information in different ways in different scenes (Retinex and ACE are examples). These scene-dependent algorithms mimic human vision, and will the subject of Section F.

1.8   Summary

The goal of this text is to present an interdisciplinary analysis of HDR imaging. It will review the work of artists, scientists, and engineers that all combine to further our understanding of the best way to reproduce the high-dynamic-range world in a manner appropriate for our human visual system.

1.9   References

Albers J (1963) *Interaction of Color*, New Haven; London, Yale University Press.

Chevreul ME (1839) *De la loi du contraste simultané des couleurs;* (1854) *The Principles of Harmony and Contrast of Colours*, Reinhold. New York, (1967).

da Vinci L (1939) *The Notebooks of Leonardo da Vinci*, MacCurdy E ed, Braziller, New York, 921.

Elder JM (1905) *History of Photography*, trans Epstein E, Dover, New York, 1978.

Hering E (1872) *Outline of a Theory of Light Sense*, trans Hurvich LM & Jameson D, Harvard University Press, Cambridge, 1964.

Hubel D & Wiesel T (2005) *Brain and Visual Perception*, New York: Oxford University Press.

McCann J (2005) Rendering High-Dynamic Range Images: Algorithms that Mimic Human Vision, *Maui: Proc. AMOS Technical Conference*, 19–28.

Mees CEK (1920) *The Fundamentals of Photography*, Eastman Kodak, Rochester.

Rumford, C (1875) *The Complete Works of Count Rumford*, Vol. IV, The American Academy of Arts and Sciences, Boston, Mass., 51–2.

von Goethe JW (1810) *Theory of Colours*, trans. Charles Lock Eastlake, Cambridge, MA: MIT Press, 1982.

Wedgwood T & Davy H (1802) *An Account of a Method of Copying Paintings upon Glass and of Making Profiles by the Agency of Light upon Nitrate of Silver*, J. Royal Institution, London I,170.

Young T (1802) *The Bakerian Lecture. On the Theory of Light and Colour*, Phil. Trans, Royal Society of London, **92**: 12–48.

2

HDR Tools and Definitions

2.1 Topics

This chapter introduces the definitions and vocabulary used in the text. Since we include many disciplines that use terms in different ways, we need to assemble here the definitions and usage of terms found in the book. The text begins by defining dynamic range. Further, we describe the vocabulary of artists, the measurement of light using physics and the measurement of appearance using psychophysics, as well as the terminology of image reproduction.

2.2 Introduction

The human eye is sensitive to a small window of electromagnetic radiation between 400 and 700 nanometer (nm) wavelengths (λ). There are four types of retinal receptors. After sitting in a light-free room for an hour humans can reliably detect flashes of light that deliver only four to six photons to their rods (Hecht, Shlaer & Pirenne, 1938). Snow on a mountaintop sends to the eye 100 million times more photons. Human visual receptors can respond to a range of light greater than 10 log-units.

What makes human vision so fascinating is the great number of paradoxes found in the study of imaging. The *Oxford American Dictionary* has a definition of paradox that reads:

> "a seemingly absurd or self-contradictory statement or proposition that when investigated or explained may prove to be well founded or true."

For example, our retinal receptors can respond to a range of light over ten orders of magnitude. It is also true that we can only differentiate details in most scenes over a range of two to three log units. Both statements are accurate, yet they seem to contradict each other. Why do we need to have such a wide range of light response when we have limited response to light from scenes? This text describes the further inspection needed to resolve this and many other paradoxes of human vision.

Human vision would be much simpler, if it behaved the same as photographic film. Film has a fixed, unique response to the number of photons falling on a small region of film (photons/area). Human

The Art and Science of HDR Imaging, First Edition. John J. McCann, Alessandro Rizzi.
© 2012 John Wiley & Sons, Ltd. Published 2012 by John Wiley & Sons, Ltd.

visual appearances are more complicated than that. What we see depends on the relationships of the light on the retina from all parts of the image.

2.3   Pixels

In the 19th century, Hering and Helmholtz discussed their visual test targets as image segments, such as, test area, surround, and background. Nevertheless, that is not how we work with scenes today. We have the technology to study each scene as arrays of millions of pixels. When we evaluate HDR images and models we are able to evaluate our hypotheses using all the information, namely all the pixels. Evaluating the influence of areas, by grouping together hundreds or thousands of pixels, was all that Hering could do. Today, with computer technology, we have to evaluate all these pixels independently in our analysis.

2.4   Dynamic Ranges

Dynamic range describes the useful range of light. A scene with a *dynamic range* of 1000:1 means that the maximum light coming from a segment of a scene is one thousand times more light than the minimum. Dynamic range has to be measured with a light meter. We cannot calculate the dynamic range from the number of bits assigned to a digital record. The number of bits describes the number of quantization levels, but not the range. Think of the scene range as the length of a *salame* (Italian for salami). They come in a variety of lengths (dynamic ranges). The number of bits is the equivalent of the number of slices used to serve the salame (Italian for salami). If one wants to calculate the range of the scene, or the length of the salame, you need to have precise measurements of the light range of each quantization level, or the thickness of each slice. Neither digital cameras, nor delicatessens provide this information for each slice.

Any system, photographic or visual, has a minimum response. Below that amount of light there is no change in response with decrease in light. Similarly, there is an amount of light that generates its maximum response. More light gives the same response. System dynamic range is the ratio of the maximum to minimum light responses. The system's dynamic range has to be measured to find the asymptotic minimum and maximum values. It cannot be calculated from individual physical limits, such as signal-to-noise of the sensor, or number of bits in the digitizer. All the properties of the camera or visual system, including optics and signal processing, play a role in determining the maximum and minimum system responses. Hence, the range limits of a real system have to be measured using the entire system.

2.4.1   Dynamic Range of Light in Scenes

Figure 2.1 is a photograph taken in sunlight of a block of wood with white, several grays, and black paints. The range of reflectances measured 30:1, or 1.5 log units.

We put the painted board in the middle of a field, in front of a garage in partial shade, with a second reflectance target inside the garage in deep shadow. We measured the light from white and black paints in sun and in deep shadow. The range of this scene is 21 900:1 or 4.3 log units.

This is an unusual scene, but it illustrates well the fact that high dynamic range images are the result of variable illumination falling on different surfaces. The range of reflectances is only 30:1 (1.5 log units) because it is limited by front-surface reflections. Regardless of the amount of light absorbers in the paint, a few percent of the light is reflected from the surface depending on the refractive indicies of the air and the media.

White Paint	Black Paint	Range
21,900 cd/m²	725 cd/m²	30:1 1.5 log units

Figure 2.1   Paint samples covering the range of reflectances.

By comparison the range of illumination is very large. If you aim a spot meter at a cloud in front of the sun it reads 179 000 cd/m². Turning around, the white paint in the sunlight is 21 900 cd/m², and the white in the garage is 1369 times darker (2.8 log units) than the white on the block in the field.

The photographs in Figure 2.2 were made using variable exposure and camera position and fixed sunlight white balance. In other words, these are conventional photographs, with exposure as the variable. The upper-left photograph is the best exposure for the painted block in the meadow. The garage in partial shade is dark. More exposure for the entire scene renders the sky and trees properly, but overexposes the painted block in the field. Still more exposure lightens the sky, trees and garage, but does not improve the information about things in the garage.

In the second row of photos taken in front of the garage, we can just barely see that there are things inside the garage in the image that overexposes its exterior. The bottom row shows the gray scale and toy vehicle not visible in the other pictures. The light entering the garage is reflected from the meadow, hence the yellow-green cast of the gray paints. We will not attempt to synthesize the best HDR image of this scene. We do not know enough yet. We use these pictures to illustrate the range of light in this very-high dynamic range scene.

2.4.2   Dynamic Range of Vision

We can take gray scale paint samples into a deep and winding light-free cave. As we walk further and further into the cave, the radiance of the black will decrease to approach zero. If we go slowly into that cave we will dark adapt, because our rod and cone sensors will become more sensitive to light. We reach the asymptotic limit of sensitivity after 30 minutes in the complete absence of light. If we move back towards the entrance of the cave we will go from no light to detection threshold. At rod absolute threshold we can detect four to six photons (Hecht et al., 1938). That is one extreme end of the ten log units range of light we can detect. We cannot see anything, all we can do is detect the presence of light on a paint sample. Increasing the illumination does two things: it allows us to see ill-defined forms or shapes, and it causes the rod vision to light adapt, thus raising detection threshold. Unlike film with fixed sensitivity to light, vision both dark adapts to gain sensitivity, and light adapts to lose it. Further increases in illumination makes shapes clearer, and then sharper, and then colorful, and then bright, and then dazzling.

The other end of the 10 log unit range is snow on top of a mountain. Light adaptation, caused by bleached photopigment, sends a signal out of the retina until the rods and cones have regenerated that photopigment (Alpern & Campbell, 1963; McCann, 1964). Mountaintop snow bleaches so much

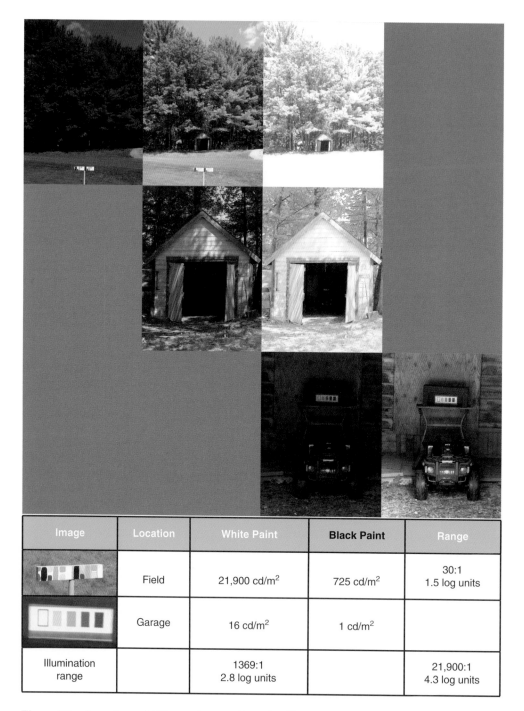

Image	Location	White Paint	Black Paint	Range
	Field	21,900 cd/m²	725 cd/m²	30:1 1.5 log units
	Garage	16 cd/m²	1 cd/m²	
Illumination range		1369:1 2.8 log units		21,900:1 4.3 log units

Figure 2.2 A montage of different photographs with different exposures to illustrate two test targets in bright sun and deep shade. From the white in the sun to the black in the garage the dynamic range is 4.3 log units.

photopigment that it takes more than 30 minutes in the dark to approach absolute threshold again. Just because the eye can detect light over this 10 log unit range of light, does not mean that we can see scene details over that range. The balance of light and dark adaptation is scene dependent and sets the range limits of we see at a given moment.

There is a second optical property of the human eye that limits what we see. Intraocular glare reduce the range of light on the retina to less than four log units at best. For most scenes the limit is between two and three log units. The range of light response we measure with spots of light, under different adaptation conditions has very little to do with what we see in real-life scenes.

We should not confuse the range of light possible in scenes in the world with the range of light we can see, or capture with a camera. The range of light on a camera's image plane is a small fraction of the physical range of light in HDR scenes. In Section B we describe careful measurements of the limits recorded by cameras and human vision. These distinct definitions of *scene range* (measurement of light in the world) and *image of the scene range* (measured and calculated light formed by optical system, such as camera's and the eye's optics) are an essential element of this text.

2.5   Measuring Light

We need to describe the major terms used to describe the measurement of light. We use *radiometry* to measure electromagnetic radiation (physics); and *photometry* to standardize human response to light (psychophysics). We have standard procedures to measure the light coming from all parts of the scene, and the light coming from a particular segment of the scene.

2.5.1   Radiometry – Measuring Electromagnetic Radiation

Radiometry is a part of physics that describes standards for measuring electromagnetic radiation. Photons with different wavelengths have different energies. From Planck's Law, we can calculate that a single 555 nanometer (nm) photon's energy equals 3.6*10 to the power–19 joules (watt *seconds).

Irradiance is the measure of the energy from the number of photons continuously falling on an area. Irradiance meters measure [watts/cm^2] using photosensitive electronic elements behind a diffuser integrating the light falling on the meter. Photographers use similar *incident* light meters to measure the light falling on the scene by standing in front of the subject, and aiming the meter at the light coming from all parts of the scene.

Radiance is the measurement of radiation coming continuously from a particular object to a particular point in space. We use a telescope with a calibrated light sensor in the image plane to measure the radiance. Such tele-radiometers measure photons per area per angle [watts/(cm^2*sr)]. Ansel Adam's Zone System described techniques to capture and print the entire range of all scenes. He measured the light from clouds, sky, rocks, trees and shadows using a spotmeter, a camera-like instrument that forms an image of a small scene segment on the light-sensitive element. By replacing the average of the entire scene with individual values for maximum light, minimum light and objects of interest, the photographer can control the image making processes to render his visualization of the scene (Chapter 6).

Photographic light meters, both incident and spotmeters, are familiar instruments, but they have the same spectral sensitivities as films. These photographic meters collect light in the same way, but do not measure radiance and irradiance values [watts/(cm^2*sr)] and [watts/cm^2] because of their spectral response. They provide exposure values (EV) for setting exposure times and lens aperture size.

2.5.2   Photometry – Measuring Visible Light

Human see a narrow window of electromagnetic radiation. If we want to correlate appearance with a physical measurement of light, we need to know the spectral sensitivity, and the characteristics of human vision's response to light.

Photometry, a part of psychophysics, is radiometry adjusted for the wavelength sensitivity standards for human vision. *Photopic Luminosity Standard (V_λ)* is the nominal sensitivity of the eye to different wavelengths. The peak sensitivity is at 555 nm (100%); half-height (50%) at 510 and 610 nm; (3%) at 446 and 672 nm. With the candela/meter^2 [cd/m^2], the standard of luminous radiance is 1/683 watt per area per angle at 555 nm. This standard is the basis of lumens per watt used in evaluating the electrical efficiency of lamps, namely [Lumen = (candela*angle)].

To convert the radiometric units (physics), *irradiance* and *radiance*, to photometric units (psychophysics) of *illuminance* and *luminance*, we multiply the radiometric values at each wavelength by the CIE standard (V_λ) curve value of those wavelengths. Illuminance is the integral of [irradiance$_\lambda$* V_λ]; luminance is the integral [radiance$_\lambda$* V_λ].

Prior to 1930, the term *brightness* was defined as both the physical measurement of light, and the psychophysical measurement of the appearance of light. The term luminance was introduced to reduce the confusion caused by the two usages of brightness with different meanings. (Chapter 5)

It is important to understand that Photometry is a standard for engineering purposes. It does not always predict visual appearance. It is well known that very high chroma stimuli do not appear to have the same brightness as white when they are adjusted to have equal measured luminances. This paradox is called the Helmholtz-Kohlrausch Effect (Wyszecki & Stiles, 1982). Further, this luminosity standard does not include the response of blue cones (Gegenfurtner et al. 1999). It is an important standard for illumination engineers, but does not predict the brightness appearance of highly colored stimuli.

2.6 Measuring Color Spaces

The V_λ achromatic standard converts radiometric light units to photometric measures of human responsiveness. Color requires colorimetric standards. V_λ was defined by measuring human response to certain stimuli. Colorimetric standards do the same, by asking observers different questions. There are two important experiments used in measuring color. Although both are very commonly used in the study of vision they are very different color spaces, because they are based on very different psychophysical questions.

One question is:

> Do two colors match?

A different question is:

> What is the color appearance of the match?

What makes color vision so interesting is that, although we may know that two mixtures of light match, we do not know the color appearances of those matches. (Land, 1964; Wyszecki, 1973). Using a color matching space is a bad idea when studying appearance. These paradoxical properties make it important to understand color spaces and their intended uses. The two most commonly used color spaces are:

1. CIE XYZ *Color Matching Functions* (CMF) psychophysically measure the color response in the cone receptors (Wright, 1987). This space is derived from data of observers who adjusted two adjacent areas until they matched (Wyszecki & Stiles, 1982).
2. Munsell's *Uniform Color Space* (UCS) asks observers to estimate the distances between hues, chroma, and lightnesses. This color space characterizes colors as they appear at the end of the color processing mechanism. (Munsell, 1905; Newhall et al., 1943).

2.6.1   Color Matching Functions

Colorimetry is the science of measuring and predicting when two patches of light will match. Given the spectrum of a patch of light, colorimetry can calculate whether it will match another patch of light. Here we ask observers to match a single wavelength (left half circle) by adjusting the amounts of red, green, and blue light (right half circle). The 1931 CIE standard describes three spectral responses that are consistent with a large pool of color matching data documented by Stockman and Sharpe (2010).

Just as with V_λ, we calculate the integral of

$$[X = radiance_\lambda * \overline{x}_\lambda], [Y = radiance_\lambda * \overline{y}_\lambda], [Z = radiance_\lambda * \overline{z}_\lambda] \qquad (2.1)$$

with normalization to Y. These XYZ values are very effective in calculating when two stimuli will match. However, as Gunter Wyszecki (1973) warned us,

> "the tristimulus values and thus the chromaticity of a color stimulus do not offer any direct clues as to what color perception will be perceived."

Additional variables, such as the other colors in the field of view, the state of adaptation of the receptors, and spatial relationships in each scene determine the color appearance. In order to measure appearance one needs to match a test color to a sample in a library of standard colors in a constant complex scene. (McCann, McKee & Taylor, 1976)

XYZ forms a cube much like Figure 1.1. We often see this data replotted in cylindrical polar space. Here, Y is the vertical axis with *achromatic* (grays) at the center. Using equations 2.2 and 2.3 we calculate the *chromaticities* (x,y), or distances from achromatic axis, in a chroma plane with constant Y.

$$x = \frac{X}{X+Y+Z} \qquad (2.2)$$

$$y = \frac{Y}{X+Y+Z} \qquad (2.3)$$

2.6.2   Uniform Color Spaces

The appearance of color space is standardized in the Munsell Book of Color, (Newhall et al., 1943) and other *Uniform Color Spaces (UCS)*. In these experiments, observers equated the visual distance between colored samples in hue, chroma and lightness. Munsell's experiments assigned colored papers to each uniformly spaced color cylinder, with white to black as the centerline of the cylinder. Unlike color matching functions that are measured in a no-light surround, Munsell designations are chosen for viewing in a real world setting. Furthermore, they measured the complete range of appearances in a 3-D color space. Physical measurements of papers in the Munsell Book provide a measurement of the color space of appearances in complex scenes.

Munsell Color has three parameters in a cylindrical polar coordinate space. Lightness (L) is the vertical component with white at the top and black at the bottom. Hue forms a circle: red yellow, green, blue, purple, red. Chroma is the distance from the white black axis caused by selective spectral colorants. Observers selected papers for each of these positions in this color space using equal-spacing criteria.

Figure 2.3 plots a side and top views of this 3-D color space using 10 of the 40 possible hue pages. It plots all the real chips in the glossy Munsell Book collection of the 5.0 Hues pages (5R, 5YR, 5Y, . . . 5RP). In this plot, the white paper has been scaled to 100 and hue and chroma have been transformed into rectangular coordinates of R-BG and Y-B, with 2.5R hue falling at 0°. The (left) top view shows the regular spacing of the papers around the achromatic center. The (middle) side view

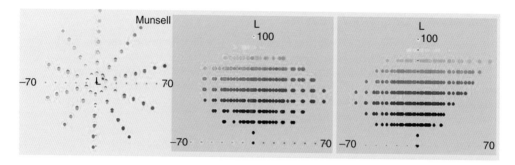

Figure 2.3   The Munsell Space locations of 10 Hue pages from the gloss samples: (left) shows top view; (middle) side view for 2.5R/2.5BG axis shows; view; (right) side view for 7.5Y/ 7.5PB axis.

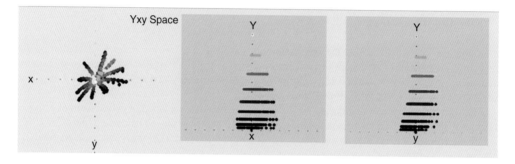

Figure 2.4   The 10 pages in Yxy colorimetric space: (left) shows the top view; (middle) same side view for x axis; (right) show side view for y axis.

shows the shape of the space along the red/blue green axis that was determined by the spectra of real papers. The (right) side view shows the color samples along the yellow/blue axis.

Given this set of real regularly spaced papers, we can measure their spectral radiances and calculate their relative *XYZ*, and *Yxy* values. Figure 2.4 plots the *Yxy* values scaled so that the maximum and minimum *Y* achromatic values are superimposed on the Munsell Lightness max/min values in Figure 2.3. (Rotating 3-D plots in Quicktime® movie files are available on the Wiley web site http:// www.wiley.com/go/mccannhdr for a better view of Figures 2.3 and 2.4 of the Munsell Book in Uniform Color Space (UCS) and Color Matching Function (CMF) space.)

In a sense we are looking backwards at the neural processes that convert cone responses (CMF) to color appearances (UCS). We want to analyze the conversion process, so we begin with the paper samples that are uniformly spaced in appearance and measure their XYZ values. The colorimetric plot (Figure 2.4) shows the color space of the XYZ responses that are transformed by color vision to UCS Munsell Space (Figure 2.3). When we plot the equally-spaced appearances of Munsell Color chips in the colorimetric *Yxy* space, we find unequal distances. In order to convert *XYZ* values to a Uniform Color Space we need a complex, non-linear transform. In other words, color-matching standards (CMF) do not have a simple correlation with appearance (UCS). The equal-spaced lightnesses are expanded in light grays, and compressed in dark grays. All chroma are significantly compressed. The hue planes are not equally spaced around the circle.

We will explore this paradox in detail throughout the text. For now, it is sufficient to say that the information we learn from color matching data (CMF) represents the spectral response of the rods and cones in the retina, but that does not represent the color space of human color appearances. Vision transforms cone responses into a very different shape of color space because of a number of different spatial mechanisms.

2.6.3   Early Pixel-based Color Matches Followed by Neural Spatial Interactions

There is an extremely large literature on colorimetry. There are a number of color spaces, color standards, and color prediction systems based on the *Color Matching Function*s that restrict their calculations to using input measurements from single scene pixels. This body of work makes it possible to predict whether two stimuli will match and not match with change in illumination. It describes the behavior of *metamers*, that is stimuli that are spectrally different, but visually identical. These *pixel-based* color systems are described in great detail in many fine texts, many of them are in the IS&T Wiley series. See Wyszecki and Stiles (1982), Hunt (2004), Fairchild (1998), Moroney et al. (2002), Morovic (2008), Ohta & Robinson (2005), Shanda (2007), and Green (2010). These texts cover in great depth: colorimetry, approximate UCS calculations, pixel-based color appearance models (CIECAM), CIE standards, and International Color Consortium device profiles. The CIECAM color appearance model uses radiance measurements of the surfaces in a scene and the illumination falling on it. Although it only uses these pixel-based measurement values in the calculation, it uses four additional, operator selected, constants to introduce scene dependent properties. We will refer to these sources, rather than duplicate their description and analysis.

WD Wright, the noted color scientist whose measurements became the basis of the CIE 1931 Colorimetry Standard, argues that:

> "Where does colorimetry end and appearance science begin? An interesting question. My short answer would be that colorimetry ends once the light has been absorbed by the colour receptors in the retina and that appearance science begins as the signals from the receptors start their journey to the visual cortex. To elaborate a little, tristimulus colour matching is governed solely by the spectral sensitivity curves of the red-, green-, and blue-cone receptors (if we may be allowed to call them that), whereas the appearance of colours is influenced by all the coding of the signals that takes place along the visual pathway, not to mention the interpretation of the signals once they arrive in the visual cortex." (Wright, 1987)

The spectral sensitivities of the cones are an essential description of the input to vision. However, as pixel-based measures of the scene information, they are insufficient to understand the entire visual process. Almost all of the non-colorimetric research in vision in the second-half of the 20th century studied the spatial interaction of retinal signals that synthesize appearance (McCann, 2005). Much of this text is devoted to the review of the studies of spatial interactions that follow the cone's quanta catch.

2.7   Image Reproduction

Real scenes have the greatest range of light. Light emitting displays (TVs) have much less range, yet more than reflective prints. Dynamic ranges characterize the light reproduction values between the most and least response for each RGB channel, for each media. Color saturation characterizes the entire range of chroma of the medium's *colorants*. Together range and chroma color gamut determine all possible spectra of the medium. With greater color saturation of the dyes and filters, we find larger color gamuts. However, these pure colors are just three, or four, points in the entire color space. While the dyes and filters control the extreme boundaries, the quality of the reproduction depends on how the medium presents the scene information in the interior of the color space. The term *tone scale* is used to describe how the medium processing alters achromatic samples in the interior of the original's color space in the reproduction.

2.7.1   Color-Forming Technologies

The simplest and oldest color technology uses opaque colorants. Here, the light is reflected from the front surface. Opaque pigments contain materials that absorb light, so as to prevent light from interacting with subsurface colorants (Figure 2.5 left).

James Clerk Maxwell's invention of additive color photography in 1861 used three black-and-white separation photographic transparencies taken with red, green, and blue filters (Maxwell, 1965). He projected the three images in superposition using the same filters used to take the pictures (Figure 2.5 center). The photographs recorded the information from the scene in each spectral region. By projecting all three records in superposition the reproduction adds the information from each waveband. White is the sum of all the light in all three. Black is the minimum transmission for all three wavebands. All other colors are additive depending on the relative transmissions in the RGB records.

Subtractive color reproductions, first demonstrated by du Hauron, use the same RGB color separation information as additive photography. Here colors are controlled by multiple absorptions (transparent dyes on top of other dyes) (Figure 2.5, right). Each transparent subtractive dye absorbs one-third of the visible spectrum, namely, minus-red (cyan), minus-green (magenta), and minus-blue (yellow). Each subtractive dye modulates the R, G, or B information.

Figure 2.6 illustrates the transmission and absorption of light in a cross-section of a hypothetical three-layer color reflection print. Figure 2.6 shows the following:

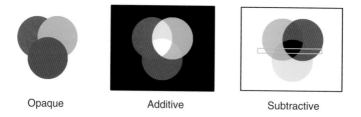

Opaque   Additive   Subtractive

Figure 2.5   Different colorant systems: opaque, additive and subtractive.

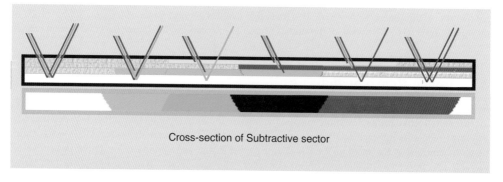

Cross-section of Subtractive sector

Figure 2.6   Subtractive images (top) light penetrating transparent dye layers on a white substrate; (bottom) colors selected from Figure 2.5 right.

absence of dye makes white in the print by reflecting R, G, B light;.

absorption of red light makes cyan (green + blue);

absorption of blue and red makes green;

absorption of all three makes black.

absorption of blue and green makes red;

absorption of only green makes magenta (red + blue).

The combination of cyan, magenta, and yellow dyes make white, red, yellow, green, cyan, blue, magenta and black colors that lie on the outer gamut of the color space for that reproduction media. All the colors in the reproduction come from combinations of these colors. Most subtractive printing systems, inkjet, electrophotography, and press printing, use a fourth print ink, black (Hornak, 2002). This has practical and financial advantages. Practically, it makes achromatic blacks and grays with a single colorant, rather than controlling precisely the amounts of yellow, magenta and cyan. Economically, at least in printing, black inks are less expensive, than colored inks.

The dynamic range of printed materials is the result of the colorants, the surface of the colorants, the support media, and the surface of the support media. Although colorants may be able to selectively absorb spectral regions, the light coming to the camera and eye is the sum of the surface refections and the underlying colorants (Emmel, 2003; Hersch, 2011).

2.7.2   Spatial Additive Color in Flat-Panel Displays

Prior to the development of subtractive Kodachrome three-layer dye systems in 1935, single-exposure color films used additive, side-by-side color forming systems. These systems included Autochrome (dyed starch grains with silver halide salts), Dufay, and Joly (1895) three-color additive patterns (Land, 1977; McCann, 1998). These are the direct ancestors of three gun color television and flat-panel *Liquid Crystal Displays* (LCD). They use the same spatial additivity to display colors. Here the display has small transparent R, G and B filters. The RGB signals sent to each triplet control the amounts of R, G, and B light. This process is called additive reproduction because the side-by-side triplets of R, G, and B stripes are too small (100 triplets/inch) for the observer to resolve at normal viewing distance. These stripes get blurred by the eye to make spatially additive reproductions. Look at an LCD computer display, or LCD television with a 10x-magnifying lens to see the additive substructure.

2.7.3   Tone Scale Control of the Interior Color Space

So far, we have discussed how to form three, or four, colors on the boundary of the color gamut. All the other colors are formed by the controlled combinations of these few colors. Each technology uses different tone-scale manipulations to manage the interior of the color space.

Color photography uses three transparent dyes on transparent, or reflective, support sheets. Light activates the silver halide sensors that in turn control the amount of overlapping transparent dye at each pixel. There is very little microstructure near visual resolution threshold, as found in additive systems.

Halftone color printing is the most commonly used technique in both computer printers and commercial publications. It reduces the amount of light coming to the eye by increasing the area of light-absorbing ink. White is no dot; black is 100 % dot. Most colored halftones use yellow, magenta, cyan and black dots.

Optimizing color halftone images involves many factors. They include: the colorants' spectra, the colorants' covering power, the colorants' opacity, the paper's spectra, the paper's surface, the presence of optical brighteners (fluorescent dyes in the paper), the size and shape of the dot patterns, the dot placement relative to each other, and dot overlap (Emmel, 2003). In 1937 Neugebauer developed a model to calculate color response of printing systems (Hunt, 2004). By measuring the response of a subset of possible printed colors, one can combine the interactions of all these complex variables. These measurements include: no ink, 50%, and 100% of all inks separately, and in all combinations. In other words, by printing selected samples of all-possible combinations, we can calculate accurately the complex system responses of its three-, or four-dimensional color space.

2.7.4 Colorimetric Reproductions

Changing the color medium usually makes it impossible to substitute accurately colorimetric matches for every pixel, because media have different colorimetric gamuts. If the reproduction color space gamut is different, (the rooms in the new house are different), then exact color reproduction techniques create errors that are easy to see, because parts of the original's information are left out. (All the furniture does not fit in the corresponding, but different sized room.)

One might think that the answer to successful reproduction is to just rescale all values between maxima and minima to fit the new color space dimensions. If the reproduction space is smaller that means a lower-contrast copy. However, observers prefer enhanced reproductions to accurate renditions and dislike low-contrast renditions (Fedorovskaya, et al., 1997). Near white, near black and along the high-chroma color gamut, reproductions change more slowly than do originals (lower contrast). In the middle of the color space preferred reproductions have higher contrast. These non-linear tone-scale curves (Chapter 5) help to preserve spatial information and reduce artifacts. Reproductions capture spatial color relationships of the original and render their copies as approximate visual appearances. High-quality reproduction requires both high-chroma colorants for color saturation, and carefully crafted tone-scale transformations within the color space. The difference between a good reproduction and a bad one is how well the tone-scale alters the original to synthesize combinations of RGB colorants that appear approximately the same as the original. Good reproductions transform the entire color space and use different colorimetric stimuli. Accurate reproduction of in-gamut pixels can lead to objectionable artifacts because out of gamut colors are easily recognized.

2.8 Contrast

The unambiguous definition of words is the most important tool we have for understanding ideas. The study of images is made more complex than necessary because of many conflicts in definitions. While there is no ambiguity in the language discussing the physics of light, there is great ambiguity and multiple usage of the terms in psychophysics and image reproduction. This creates confusion and misunderstanding.

One of the most confusing words is *contrast* (Table 2.1). It has distinctly different meanings in physics, psychophysics, photography, video, and digital processing. Michelson contrast is the ratio (max−min)/(max+min) (Michelson, 1927). Photographic contrast is the slope of the response function (optical density vs. log luminance). Video contrast is the gain that adjusts the luminance of the white. Wikipedia lists 11 different definitions for display contrast. Ironically, video contrast is a measure of

Table 2.1 Contrast describes measurements of the scene (physics); retinal contrast will describe the image on the retina (glare calculation); neural contrast will describe human image processing; and simultaneous contrast will describe appearances (psychophysics).

Term	Definition
Contrast	changes in amount of light in the scene
Retinal Contrast	changes in the calculated amount of light on the retina
Neural Contrast	changes caused by neural spatial image processing
Simultaneous Contrast	changes in appearance from identical scene elements

the gain which increases the luminance of whites. Video brightness is the adjustment for black level. Video brightness does not influence the luminance of whites. Photographic and video terminology have contradictory definitions.

There isn't any easy way to replace contrast with a better word. In this text we will use the following definitions.

There are many definitions of *lightness* and *brightness* used by different authors (Blakeslee et al., 2008). They all share the common goal of understanding how vision converts the quanta catch of retinal receptors to achromatic appearances. In response to different appearances from the same quant catch, 19th century psychophysics made the distinction between changes in appearances of light, and changes in appearance in the presence of white.

The CIE (Wyszecki & Stiles, 1982) defines them as the attributes of a visual sensation in which a visual stimulus is:

> *Brightness*: appears to emit more or less light.

> *Lightness*: appears to emit more or less light in proportion to that emitted by a similarly illuminated area perceived as a "white" stimulus.

The underlying issue is whether humans see illumination in a different mode than they see the reflectance of objects' surfaces. Since Helmholtz, this idea has been used as a suggested explanation for color constancy. As we will see in Chapters 21, 29, and 32, the appearance of lightness in complex scenes is a central issue of this text. We will begin this text by using a simple operational definition of lightness. We will measure the lightness of an area in a scene by asking observers to select achromatic lightness chips in the Munsell Book that has the same apparent lightness. Analogously, we will define hue and chroma as matches to the Munsell Book.

2.9 Digital Imaging

With digital imaging we can capture and display scenes effortlessly. With this added efficiency and automatic image processing, it is essential to remember the critical importance of calibration measurements of displays used in psychology experiments. Digital values for each pixel in a photograph of a scene are not equal to, nor proportional to, the array of radiances from the scene. The automatic image processing in all cameras and all displays apply complex transforms of scene radiances to achieve pleasing images. As well, digits in computer memory are not proportional to displayed radiances. One must carefully calibrate all electronic devices, over the entire 3-D color space, to understand the relationship of RGB digits and radiance.

The outermost regions of the 3-D color space, including whites, blacks, and all high-chroma colors, are often of importance and interest in psychology experiments. Printers and displays require careful calibration of near-gamut-boundary colors, because devices have the least predictable control of light stimuli for these colors. The response of a device near the gamut boundaries has very different properties than the middle of the color space.

2.10 Summary

In film photography, there is a simple relationship between the amount of light (number of photons at each wavelength) and the film's response. The same number of photons will initiate the same response everywhere in the image, for every image, every time. In psychophysics, we observe complex responses of the human visual system to the number of photons. We observe that the same number of photons do not initiate the same response in the same image (simultaneous contrast), for every image (scene dependence), and every time (adaptation). The use of photometric standards is helpful in describing stimuli and improving communication among authors. However, in human vision there is no simple, precise, determinative relationship between measured light at a pixel and appearance. It requires considerable information on the spatial content of the image to predict appearance.

2.11 References

Alpern M & Campbell FW (1963) The behaviour of the pupil during dark adaptation. *J Physiol* **165**, 5–7 P.

Blakeslee B, Reetz D, & McCourt, ME (2008) Coming to terms with lightness and brightness: Effects of stimulus configuration and instructions on brightness and lightness judgments. *J Vision* **8**(11), 3 1–14,

Emmel P (2003) Physical models for color prediction, in *Digital Color Imaging Handbook*, Sharma G ed Editor, CRCPress, Boca Raton, 173–238.

Fairchild MD (1998) *Color Appearance Models*, Addison-Wesley, Reading.

Fedorovskaya E, de Ridder H & Blommaert FJJ (1997) Chroma variations and perceived quality of color images of natural scenes, *Color Res & Appl*, **22**, 96–110.

Gegenfurtner KR, Sharp LT & Boycott BB (1999) *Color Vision: From Genes to Perception*, Cambridge University Press, Cambridge.

Green P ed (2010) *Color Management Understanding and Using ICC Profiles*, John Wiley & Sons, Ltd/IS&T, Chichester.

Hecht S, Shlaer S & Pirenne MH (1938) Energy, Quanta and Vision, *J Gen Physiol* **25**, 819.

Hersch D (2011) http://diwww.epfl.ch/w3lsp/hersch/.

Hornak JP (2002) *Encyclopedia of Imaging Science and Technology*, John Wiley & Sons, Inc, New York.

Hunt RWG. (2004) *The Reproduction of Color*, 6th ed, JohnWiley & Sons, Ltd, Chichester, 483–485.

Joly J (1895) On a Method of Photography in Natural Colours, Transactions, Royal Dublin Soc, vi, v, June 26, in *Photography in Ireland: The Nineteenth Century*, Chandler E, ed., Edmund Burke, Dublin, 115–119.

Land EH (1964) The Retinex, *Am Scientist*, **52**, 247–64.

Land EH (1977) An Introduction to Polavision, *Photographic Sci & Eng* **2**(5), 225–36.

Maxwell JC (1965) *The Scientific Papers of James Clerk Maxwell*, ed. Niven, New York: Dover 410–44.

McCann JJ (1964) Pupil Response to Light, Senior thesis, Harvard College.

McCann JJ (1998) Color Theory and Color Imaging Systems: Past, Present and Future, *J. Imaging Sci & Technol*, **42**, 70–8.

McCann JJ (2005) Rendering High-Dynamic Range Images: Algorithms that Mimic Human Vision, *Proc. AMOS Technical Conference, Maui*, 19–28.

McCann JJ, McKee SP & Taylor T (1976) Quantitative Studies in Retinex Theory, A Comparison between Theoretical Predictions and Observer Responses to Color Mondrian Experiments, *Vision Res*, **16**, 445–58.

Michelson A (1927) *Studies in Optics*, University of Chicago Press, 41.

Moroney N, Fairchild M, Hunt R, Li C, Luo M & Newman T (2002) The CIECAM02 Color Appearance Model, *IS&T/SID Color Imaging Conf*, **10**, 23–7.

Morovic J (2008) *Color Gamut Mapping*, Wiley/IS&T, Chichester.

Munsell AH (1905). Munsell® Book of Color, Munsell Color, MacBeth Division of Kollmorgen Instrument Corporation, Baltimore, MD.

Newhall SM, Nickerson D & Judd DB (1943) Final Report of the O.S.A. Subcommittee on the Spacing of the Munsell Colors, *J Opt Soc AmOSA*, **33**(7), 385–411.

Ohta N & Robertson A (2005) *Colorimetry: Fundamentals and Applications*, John Wiley & Sons, Ltd/IS&T, Chichester.

OSA (1953) Committee on Colorimetry, Optical Society of America (1953) *The Science of Color*, Crowell, New York.

Shanda J (2007) *Colorimetry: Understanding the CIE Systems*, John Wiley & Sons, Ltd/IS&T, Chichester.

Stockman A & Sharpe LT (2010) Website http://cvision.ucsd.edu/.

Wright WD (1987) A plea to Edwin Land, *Color Res & Eng*, **12**(3), 120–2.

Wyszecki G (1973) Colorimetry, in *Color Theory and Imaging Systems*, Soc Photographic Sci & Eng, Eynard R, ed., Washington, 24–49.

Wyszecki G. & Stiles WS (1982) *Colour Science: Concepts and Methods Quantitative Data and Formulae*, 2nd ed, John Wiley & Sons, Inc, New York, 486–513.

3

HDR in Natural Scenes

3.1 Topics

The range of light in the world is the product of material reflectance and scene illumination. Much of our time is spent in rooms designed to have uniform illumination and desirable spectral content. Both factors improve the color constancy of objects. If we go outside we can appreciate the range of natural light and its effects on appearance.

3.2 Appearance in HDR and Color Constancy

Almost all of us regard high-dynamic-range (HDR) imaging as a different topic from color constancy. However, the two fields of study have considerable overlap. As discussed in Chapter 2, the visual environment created by natural scenes can have very large ranges of light, because of the independence of objects' reflectances and scenes' illuminations. In an office, or school, environment we live in a relatively uniform illumination. The architects of the building worked hard to provide each of us with uniform, desirable light, with spectral content chosen carefully to avoid shifts in the appearance. We spend much of our time in relatively low-dynamic-range of light environments. Objects appear nearly constant if we move them through that environment. If we go outside, and encounter the natural environment, we find a wider range of light, more diversity of spectral content, and significant departures from uniform illumination.

A good photograph renders appearance. Ansel Adams talked about his photographs of the southwest as portraying the "Eloquent Light". Good photographs render both the material and the illumination. The photographs in Figure 3.1 were taken at Bryce Canyon, Utah. They do a reasonable job of recording what we saw, sitting on a bench for an hour. A snow storm was approaching from the south. Bryce Canyon is at 9000 feet elevation, so the clouds were close to the ground and moving very quickly. The materials in the image were quite constant (over eons). The light changed rapidly and had both sharp edges from low clouds and gradients that were less visible. Observing the scene showed the variability of appearances from changes in illumination.

The Art and Science of HDR Imaging, First Edition. John J. McCann, Alessandro Rizzi.
© 2012 John Wiley & Sons, Ltd. Published 2012 by John Wiley & Sons, Ltd.

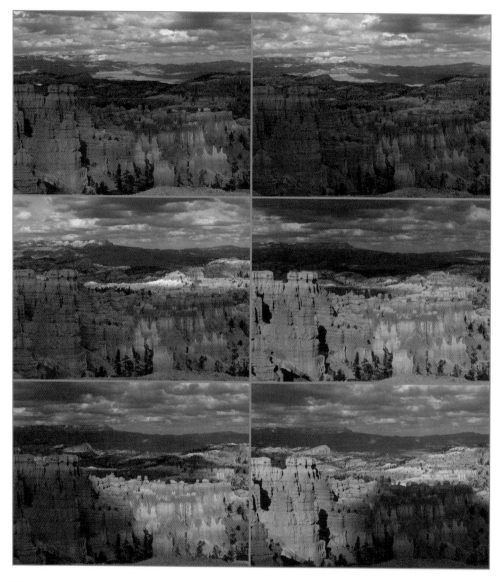

Figure 3.1 These six photographs were taken over a period of one hour. The scene is constant; only the illumination changes.

The top left photograph in Figure 3.1 shows light passing through thin clouds hitting the foreground, and bright sunlight falling on the distant bluffs in the top left. The next photograph shows the appearance a few seconds later. The overall pattern of clouds has not changed much, but the bluffs are much lighter in full sunlight and the foreground is much darker. A short time later a break in the clouds illuminates a white limestone patch in the center with full sunlight. That area has changed from a light gray to bright white. At the same time there is a dark sharp shadow on the bluffs and the foreground is much lighter.

In the fourth photograph, the dark nearly black cloud that has been moving from right to left makes the white band darker, while letting sunlight hit the foreground making it appear uniformly light. The bluffs have become nearly uniform black. With further changes in illumination, the mid-distance gets darker, and the bluffs get lighter. The foreground gets direct sunlight through a hole in the clouds on the right. It looks lighter. In the final photograph, the hole in the clouds moved making the left side lighter. The mid-distance gets lighter and the bluffs stay the same.

These observations were more dramatic and dynamic sitting on the bench at the edge of the canyon. The wind was strong enough so that the movement of the shadows was easy to observe. The appearance of each constant reflectance rock changed in slow motion. The shadows with sharp edges made the limestone surface considerably darker. The diffuse shadows that made gradients in illumination were much more difficult to detect.

Illumination is the cause of HDR imaging. As we saw in Chapter 2, the range of reflectances of materials in the world, in artists' paintings, and in a print reproduction are all 30:1 dynamic range. Uniform illumination is found in laboratories, and on a sandy beach with a cloud-free sky. However, most scenes have non-uniform illumination that dramatically increases the scene dynamic range.

Observing this scene in changing illumination raises interesting paradoxes about vision. We see that the rocks get lighter and darker as the clouds move. We know that the rocks are not changing their surface reflectance. We can easily get tangled in our words.

The rocks have constant physical reflectance surfaces.

We recognize that the rocks are constant.

We recognize that the illumination changes.

We see the appearance of the rocks changes.

What we recognize about the rocks is in conflict with what we see.

In "The Science of Color" the Committee on Colorimetry of the Optical Society of America (OSA, 1953) makes a very important distinction between sensation and perception, following the definitions introduced by the Scottish philosopher Thomas Reid in the 1820s. The Committee defines *sensation* as the "mode of mental functioning that is directly associated with stimulation of the organism". The rocks generate different sensations with changes in illumination. The rocks appear lighter and darker because we would match their appearance to different chips in the Munsell Book.

The OSA defines *perception* as the "mode of mental functioning that includes the combination of different sensations and the utilization of past experience in recognizing the objects and facts from which the present stimulation arises". The distinction centers on the roles of cognition and recognition in the perception of objects in life-like complex images. We recognize perceptions of constant rocks and variable illumination (McCann & Houston, 1983; McCann, 2000; Chapter 29).

The sensation of a rock and its perception are different. Is HDR supposed to render the accurate radiances from the scenes, the sensations it generates, or the perception of the surfaces?

Figure 3.2 shows two photographs taken at the end of the day with increasing length of shadows (top row). The bottom pair of photographs was taken the next morning slightly after sunrise (left) and an hour later, with the arrival of the snowstorm. We can all recognize that the materials of the rocks are constant. The objects in the scene do not change. Any of the reproductive media described earlier can easily reproduce the dynamic range of the rocks. None of that media can accurately reproduce the combination of the material and the dramatic changes in illumination. The range of light in these scenes is greater than the range that cameras and humans can accurately record (Section B). Nevertheless, we can see details in the shadow, at the same time as seeing structure in the clouds. Just as our visual system can render HDR sensations of the world, we can synthesize low-dynamic range renditions of visual appearances. What we see is, in fact, a synthesis of the spatial content of the scene, rather than

Figure 3.2   The photographs taken close to sunset and sunrise.

an accurate measurement of scene radiances. Successful reproductions render equivalent spatial information.

As we will see in Section B, it is not practical to capture and reproduce actual scene radiances. We need to generate the equivalent appearance of the scene. To do that we need to understand how we generate appearances.

3.3   Summary

These photographs of Bryce Canyon, Utah show the challenge of HDR imaging. They demonstrate the interplay of objects' reflectances and illumination. We see a number of different paradoxes by simply sitting in place and observing the changes in appearance.

The quanta catch on our retinas changed dramatically because the illumination changed, but we can recognize that the rocks were not changing.

The cognitive knowledge that the rocks were constant did not make the rocks appear constant. They got lighter and darker, yellower and bluer, with changes in illumination.

One approach to reproducing these HDR scenes is to accurately record and reproduce the light from the scenes, but there are limits to scene capture and reproduction.

At the other extreme, we could attempt to render the scene by calculating the surface reflectance of all the objects and remove the effects of illumination. So far this goal has not been reached for real scenes. Further, it also would be highly undesirable. If achieved, the renditions of the 10 scenes in this chapter would be identical images.

Human vision is successful at rendering these illuminations of the scene. It does it with low-dynamic-range optic nerve cells. It does not accurately report the light from the scene. It does not remove the illumination.

Appearance is the result of the spatial interactions of gradients and edges, and has little to do with isolating materials from illuminations. As we will see later in the text, the study of HDR imaging is very similar to that of color and lightness constancy. We would do very well if our HDR imaging technology could mimic what our human visual system does in generating sensations.

3.4   References

McCann J (2000) *Simultaneous Contrast and Color Constancy: Signatures of Human Image Processing*, ed. Davis, Vancouver: Oxford University Press, USA, 87–101.

McCann J & Houston K (1983) *Color Sensation, Color Perception and Mathematical Models of Color Vision, Colour Vision*, ed. Mollon & Sharpe, Academic Press, London, 535–44.

OSA Committee on Colorimetry (1953) *The Science of Color*, Optical Society of America, Washington.

4

HDR in Painting

4.1 Topics

This chapter reviews the long history of paintings of HDR scenes from the Renaissance to modern photorealism.

4.2 Introduction

As we saw in the last chapter, it is the illumination that controls the dynamic range of the scene. The reflectance range varies from maximum (white) to the minimum. Almost always, in reflected light, the minimum is a surface reflection. There is a small fraction of reflected light from an object's front surface. When it is larger than the light reflected from the underlying high-absorption colorant, the lower limit of surface reflectance is set by the surface. In general, that limit is about 3 % of the incident light. That means there are only about 1.5 log units of range controlled by objects in a scene.

Illumination can vary from no light to very high values in direct sunlight, covering many log units of range. At times, the light source can be considered a part of the scene. Paintings can successfully render HDR scenes with considerable ranges of illumination. They do this in reflective media, such as oil paints, that control only the low dynamic range of surface-limited colorants.

For many centuries painters portrayed people and objects in a style that rendered the subject without the need to reproduce the surrounding scene and lighting environment accurately. That changed in the Renaissance, first with Brunelleschi's *perspective* for rendering the geometry of the spatial environment, and later with *chiaroscuro* for rendering the high-dynamic-range illumination environment. There are many examples in painting that show how artists have succeeded in rendering high-dynamic-range (HDR) scenes in low-dynamic-range (LDR) media (McCann, 2007; McCann & Miyake, 2008; Parraman, 2010a, 2010b).

4.3 Ancient Painting

The earliest examples of painting are preserved in rock art. The Lascaux cave paintings have been dated to 13 000 to 15 000 BC. As best we can tell, these are the artistic expressions of individuals, with possible social implications (Figure 4.1).

In North America petroglyphs are dated to 3000 BC. Carvings on Newspaper Rock (Figure 4.2) began around 2000 years ago. The figures were made by abrading off the dark "desert varnish" on the

The Art and Science of HDR Imaging, First Edition. John J. McCann, Alessandro Rizzi.
© 2012 John Wiley & Sons, Ltd. Published 2012 by John Wiley & Sons, Ltd.

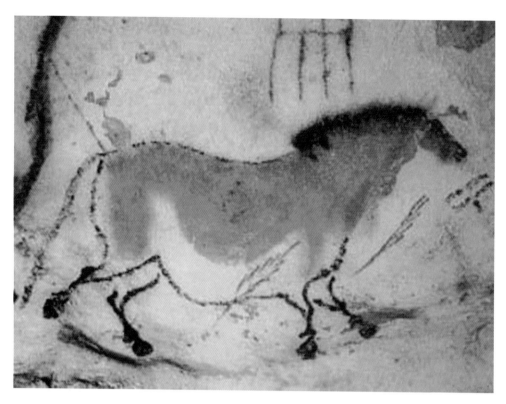

Figure 4.1   Lascaux caves painting, painted 160 centuries ago, Lascaux France.

Figure 4.2   Thirty century old Petroglyphs on "Newspaper Rock" near Canyonlands National Park, USA.

Figure 4.3 Su "Children Playing on a Winter Day", detail in Sung dynasty scroll (1130–1160), National Palace Museum, Taipei.

rock surface to reveal the lighter sandstone rocks underneath. Anasazi Indians in Chaco, New Mexico around 1000 AD used a camera-like cave with a small slit so as to project light onto a spiral petroglyph to identify summer solstice, equinox and the cycles of the moon (Solstice, 2007; Sinclair et al., 1987).

For centuries the picture makers were artists who painted them for themselves, or for noble clients. Pictures were rare; their users were few. Figure 4.3 shows a small section of a thousand year old Sung dynasty scroll (Ch'in Hsiao-I, 1986). The children are seen on a plain unpainted background. They appear to be in uniform illumination. The painting renders the objects of interest without the seemingly unimportant surroundings and illumination.

4.4 Perspective

In the Italian renaissance, art became more public. Family dynasties such as the Medici supported public art just as today benefactors support public parks and museums. This led to artists' workshops that resembled 20th century research laboratories. These workshops were well funded and well staffed and generated a considerable volume of images available to a wider range of viewers. The practice of

Figure 4.4   Giotto's "Innocent III's Dream" c1297, Basilique Assise.

portraying people and objects without an accurate rendition of the surrounding scene and lighting environment changed in the Renaissance.

Figure 4.4 show Giotto's painting in the Proto-Renaissance around 1300 AD. It illustrates Pope Innocent's dream of Saint Francis of Assisi holding up the church of San Giovanni in Laterano. This painting describes the legend, but neither renders the scene in perspective, nor with realistic illumination.

4.4.1   *Perspective in the Renaissance*

First, Brunelleschi's perspective rendered the geometry of the spatial environment. The backgrounds in paintings around the portraits began to look realistic. The scene was rendered in correct geometry.

Figure 4.5 Piero della Francesca's "Flagellation of Christ" (1455–1460), National Gallery of the Marche, Umbria.

In about 1420 Brunelleschi demonstrated the geometrical method of perspective, used today by artists, by painting the outlines of various buildings in Florence onto a mirror. When the building's outlines were continued, he noticed that all of the lines converged on the horizon line. Vasari (1972) reports he then set up a demonstration of his painting of the Baptistry of the Cathedral in Florence. He had the viewer look through a small hole on the back of the painting, facing the Baptistry. He would then set up a mirror, facing the viewer, which reflected his painting. To the viewer, the painting of the Baptistry and the Baptistry itself in the mirror were nearly indistinguishable.

Figure 4.5 shows Piero della Francesca's "Flagellation of Christ" which presents an Early Renaissance painting with both perspective drawing, and two distinct illuminations: indoor and outdoor.

4.5 Chiaroscuro

Art historians credit Leonardo da Vinci with the introduction of chiaroscuro, the painting of light and dark. His "Lady with an Ermine" (Figure 4.6) captures the illumination as well as the figures. One sees that the illumination comes from a particular direction with consistent highlights and shadows.

Michelangelo Merisi da Caravaggio's paintings, such as "The Musicians" (Figure 4.7), portray people and illumination with equal importance.

4.6 Gerritt van Honthorst (Gherardo delle Notti)

In turn Caravaggio influenced Dutch painters, among them Gerrit van Honthorst.

Figure 4.6 Leonardo da Vinci's "Lady with an Ermine", c1490, National Museum, Czartoryski Collection, Cracow.

Chiaroscuro paintings rendered the high-dynamic-range (HDR) illumination environment emitted from a single candle. Figure 4.8 shows Gerald van Honthorst's painting of Christ holding a candle. Christ has the lightest face. Joseph, further from the light, is slightly darker. The other children, progressively further from the light are progressively slightly darker.

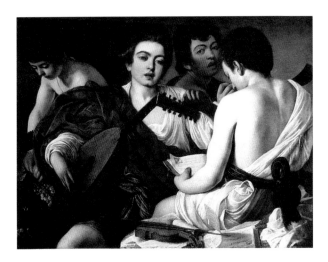

Figure 4.7   Caravaggio, "The Musicians", c1595, Metropolitan Museum of Art, New York.

Figure 4.8   Gerald van Honthorst's "The Childhood of Christ", 1620, Hermitage. Photograph © The State Hermitage Museum. Photo by Vladimir Terebenin, Leonard Kheifets, Yuri Molodkovets.

Figure 4.9   Rembrandt van Rijn, "The Company of Frans Banning Cocq and Willem van Ruytenburch," known as the "Night Watch", 1642, Rijksmuseum, Amsterdam.

4.7   Rembrandt van Vijn

Rembrandt's, "Night Watch" (1642) is an almost life size painting (363 x 437 cm) of a military company receiving orders to march (Figure 4.9). It is known for its effective use of light and shadow, and perceived motion.

4.8   John Constable

John Constable was an early 19th century English romantic painter. Although he often painted the natural scene, he almost always included churches, farmhouses and cottages, and the work of humans in its visual environment. Figure 4.10 show Salisbury Cathedral in bright sunlight, and its Bishop with wife in shade. This scene is similar to many examples of digital HDR photography today.

4.9   John Martin

As we will see in Chapter 5, this is the time period when photographers started to use different film exposures to capture HDR information. John Martin painted visualized scenes, such as "the Destruction

Figure 4.10   John Constable, "Salisbury Cathedral from the Bishop's Garden" c1825, Metropolitan Museum of Art, New York.

of Sodom and Gomorrah", (Figure 4.11) about the time that photography had become practical and popular. While photography was beginning to capture real scenes using silver halide technology, the recreation of Sodom and Gomorrah required a painter's imagination and skill. Carinna Parraman (2010b) described his paintings:

> "Martin was interested in highly apocalyptic subject matter, images included: lightening strikes, craggy hillsides, fantastic temples built into mountainous landscapes, boiling seas, shafts of light, burning clouds, fiery molten rock. The figures in the works played minor parts and were certainly overshadowed by the tempestuous landscapes."

4.10   Impressionism

The role of light is fundamental to late 19th century Impressionism. Monet, Renoir, Degas, Pissarro, Sisley, Morisot and colleagues rendered onto canvas what they saw. These artists banded together in 1874 to display their works in the "First Impressionist Exhibition", independent of the official Salon of the times (Tucker, 1998).

Figure 4.12 shows Monet's "Camille and Jean on a Hill, 1875". It is a portrait of his wife and child, the sky and clouds, the sunlight and shadow.

Figure 4.11 John Martin, "The Destruction of Sodom and Gomorrah", 1852, Laing Art Gallery, Newcastle on Tyne.

Figure 4.12 Claude Monet's "Camille and Jean on a Hill", 1875 National Gallery Washington, Mellon Collection.

Figure 4.13   Richard Estes, "Downtown – Reflections, 2001", color woodcut, Marlborough Galleries.
© Richard Estes, courtesy Marlborough Gallery, New York.

4.11   Photorealism

Richard Estes is an American painter who works from photographs to render HDR scenes in paint and
other media. His work is known for its HDR rendering and minute realistic detail. Figure 4.13 shows
a color woodcut of sunlight, shade, and reflections.

4.12   Summary

The examples of pre-Chiaroscuro painting, shown above, render the objects in the scenes. There is
minimal attention in the painting to the illumination. The painter removed the visual effects of lighting.
After Renaissance paintings introduced chiaroscuro techniques, illumination became important. In fact,
in the 17th century it became a challenge, as seen in the "Childhood of Christ". These paintings successfully rendered scenes with dynamic ranges greater than the physical dynamic range of oil paints
and prints. These images are just of few examples of HDR scenes rendered by painters in the low-
dynamic-range of reflective media. Over the centuries, chiaroscuro techniques were used in many
movements and genres of painting. These extremely successful examples show that HDR scene reproduction is possible. They also show how much the rendition of illumination adds to paintings. Imagine
for a moment what Figures 4.6 through 4.13 would look like if we removed all trace of illumination.
Imagine them as accurate renditions of the object's reflectances. Frightening, isn't it?

4.13   References

Hsiao-i C (1986) *The National Palace Museum in Photographs*, The National Palace Museum, Taipei.

McCann JJ (2007) Art, Science, and Appearance in HDR Images, *J Soc Info Display*, **15**(9), 709–19.

McCann JJ & Miyake Y (2008) The interaction of Art, Technology and Customers in Picture Making, *IEICE Transactions*, **91-A**(6), 1369–82.

Solstice (2007) http://www.solsticeproject.org/research.html.

Sinclair R, Sofaer A, McCann JJ & McCann JJ Jr (1987) *Marking of Lunar Major Standstill at the Three-Slab Site on Fajada Butte*, American Astronomical Society, 17 Annual, Austin.

Parraman C (2010a) The development of alternative colour systems for inkjet printing, PhD. thesis, School of Creative Arts, University of the West of England, Bristol. ch 2.

Parraman C (2010b) The drama of illumination: artist's approaches to the creation of HDR in paintings and prints, Proc SPIE 7527, 75270U.

Tucker P (1998) *Claude Monet: Life and Art*, Yale University Press, New Haven.

Vasari G (1972) *The Lives of the Artists*, Simon and Schuster, New York.

5

HDR in Film Photography

5.1 Topics

Examples of HDR photography date back 160 years. This chapter reviews the evolution of many approaches, including multiple exposures, and tone-scale functions in rendering HDR scenes. HDR scenes presented severe problems for early landscape photographers. Multiple exposure techniques for rendering HDR scenes go back to the earliest days of negative-positive photography. Early photography had to deal with the great range of light in the world using home-made emulsions, controlled by craft, rather than scientific measurements. This chapter covers the history of HDR photography from 1830s to the 1980s with the birth of digital photography.

5.2 Introduction

There are many excellent histories of early photography that describe the rich and varied story of writing images with light. A particularly good summary is found in Mees (1937) based on the Christmas Lectures given at the Royal Institution of Great Britain. It summarizes the history from 1727, when Schultze experimented with light-sensitive silver salts, through Thomas Wedgewood's and Sir Humphrey Davy's work on silver nitrates, and through the public and private studies of Niepce, Daguerre and Fox Talbot, including Sir John Herschel's major contribution of hypo, the silver-salt clearing agent. A detailed history, including ancient Greek references to light sensitive matter and descriptions of camera obscura, can be found in Elder and Epsteam (1932). Other histories are: Newhall (1982), Scharf (1996, 2000) and Frizot (1998).

In the late 1830s, silver halide photography took a great leap forward from the research on light sensitive material to the development of practical photographic systems. The key advance was Daguerre's discovery of silver development by mercury vapor. In 1839 Daguerre made public disclosure of his technique to the Academie des Sciences, Paris in exchange for a 6000 francs per annum for life (4000 francs for M. Niepce) from the French government.

Today's negative-positive photography descends from the 1835 experiments with Fox Talbot's Calotype process (Scharf, 1996, 2000). In the Daguerreotype the silver plate was developed to a positive image with fuming mercury. With Fox Talbot's Calotype the silver salts on paper were developed

The Art and Science of HDR Imaging, First Edition. John J. McCann, Alessandro Rizzi.
© 2012 John Wiley & Sons, Ltd. Published 2012 by John Wiley & Sons, Ltd.

in a water bath to a negative image. This negative, when printed on a second AgCl paper, made the positive print when developed. In 1851 Scott Archer described a wet collodion process with which photographers made their silver halide negatives at the time of exposure. Landscape photographers carried their darkrooms with them to process negatives in the field. By the 1850s the fascination with making pictures led to societies for the discussion of techniques. An excellent example is the 1854 *Journal of the Photographic Society of London* containing The Transactions of the Society and a general Record of Photographic Art *and Science* (Henfrey, 1976). It, along with many other documents in the collection of the Royal Photographic Society of Great Britain, provides a time capsule view of life in the 1850s.

5.3   Multiple Exposures in the 1850s

Photographers made images of outdoor scenes as soon as it was possible, and well before it was practical. When confronted with the extremely large dynamic range of outdoor scenes, they improvised with multiple exposures.

5.3.1   Edouard Baldus

Baldus was trained as a painter, but became an early member of Benito de Monfort's Société de Héliographique in France in 1851. He extended Talbot's calotype process by replacing wax with gelatin in the negative. He made the print shown in Figure 5.1 in 1853 using 10 negatives (Flukinger, 2010).

Figure 5.1   Baldus's "Cloisters of the Church of St. Trophime, Arles", 1853, University of Texas, Austin, made from 10 paper negatives.

5.4 HP Robinson

Henry Peach Robinson was a remarkable photographer, teacher and role model. Everyone knows Lewis Carroll's book, *Alice in Wonderland*. Fewer people are familiar with him as "The Reverend Charles Lutwidge Dodgson", Anglican clergyman and accomplished lecturer in mathematics at Oxford (1856–81). Still fewer are aware of his portrait photography. In Gernsheim's (1950) book *Lewis Carroll Photographer* he reports that Lewis Carroll admired the photographs of HP Robinson.

Robinson's work is remarkable because it remains of great interest today for two completely different reasons. First, it competed directly with painters in the creation of fine art. It went beyond scene reproduction to creating images designed to evoke an emotional response. Second, to be truly competitive with artist's paintings, it had to overcome all the technical limitations of image rendering. It had to solve all the problems of today's HDR imaging.

Figure 5.2 shows HP Robinson's 1858 composite photographic print "Fading Away" made using five differently exposed glass negatives (Mulligan and Wooters,1999). This dramatic still life was staged using actors.

This multiple negative process is described in detail in the Robinson and Abney (1881) book, *The Art and Practice of Silver Printing*. The negative capture and positive print are very important to this process. When the photographer developed his negative of the scene exposed for the areas with the most light, the well-exposed areas were darkened with developed silver. The unexposed areas in the negative, the parts of the scene with much less light, were clear glass. That meant that the photographer could take a second, longer exposure to record details in the shadows, develop it and superimpose the two negatives. This is just like layers in Photoshop®, only 150 years earlier. The shadow details were added through the nearly clear glass of the first negative. This sandwiching of negatives combined with cutting and juxtaposing different negatives of the same scene produced HDR renditions that are as good as today's digital techniques. With insensitive AgX print paper, these prints were made in sunlight, a good source of collimated light.

Figure 5.2 HP Robinson's "Fading Away", 1858, made from five combined negatives.

Figure 5.3 HP Robinson's "When the Day's Work is Done", 1877, Princeton University Art Museum, made from six negatives and three exposures.

Figure 5.3 shows HP Robinson's 1877 print "When the Day's Work is Done". It is the combination of six negatives made by six separate camera exposures. The process begins with a rough sketch, followed by a detailed actual size (32 x 22 inches) sketch of the final image. Two separate photographs were made of the left side (white wall), and the right side (black shadow wall). The glass negatives were scored with a diamond, cut and butted together and mounted to a larger clear glass plate. Robinson describes this combination of two negatives on glass as "one large negative of the interior of the cottage, into which it would be comparatively easy to put almost anything". The third negative printed contained the old man, the table and chair with the matting under his feet. The man was photographed against a black background. The edge of the negative ran along the square back of the woman's chair and up the wall. On the other side the joint runs along the table down the leg and across the floor at the edge between the carpets. The fourth photograph was the old lady. The fifth photograph was the group of baskets in the right in the corner, and the sixth was the view of the village through the window. The old man and the window were mounted on a second large glass plate. The old woman and the corner baskets were mounted on a third glass plate. Robinson used marks on the glass to register the three combined negative plates.

HP Robinson used photographic images to generate emotions. These photographs were intended to be used as fine-art paintings were used in Victorian Britain. The intent was not to capture the real scene and reproduce it, rather the intent was to synthesize images that generate feelings of sadness, or nostalgia.

5.5 Hurter and Driffield-Scientific Calibration of AgX Film Sensitivity

Robinson's techniques were empirical. In the 1870s and 1880s Hurter and Driffield established the field of photographic sensitometry (Hurter & Driffield, 1974). They measured the response of the AgX film

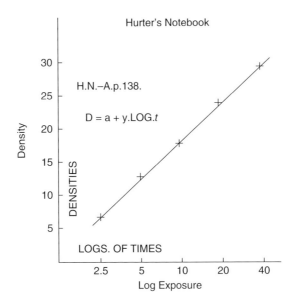

Figure 5.4 Driffield's plot of log exposure times vs. density.

(density) to exposure to light. Hurter's notebook B plots log time of exposure vs. *optical density* and finds a straight line fit for the data. He wrote:

"If the experiments be made in which the times of exposure is prolonged in geometrical progression, the density of the plate is found to be in arithmetic progression" (Hurter & Driffield, 1974).

Figure 5.4 is a recreation of Hurter's plot of log exposure times vs optical density. This is the foundation of the science of photography. It showed that the log of the amount of light (number of photons) had a linear relationship to the log transmission of the negative (density).

The horizontal axis plots the time (logarithmically spaced) covering a range of exposures of 16:1. This is a negative image so with increased exposure the density increases (transmission decreases). Whites are bottom left; and blacks are top right. The slope of the curve *y* varies with the chemical composition of film and developer, as well as the time of development. This result of the analysis of response to light is the basis of today's *camera response functions* and *input/output lookup tables (LUTS)*.

This plot was a watershed event in the history of the science of photography and the engineering of picture making. The work that went into making instruments that could control the light falling on the film was rewarded by the result that the film's response was now proven to be the logarithm of the exposing light. In 1888, this was a remarkable accomplishment.

Hurter and Driffield's (1890) first publication describes the characterization of silver halide films as the plots of density vs. exposure. This work was quickly recognized and became known as the "*H&D curve*". English film manufacture's marked their product with H&D ratings. Hurter and Driffield wrote a long series of research reports starting in 1881, and continuing through 1903 that studied all aspects of film sensitometry. In addition to measuring films, these journal articles include experimentation with instrumentation, techniques for measuring density, and the accuracy of grease spot photometers. The final paper by Driffield, after Hurter's death, entitled "The Hurter and Driffield System: Being a Brief Account of their Photo-Chemical Investigations and Method of Speed Determination" (1903) is fascinating for two very different reasons. First, it is beautiful science from a time in which scientists had to create their own tools of analysis. Second, it is a time capsule for imaging. It makes apparent the remarkable progress in imaging over the past 100 years.

5.6   Sheppard and Mees

Sir William de W Abney's (1900) book *Instruction in Photography* became a popular reference. It was originally written in the mid 1880s with a dozen new editions over the next 20 years. Abney describes a number of measurements recording film response data for lantern slide transmission vs. exposure. In 1900 CEK Mees became an undergraduate at University College London. He met another student SE Sheppard who shared many of his interests. The two of them worked under Sir William Ramsay, a physical chemist (Mees, 1956). They were both interested in photography, and found that equations in Captain Abney's Instructions in Photography "did not add up". Training as physical chemists they wanted to know the kinetics of the AgX development reaction. They then read Hurter and Driffield (1890) *Photochemical Investigations and a New Method of Determination of Sensitivities of Photographic Plates*. They had found a "model for attack on the nature of the photographic process" (Mees, 1956). They repeated Hurter and Driffields's experiments with newer apparatus including an acetylene burner as a standard light source. Mees modestly described this undergraduate research by saying: "We made little progress beyond what Hurter and Driffield had put on the record" (Mees, 1956). Nevertheless it earned them their B.Sc. degree and first scientific article, published in the Photographic Journal at the age of 21. Sheppard and Mees continued their collaboration earning their doctorates in 1906, and the publication of this work is found in a book *Investigations of the Theory of the Photographic Process* (Sheppard and Mees, 1907).

5.7   19th Century – Professional Amateur Photography

The frontispiece of Mees's (1937) book *Photography* showed an iconic image "Coating a gelatin dry plate by hand". It illustrates beautifully one of many steps needed to make a picture; many of these steps were performed in complete darkness. Nineteenth century photography started with a few scientists experimenting with light changing matter. It ended with the beginnings of the massive industrialization of photography. Over most of the 18th century photography was developed by thousands of individuals and small groups who painstakingly made each photograph from start to finish. Photography required the commitment of time, and the investment in facilities and equipment, beyond just cameras and lenses. By the end of the century there were innumerable small businesses, in unique niches, finding independent solutions to common problems (Gernsheim, 1950).

5.8   20th Century – Corporate Photography

In 1900 Kodak introduced the Brownie Camera. It cost $1 ($0.15 for film). The ads read "You press the Button, We do the rest" (Mees, 1937). It was the corporate organization that put simultaneous advertisements in every US newspaper that changed photography. It universalized the photographer. Up until then photography was a small industry occupation, or a serious avocation. The new "Brownie" photographer clicked the shutter, mailed the camera to Kodak, and received by mail their photos, and their camera loaded with fresh negatives. All the dangerous chemicals, and the bulky dark tents were taken away by a large corporation. Eastman introduced convenience into photography. In 1905 Kodak published a 190 page, hard-covered book, "The Modern Way in Picture Making" (Eastman Kodak, 1905). In the preface, it said,

"In its compilation we have endeavored to cover fully and clearly every point on which he should have information. With equal care we have avoided useless discussion of theory and have given no space to topics that would not appeal to those who take pictures for the love of it."

EASTMAN KODAK CO.

This states quite clearly the change from 19th century skilled amateur hobbyist, interested in understanding the processes, to the new 20th century consumer, who was assumed to be disinterested. Just after the introduction of the Brownie, George Eastman went on a different quest, an industrial research lab. It took several years, but in 1912 he persuaded CEK Mees to move from London to Rochester, and become Kodak's Director of Research. Mees's condition was that Eastman buy his present employer Wratten and Wainwright (Mees, 1956). Mees, although regarded by some as a chemist, became a charter member of the Optical Society of America in 1916.

5.9   20th Century Control of Dynamic Range

In Mees's (1920) *The Fundamentals of Photography*, he shows an example of a print made from multiple negatives with different exposures. This example of multiple exposures was not so much an artistic technique, as it was a demonstration of an improvement in image quality. The example showed that it was possible to improve a photograph using multiple exposures, but that process required great skill and was very time consuming. It did not fit the Kodak goal of producing the best image for the customer. Mees turned to the science of silver halide response and the engineering of film to find a better solution.

5.9.1   The Tone Scale Curve

In Mees (1920) *The Fundamentals of Photography* (first edition), he described the reproduction of Light and Shade in (Chapter VII). The description starts by showing the reproduction of a cube using two, three, four, five and six tones (Figure 5.5). Mees used "tones" to explain in simple terms the progression from white to black in a photo. His chapter gives a detailed discussion of how the shape of the photographic tone-scale curve affects the photographic print.

Mees introduced the term "tone-scale" to avoid scientific jargon. Figure 5.5 shows illustrations from Mees's chapter "Reproduction of Light and Shade". Although Mees was exemplary in publishing his fine research in scientific journals, and in supporting published industrial research, he followed the Kodak philosophy of instructing a new kind of customer. He used tone-reproduction curve to replace the term H&D curve in his and subsequent Kodak publications. In fact, the term *tone scale characteristic curve* is used today in digital high-dynamic-range imaging. In today's terminology, a *tone scale*

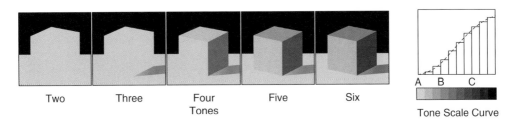

Figure 5.5   Mees's tone scale response to different intensities of light.

curve is the equivalent of a one-dimensional *lookup table* (*LUT*) that converts light on the camera's image plane to the camera's output digits. It modifies the quanta catch to the desired camera response.

5.9.2   The Physics Behind the Tone Scale Curve

Mees, as director of Research at Kodak for half a century, led the development of negative films that can capture a greater range of luminances than possible on camera image planes for the vast majority of scenes (McCann & Rizzi, 2007). This film design was the result of extensive photographic research (Mees, 1956, 1961, Mees & James 1966).

When the Kodak Research Laboratory opened in 1913 it had a staff of 20 people. They studied optics in physics and colloidal, physical, and organic chemistry. PG Nutting, from the National Bureau of Standards, Washington, was Kodak's first chief physicist. He brought with him LA Jones, who succeeded him, and was chief physicist from 1917 to 1954.

Mees (1961) describes LA Jones' work on both physics and psychophysics that defined the optimal *tone reproduction curve* for films used in the second half of the 20th century.

"The theory of photographic tone reproduction attracted the interest of Jones in the early years of the Kodak Research Laboratories, and in 1920 he stated the concepts on which the theory of photographic reproduction is based. According to Jones, the object of the photographic process is to effect a reproduction in which the scale of tones in the finished print corresponds as closely as possible to the subject, or in which the scale of tones is modified in a known manner, either from necessity, owing to the limitations of the process, or in order to produce the desired effect. The primary object in making a photograph is not in most cases the exact reproduction of brightness and contrast of the original but rather the production on a flat surface of variations in reflecting power such that the impression obtained when this surface is viewed is as nearly as possible identical to that produced in the mind of the observer when viewing the object photographed. . . . It is clearly seen that a differentiation must be made between the two phases of the problem, which, for convenience, may be termed objective and subjective phases. The former deals with the reproduction of the objective brightness and contrast of the original, these factors being measurable by the usual method of photometry. The latter, dealing with the subjective evaluation of physical brightness and contrast, requires the application of those psychophysical laws which determine the subjective impressions produced in the observer." (Mees, 1961, page 71)

The Kodak Research Laboratory set out to continue the objective described by Driffield in 1903. Hurter and Driffield felt that the art in photography ended at the moment the light hit the film. Their goal was to raise technical photography from an empirical art to a quantitative science (Mees, 1961, page 70). However, by 1920 the Kodak lab's measurements showed that the physics of light and the chemistry of film were not sufficient to make good reproductions of all scenes. The psychophysics of appearance played an important role in making pictures. Accurate reproductions of the scene radiances were not the goal of photography. Jones and Nelson's study of prints concluded that exact, or even proportional reproduction, is not essential to achieving excellent photographic results. They said that exact reproduction was desirable in the mid-tone region and that lower slope reproduction was desirable in the highlight and shadow regions (Mees, 1961, page 74).

5.9.3   Jones and Condit – Range of Light in Scenes

The Jones and Condit (1941) study of 128 outdoor scenes provided two important benchmarks in HDR imaging. First, it compared photographic images and spot photometer (Luckiesh and Taylor, 1937) measurements from scenes. It characterized these images into three homogeneous groups of scene: Illumination characteristics (sunlit, haze, light clouds, heavy clouds), Viewing aspects (front, cross and back lighting) and Spatial distributions (distant, remote, near-by, close-up). In all there are 17 classifications. The minimum dynamic range ratio was 27:1; the maximum was 750:1; and the average was 160:1. These measurements did not include specular highlights.

Second, and more relevant here, Jones and Condit devoted a significant portion of the paper to the careful analysis of flare in the image falling on the camera's image plane. They calculated a flare factor for each image. Intentionally, they used a Bausch and Lomb Vila Protar 8-element lens cemented into two-components in a Speed Graphic camera with a 5 x 7 inch image size to reduce flare. They explained that small image sizes in typical cameras (by 1940s standards) had a much larger glare problem. By comparison, using 1/4 inch electronic sensors (cell phones), and 9-element zoom lenses (digital cameras) generates even more flare than measured in the Jones and Condit study.

Jones and Condit also measured the film's limit of response to light. At some maximal exposure the film stops getting darker with increases in exposure. As well, at some minimal exposure the film stops getting less dense with decreases in exposure. The limit of film response to no light exposure is called the fog level of the film. This is the equivalent of the various noise limits in blacks in CCD and CMOS sensors. Jones and Condit showed that the film's fog-limit range was significantly greater, than the camera flare-limit range. Although many papers discuss digital camera noise limits, few discuss flare limits. Flare, not sensor signal-to-noise ratios, determines the limit of usable dynamic range in cameras.

Over the years the science of silver-halide imaging improved rapidly. Mees established standards for high-dynamic range image capture on the negative, and high-slope rendering on prints. Negatives are designed to capture all the information in any scene. The negative response function changes very slowly with change in exposure (low-slope film). This property translates into capturing a wide range of scene luminances. Further, it relaxes the requirements for cameras to make accurate film exposures. Once the scene is captured and the negative is developed, the final print can be made under optimal conditions at the photofinishing facility. The print paper has a rapid S-shaped nonlinear response to light (high slope). The resulting positive print is higher in contrast than the original scene. The loss of scene detail in shadows and highlights occurs in printing the negative onto high-contrast print paper.

5.9.4   Color Film

Jones's work on the dynamic range of scenes led to a single *tone reproduction curve* for color prints. It was used in all manufacturers' color films for the second half of the 20th century. Innumerable experiments in measuring users' print preferences led the massive amateur color print market to use a single tone-scale-system response. Even digital camera/printer systems mimicked this function (McCann, 1996). It is important to note that this *tone scale function* is not slope 1.0. It does not reproduce the scene accurately. It compresses the luminances in both whites and blacks, enhances the mid-tones (increased color saturation) and renders only some dark edges and light skin-tone details accurately with slope 1.0. This universal tone-reproduction curve, for all scenes, was very successful. It became the standard for the trillions of color prints made in the second half of the 20th century.

5.9.5   LA Jones

LA Jones made many contributions to vision science and optics, as well as to photography. In 1917 LA Jones studied the visibility threshold of ships on the horizon, in response to a request from the US Navy (Mees, 1961). The best color for the northern portion of the danger zone from German submarines was omega gray (a grayish blue-green reflecting power of 40%). This study led to the adoption of *omega gray paint*, still used today, to minimize the visual contrast of navy ships. He played a key role on an advisory committee at the University of Rochester developing their center for optics. In 1927, Jones, along with TR Wilkin (University of Rochester) and WB Rayton (Bausch & Lomb), drafted the original proposal and curriculum of the Institute of Applied Optics, University of Rochester.

Jones was on an Optical Society of America Committee on Colorimetry (1922), chaired by LT Troland, 1922). In 1933, he was appointed the Chairman of the group and asked to update recent progress. They published a preliminary report (Jones, 1937). The term *luminance* was introduced by this report. Jones (OSA, 1953) wrote:

> "Of particular interest was the Committee's decision to recommend that the word 'luminosity' be substituted for 'visibility' with reference to the ratio of luminous flux to radiant flux. . . . A somewhat more radical change, but in the opinion of the Committee no less desirable, was the reservation of the term 'brightness' as the name for the sensory attribute correlated with the photometric quantity to which is now assigned the term 'luminance'."

The final report was published as "OSA, The Science of Color" (1953), 20 years after the formation of the committee. In 1943 Jones received the OSA Ives Medal.

5.9.6 Color Measurement vs. Color Photography

Kodak and Fuji make two classes of positive transparency films. Transparencies are simpler to analyze because the exposed film becomes the finished image. One Kodak film is for photography [(Ektachrome E200•E-28 (Kodak, 2005)], the other is for copying photographic transparencies [Ektachrome EDUPE • E-2529 (Kodak, 2004)]. If we plot the green record log exposure vs. optical density for these two films we see two very different responses to light. Figure 5.6 (left) plots the response to light for Ektachrome Slide Dupe film EDUPE in red; and Ektachrome camera film E200 in blue. The camera film curve shows the S-shaped film response designed for making pictures that customers like. The camera film is more sensitive to light (faster) because the blue curve is shifted to the left with less lux-seconds of exposure. Cameras for this film can use smaller apertures and shorter exposure times for better performance. The duplication film (red) is slower and has a straight-line response.

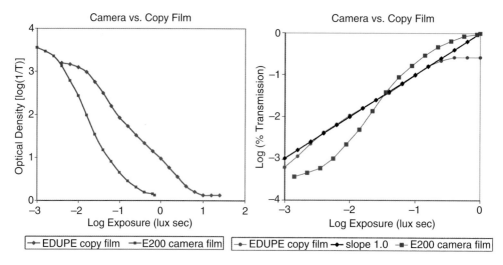

Figure 5.6 (left) The green response to light for two Kodak Ektachrome films: camera use (blue) and copy (red); (right) replots the data normalized to middle gray.

Figure 5.6 (right) plots the same data on a more conventional log scale used in photography today. It also add a slope 1.0 curve (black) that illustrates a perfect reproduction over a range of 3.0 log units. Whites are found in the top right; blacks in the bottom left. The curves have been normalized so that their middle grays (18 % transmission) were superimposed.

Obviously, copy film attempts to be an exact reproduction. The physical proof of a perfect copy is a measurement of slope 1.0 plot of original optical density equal to copy's optical density. Over its entire dynamic range of 3.0 log units the Ektachrome EDUPE (red curve), has a response very close to slope 1.0. It is an excellent reproduction of the scene, as required of a film designed to make copies that are accurate reproductions of originals.

The S-shaped camera film curve (Figure 5.6 right-blue) has the same 3.0 log unit exposure dynamic range. Only in mid tones, does it have a straight line (linear) function with a range of 1.2 log units. At higher and lower exposures the response is non-linear, passing through slope 1.0 to a low-slope reproduction of white and black regions. The low-slope regions are needed to extend the range to 3.0 log units, and to render scene details near white and black. The low-slope function compresses the scene's contrast in its rendition.

The high-slope mid-tone region expands the scene's contrast. The differences between grays are amplified. That also means the chroma of colored objects is amplified. Gray objects generate the same response to light in each of the RGB color separations. Colored objects have different response in different separations. For red objects, the R separation has more response than the G separation, etc. High-slope response increases the differences in the color separation records, thus increasing chroma.

With transparency films the 3.0 dynamic range of sensitivity to light is matched by the 3.0 log unit range of dye density. That is not the case for prints. The surface reflections (Chapter 2) limit the range to roughly 1.5 log units in normal viewing conditions. For prints the problem becomes rendering the more than 3.0 log unit world in a 1.5 log unit media. The trillions of color prints made in the past 60 years used the S-shaped tone scale curve to make this HDR transformation. Figure 5.7 illustrates how camera films render HDR scenes using compression at the extreme exposures and expansion of mid-tones.

The plots in Figures 5.6 and 5.7 show that commercial camera films are not designed to reproduce scenes exactly. Instead, they are designed according to the Mees-Jones theory that used pixel-based *tone reproduction curves* to make the pictures that people want. The corollary is that people do not want accurate reproductions.

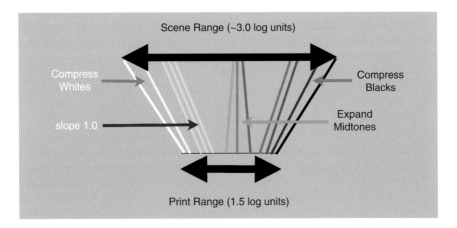

Figure 5.7 How the S-shaped camera film renders a 3.0 log unit scene. The highlights and shadows are compressed while the mid-tones are expanded.

5.9.7   HDR Pseudocolor Measurement – Wyckoff

Charles Wyckoff (1961, 1962) developed a multi-layered color film with different ASA panchromatic sensitivities (speeds) in each set of color forming layers. The film was used to photograph the detonation of nuclear explosions. The bottom, yellow color separation, layer had ASA = 2 and the top, cyan layer ASA = 600. This made a single exposure pseudocolor film able to measure a dynamic range of 8 log units.

5.10   Other Silver-Halide Stories

This brief chapter barely covers the story of HDR aspects of silver-halide photography. There are many other stories that can be found in the literature on the development of color photography (Wall, 1925; Friedman, 1944; Mees, 1961). There are optical masking systems in color negatives that correct for unwanted spectral absorptions. There are chemical chroma-enhancing systems built into the color specific film layers, so that colored objects have a different H&D curve than gray ones. These interimage effects are caused by *Development Inhibitor Release* (DIR) couplers (Krause, 1989). There are also Anchimeric (DIAR) versions that affect more distant regions because they have a built-in time delay for spatial diffusion. These compounds affect the shape of the dye cloud around a developed silver grain, and enhance image sharpness (Jarvis, 1984). Although these spatial reactions have a large effect on granularity and apparent sharpness, the actual spatial distances are a very small fraction of the image's dimensions.

Instant Photography is another parallel track of image formation chemistry (Land, 1977; Land et al, 1977; Walworth & Mervis, 1989; Land, 1993). Edwin Land extended considerably George Eastman's idea that the photographer should take the picture and photographic company should do the rest. He delivered the picture instantly. He put the film processing factory in the camera. Charles and Ray Eames (1972) made a documentary about Instant Photography that can be found on the internet.

Land played a top secret role in the development of the first Charge Coupled Devices (CCD's) at the Bell Labs. It was developed so that electronics could replace film in satellite cameras for aerial recognizance. Land was a Special Advisor for Foreign Intelligence to Presidents Eisenhower, Kennedy, and Johnson, and played a key role in the research, development and implementation of satellite imaging.

Land understood the limits of systems with a fixed H&D curve. There was no choice in instant chemical photography; the film's response function to light was fixed in the factory. It could not respond to scene content. Land was quick to realize that vision was different (Chapter 26). He became fascinated with color constancy (Chapter 28), and models for calculating apparent lightness (Chapter 32). Land's interest in electronic imaging and the application of models of vision to scene rendition has outlived Polaroid, his company that made Instant Photography.

5.11   Summary

This chapter compresses the 16 decades of HDR imaging photography. At first films had very limited dynamic ranges, so multiple exposures were necessary. With intensive research dynamic ranges increased and glare became the limit of the range of luminances that can be measured accurately by a camera. For decades, black & white and color negatives were designed to capture the entire range of light falling on cameras' image planes from almost every scene using single exposures. Multiple exposures were unnecessary. The problem was not scene capture, rather scene rendering on the print.

As early as 1920, theories of photographic reproduction used a balance of accurate physical reproduction of light and subjective substitution of equivalent, or desirable appearances. Jones showed that the range of scenes far exceeded the range of prints. He showed that the actual reproduction of the original was not important. Rather, the photograph should synthesize a desirable image of the original. That apparent image had to have different physical measurements of light coming to the eye. The only tool available to film photography was to control the tone scale curve. Since film could not respond in a different manner to different widely-spaced portions of an image, it had to use the best compromise – a universal tone reproduction curve.

5.12   References

Abney W (1900) *Instruction in Photography*, 10th ed, Sampson Low, Marston & Co, London.

Eames C & Eames R (1972) SX-70, http://www.youtube.com/watch?v=5jaiq_ZZ_eM.

Eastman Kodak (1905) *The Modern Way in Picture Making*, Eastman Kodak Co, Rochester.

Elder J & Epsteam E (1932) *History of Photography*, 4th ed., Dover Publications Inc, New York.

Flukinger R (2010) *The Gernsheim Collection: Harry Ransom Center*, University of Texas Press, Austin.

Friedman JS (1944) *History of Color Photography, The* American Photo Publishing Co, Boston.

Frizot M (1998) *A New History of Photography*, Konemann, Koln.

Gernsheim H (1950) *Lewis Carroll Photographer*, Chanticleer Press Inc, New York.

Henfrey A (1976) *The Journal of the Photographic Society of London: Containing The Transactions of the Society and a General Record of Photographic Art and Science*, **1** (facsimile), Argus Press, Watford Herts.

Hurter F & Driffield V (1890) Photochemical Investigations and a New Method of Determination of Sensitivities of Photographic Plates, *J Soc of Chem Industry*, 31ST May.

Hurter F & Driffield V (1903) The Hurter and Driffield System: Being a Brief Account of their Photo-Chemical Investigations and Method of Speed Determination, Photo. *Miniature*, 56.

Hurter F & Driffield V (1974) *The Photographic Researches of Ferdinand Hurter & Vero C. Driffield*, Ferguson WB ed, Morgan and Morgan, Dobbs Ferry.

Jarvis JR (1984) The Calculation of Sharpness Parameters for Colour Negative Materials Incorporating DIR Coupler, *Proc RPS, Symposium Photo & Electr Image Quality*, 10–20.

Jones LA (1937) Colorimetry: Preliminary Draft of a Report on Nomenclature and Definitions, *J Opt Soc Am*, **27**(6), 207–11.

Jones LA (1944) Psychophysics and Photography (Ives Medal Address), *J Opt Soc Am*, **34**(2), 66–88.

Jones L A & Condit H (1941) The Brightness Scale of Exterior Scenes and the Computation of Correct Photographic Exposure, *J Opt Soc Am*, **31**, 651–78.

Kodak (2004) http://www.kodak.com/global/en/professional/support/techPubs/e130/e130.pdf.

Kodak (2005) http://www.kodak.com/global/en/professional/support/techPubs/e28/e28.pdf.

Krause P (1989) *Color Photography, in Neblette's Imaging Processes and Materials*, 8th ed, Sturge J, Walworth V & Shepp A, Van Nostrand Reinhold, New York.

Land EH (1977) An Introduction to Polavision, *Photographic Sci & Eng* **2**(5), 225–36.

Land EH (1993) *Edwin H. Land's Essays*, McCann M ed, IS&T, Springfield.

Land EH, Rogers HG & Walworth VK (1977) One-Step Photography in *Nablette's Handbook of Photography and Reprography*, 7th ed., 258–330.

Luckiesh M & Taylor A (1937) A Brightness Meter, *J Opt Soc Am*, **27**, 132.

McCann J (1996) Color Imaging Systems and Color Theory: Past, Present and Future, *Proc. of SPIE*, **3299**, 36–46.

McCann J & Rizzi A (2007) Camera and visual veiling glare in HDR images, *J Soc Info Display*, **5**(9), 721–30.

Mees C (1920) *The Fundamentals of Photography*, The Macmillian Co, New York.

Mees C (1937) *Photography*, The Macmillian Co, New York.

Mees C (1956) An Address to the Senior Staff of the Kodak Research Laboratories, Rochester: Kodak Research Laboratory.

Mees C (1961) *From Dry Plates to Ektachrome Film: a Story of Photographic Research*, Ziff-Davis Pub, New York.

Mees C & James T (1966) *The Theory of the Photographic Process*, 3rd ed., Macmillan, New York.

Mulligan T & Wooters D (1999) *Photography from 1839 to Today*, George Eastman House, Rochester, NY.

Newhall B (1982) *The History of Photography*, Little Brown & Co, Boston.

OSA Committee on Colorimetry (1922) Report of Committee on Colorimetry for 1921–22, *JOSA & Review of Scientific Instruments*, **6**, 527.

OSA (1953) *The Science of Color* L.A, Jones Chair, Crowell, New York.

Robinson HP & Abney R (1881) *The Arts and Practice of Silver Printing, The American Edition*, E & H.T. Anthony & Co, New York.

Scharf L (1996) *Records of the Dawn of Photography: Talbot's Notebooks P&Q*, University of Cambridge Press, Cambridge.

Scharf L (2000) *The Photographic Art of William Henry Talbot*, Princeton University Press, Princeton.

Sheppard C & Mees C (1907) *Investigations of the Theory of the Photographic Process*, Longmans, Green, and Co, London.

Troland LT (1922) Report of Committee on Colorimetry for 1920–21, *J Opt Soc Am*, **6**(6), 527–91.

Wall EJ (1925) *The History of Three-Color Photography*, American Photographic Publishing, Boston.

Walworth VK & Mervis SH (1989) Instant Photography and Related Processes, in *Neblette's Imaging Processes and Materials*, 8th ed, Sturge J, Walworth V & Shepp A, Van Nostrand Reinhold, New York, 181–225.

Wyckoff C (1961) An Experimental Extended Response Film, Tech Rep No. B31, Edgerton Germeshauser and Grier, Inc, Boston.

Wyckoff C (1962) An Experimental Extended Response Film, SPIE Newsletter.

6

The Ansel Adams Zone System

6.1 Topics

*Ansel Adams' Zone System integrated image science with the practical needs of creative pho-
tographers. While amateurs were quite happy with the compromises found in commercial
films, artists needed tools to control the aesthetic content of their images. Adams' technique
began with a complete understanding of the tone scale characteristics of film and processing;
then incorporated visualization of the final print before exposure; and added spatial image
processing to compress the world's great dynamic range into the limited range of prints.
Adams' teaching of the Zone System, illustrated by his extraordinary images, is a high point
in HDR imaging.*

6.2 Introduction

We tend to think of everyday, consumer, silver-halide photography as a fixed window of scene capture
with a limited, standard range of response. This description of photography was certainly true, between
1950 and 2000, for instant films and negatives processed at the drugstore. These systems had fixed
dynamic range and fixed tone-scale response to light. All pixels in the film have the same response to
light, so the same light exposure to different pixels was rendered as the same film density.

Ansel Adams, along with Fred Archer, formulated the Zone System, starting about 1940. What
Adams called the *Zone Scale,* in the first step of his process, is very similar to what Mees and Jones
called the *Tone Scale*. It was earlier than the trillions of consumer photos in the second half of the 20th
century, yet it was more sophisticated than today's digital techniques. This chapter describes the chemical mechanisms of the zone system in the parlance of digital image processing. It will describe the
Zone System's chemical techniques for image synthesis. It also discusses dodging and burning techniques to fit the HDR scene into the LDR print.

6.3 Compressing the HDR World into the LDR Print

If you measure the range of light coming from an Ansel Adams print on a wall in a room, you get a
very small ratio of maximum to minimum. For example, the luminance of the white clouds in the center

The Art and Science of HDR Imaging, First Edition. John J. McCann, Alessandro Rizzi.
© 2012 John Wiley & Sons, Ltd. Published 2012 by John Wiley & Sons, Ltd.

of "Moonrise at Hernandez" is 17.1 cd/m^2; and the black sky is 0.15 cd/m^2. It has a dynamic range of only 114:1. Clearly, an Ansel Adams print does not reproduce scene luminances, pixel by pixel. It renders the artist's intent of what he saw when looking at the scene. How did he do it?

For Ansel Adams there was an orderly and disciplined approach to making photographs. (Adams, 1967, 1981, 1983a, 1957, 1956, 1978, 1983b). First, he visualized the final print that the scene will eventually produce. Second, he measured the scene's dynamic range and the intermediate light coming from key features. Ansel used these scene luminances to determine the technique for capturing the scene information in the negative. Third, he used dodging and burning in exposing the print. He used this manual spatial image processing to modify the scene information captured in the negative, so as to generate the desired rendition consistent with his visualization.

6.4   Visualization

In Ansel's lectures and in his book *Examples: The Making of 40 Photographs* (Adams, 1983b), he described his first use of visualization in making the image "Monolith, The Face of Half Dome", 1927. He had hiked halfway up the plateau above Yosemite Valley with his wife and three friends, carrying his . . . "Korona 6 1/2 × 8 1/2 view camera with two lenses, two filters, a rather heavy wooden tripod, and twelve Wratten Panchromatic glass plates." (Adams, 1983b) The left image in Figure 6.1, after one of Ansel's lecture slides, shows the image taken with a yellow filter. After exposing that negative, Ansel had one negative left. He stopped and visualized how he expected the print to look with the Half Dome against the sky. He "visualized" that the sky photographed with the yellow filter would be brighter than the rock and would detract from the image he wanted to make. He decided to take the last negative

Figure 6.1   Pre- and post-visualization made from Ansel's lecture slides. © The Ansel Adams Publishing Rights Trust.

with a red filter that would render the sky darker than the rock (Figure 6.1 right). He often said that he ran down the trail to get to the darkroom to see how well his visualization worked.

6.5   Scene Capture

The Zone System imposed the discipline of visualizing the final image by assigning image segments to different final print zones prior to exposing the negative. Adams was a professional pianist. He often described the negative as the analog of the musical score and the print as the performance. It was essential that the negative captured all the information in the scene, and that the printing process rendered this information in the print.

Although the terminology has changed over the years with the transition from silver halide to digital photography, the problem of successfully capturing scene information in an efficient manner remains the same. Discussions of the "range of the scene" in film circles involved techniques to preserve details in the highlights and shadows. Range problems were described as burning out facial skin-tones in the toe of the response function (tone-scale curve), or the lack of definition in the shoulder. Today, these issues are described as having enough bits to represent the scene.

6.5.1   Assigning Scene Luminances to Zones in the Print

For Adams it was essential that the exposed negative captured all the details in the highlights and the shadows. This is the usable information needed for an optimal final print. Adams describes in detail the techniques for exposing and developing negative to, not only capture the usable range, but also to optimize how the scene information is positioned on the tone scale.

Adams used a spot photometer to measure the luminances of image segments and assigned them to zones in the scale from white to black in the final photographic print. In "The Negative" Adams (1981) used the photograph of "Silverton Colorado" (Figure 6.2) to illustrate the relationship of light meter readings that are transformed into print values.

The spot meter readings established the midpoint of the exposure range (Zone V) and the relative amount of light from the maximum and minimum luminances. The process would transfer the exposure Zone V to the print value of middle-gray density. The max and min readings determined whether the range of the scene fit the normal exposure/development process. If the range of the scene was greater or smaller, then changes in exposure and development could compensate.

6.5.2   Zone System: Interplay of Exposure and Development

Photographic contrast is the rate of change of density vs. exposure. In the negative, the values are controlled by both development and exposure. The zone system provided the necessary information to select appropriate exposure and processing for each scene's dynamic range.

Ansel rehearsed photography the way he practiced on the piano. He knew his negative's response to light and dynamic range. His light measurements of the scene determined whether the negative's range should be expanded (lower slope and higher dynamic range), or contracted (higher slope and lower dynamic range). Both the exposure and subsequent development were optimized before taking the picture. Adams described his exposure and development techniques:

"It is important to understand that the primary effect of expansion or contraction is in the higher values, with much less alteration of the low values. This fact leads to an important principle underlying exposure and development control: the low values (shadow areas) are controlled

Figure 6.2 Adams' example of zones found in a scene. Areas in the print are shown to correspond to zones on the top of the print. © The Ansel Adams Publishing Rights Trust.

primarily by exposure, while the high values (light areas) are controlled by exposure and development." (Adams, 1981, page 72)

A familiar way to describe the above is to plot the Tone Response (Net Density) vs. Zone in the scene (Figure 6.3). We see the plots of the negative's density values with under development (N−1), normal, and over development (N+1 and N+2). In this manner, increased exposure and reduced development compresses high-dynamic range scenes into the normal tone-scale range of the negative/positive print system. Alternatively, reduced exposure and increased development expands low-dynamic range scenes.

The range of densities in the negative is equivalent to the number of gray levels in a digital image. When performing calculations with limited precision one has to be careful not to exceed the fixed digital range. With linear digitization of HDR scenes, exceeding the limits is easy. However, if one uses logarithmic encoding of HDR scenes, then the range is not limited by the number of bits. Just as one should optimize the encoding of the range of bits available to perform a calculation, Adams optimized the gray levels in the negative by controlling exposure and determining development before taking the picture. This is equivalent to tuning the A/D converter in a digital camera before exposure.

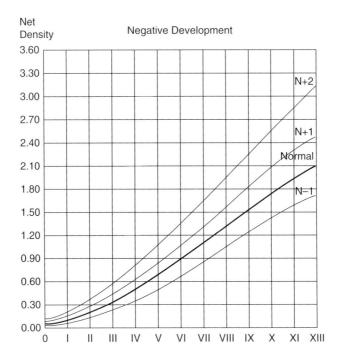

Figure 6.3 Adams's Film Test Data for Kodak Tri-X Professional (Adams, "The Negative", page 247). © The Ansel Adams Publishing Rights Trust.

Figure 6.4 shows the effects of different silver halide papers and development. It is easy to see their similarity to digital tone scale plots.

It is possible to record in a single exposure all the usable scene information in scenes using film negatives. By controlling the exposure and chemical processing of the negative and print Adams optimized the use of the captured dynamic range.

6.5.3 Compressing the HDR Scene into the LDR Print – Spatial Image Processing

Print paper responds to specific amounts of light. Each tone in the final image is associated with a unique exposure. In photography the tools to render the negative are the exposure, the type of printer (contact, diffuse or projection illumination), the contrast of printing paper, print development, and the surface of the paper. The photographer has to convert the signal stored in the negative to the desired appearance of the print.

We have already seen that the rendering of HDR scenes is not done by reproducing the luminances of the scene. What is the reason that the Zone System and HDR spatial processing make better pictures?

The answer is that the dynamic range is less important than how one handles the information between maximum and minimum. It is the encoding of the tone-scale values between white and black that controls the quality of the rendition. The limitations we discussed above, namely keeping details in the highlights and shadows have little to do with the range of the sensor. Instead, it has to do with how one encodes the tone values within the range. The highlights get "burned out" because the spatial

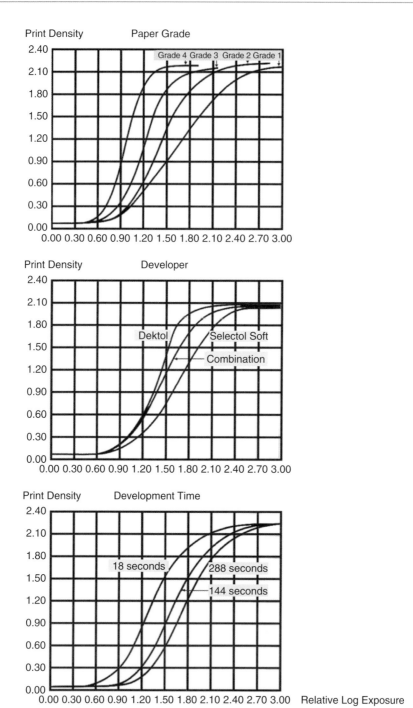

Figure 6.4 Tone scale response: paper grade, developer and times of print development (Adams, "The Print", pages 198–200). © The Ansel Adams Publishing Rights Trust.

information captured in the negative is too low in contrast for what the print paper needs to render it properly. Negatives can capture the range of the scene. The problem is fitting that information into the low-range print.

6.6 "Performing the Score"

Ansel's favorite analogy was that the negative was the musical score, and making the print was the performance. So far we have described all the technical details of making negatives that optimally record the range of light in the scene. We have described Ansel as a skilled recording specialist for light. In making the print Ansel's role changed dramatically. He became the performer of the score in the negative. He became the interpreter of the optimally stored information. The tone-scale functions that he studied and mastered were necessary, but not sufficient to render his visualization of the scene. Spatial manipulation of light projected from the negative onto the print paper was needed to successfully render the visualization. Each scene required a unique spatial intervention.

6.6.1 Dodging and Burning

The next stage was to control the amount of exposure for each local part of the image (dodge and burn) to render all the desired information from a high-dynamic-range scene into a low-dynamic-range print. This process starts with a preliminary test print using uniform exposure. Examination of the print identifies the areas with under-exposed whites and over-exposed blacks that have lost spatial detail. *Dodging* refers to holding back exposure from areas that are too dark. Less exposure lightens this local region of the negative-acting print paper and gives better detail rendition in the blacks. *Burning* refers to locally increasing the exposure to make an area darker. This is a local spatial manipulation of the image. Not only can these techniques preserve detail in high and low exposures, they can be used to assign a desired tone value to any scene element.

Dodging, withholding light from the print paper, is often done with a circle of black opaque paper disk (several inches in diameter) on a wire, or stick. It looks like a lollypop. The lens of the negative enlarger has a small aperture, so as to have a long exposure time. The operator holds the disk above the print paper so that the shadow it creates is out of focus. The operator moves the opaque paper in slow, smooth motions. This moving, out of focus mask makes the area under the disk lighter, but leaves no visible image of the disk. A brief exposure of a static mask would make sharp edges in the print that could be easily seen.

In image processing parlance, dodging introduces gradients in luminance, thus changing the low-spatial frequency content of a portion of the scene. These low-slope gradients are below visual threshold, and hence not visible. The effect of the reduced light makes the area of interest lighter. That means that the details in the negative of black scene areas become visible in the print.

Burning is the opposite process for making white scene details visible in the print. It makes the area of interest darker. Here the operator uses a circular hole (several inches in diameter) in a black opaque paper. Again, the image of the circle is out of focus on the print paper. Here the increase in light darkens the area of interest. In highlight areas of the scene, this added exposure makes the details in whites more visible.

Adams described the dodging and burning process in detail (Adams, 1981, pages 102–116) and wrote about using it in many of his most famous images (Adams, 1983b) He executed remarkable control in being able to reproducibly manipulate his printing exposures.

Adams' final print was a record of his visualization of his desired image, not a simple record of the radiances from the scene. The Ansel Adams Zone System combined the chemical achievements of capturing wide ranges of luminances in the negative with dodging and burning to synthesize his aesthetic intent.

6.7   Moonrise, Hernandez

Adams (1983b, page 41) wrote that "Moonrise, Hernandez, New Mexico, 1941" was his most popular single image. Adams describes the details of making this image. He recounts: seeing the scene, visualizing the beauty of the light on the crosses in the foreground, stopping the car, unpacking the camera, getting on the platform on the roof of the car, setting up the camera, focusing, estimating the exposure based on the luminance of the moon, and taking the picture. Figure 6.5 shows Ansel's "Moonrise" print.

Figure 6.6 shows the "straight print" of the Ansel Adams negative taken of the moonrise scene. This print was made by John Sexton using the Adams negative with uniform illumination, without dodging and burning. It shows the information recorded on the negative. John Sexton made this print to illustrate the major role of dodging and burning in Ansel's final print.

The Adams print shows that the relative spatial information recorded in the negative can be rendered by selective dodging and burning. Adams (1983b, page 42) describes the details of his transformation of a low-range scene capture to this most remarkable print.

The Adams print shows that:

Burning darkened the sky from middle gray to black.

Dodging lightened the buildings and graveyard.

Dodging the graveyard crosses made them white in the print.

6.8   Apparent vs. Physical Contrast

Ansel (1978, page 178) showed a very powerful example of the difference between actual dynamic range and apparent dynamic range in Polaroid Land Photography.

Adams described the pre-exposure technique to see more details in shadows. He uniformly pre-exposed the film to a very low light level. Figure 6.7 shows: (top left) control of film response to HDR scene; (bottom left) pre-exposure; (right) pre-exposure plus scene. The control photograph has large areas of uniform black that are darker than the range of the film. The pre-exposure raised the film response to its threshold everywhere in the print. Additional amounts of light from the darkest areas in the scene filled in the details in the shadows. This technique adds light everywhere in the image, and hence lowers the actual dynamic range of the film. Nevertheless, it increases the apparent range of the print. This is an excellent example of why preserving local spatial information is the key to successful HDR imaging.

6.9   Summary

In 1940 Ansel Adams first described the Zone System for photographic exposure, development and printing. It described three sets of procedures: first, for measuring scene radiances; second, for controlling negative exposure and development to capture the entire scene range in single exposures, and third, spatial control of exposure to render the high-range negative into the low-range print.

Adams used a spotmeter to measure the luminances of image segments and assigned them to zones in the scale from white to black in the final photographic print. The Zone System imposed the discipline of visualizing the final image by assigning image segments to different final print zones prior to exposing the negative. Adams was a professional classical pianist. He often described the negative as the analog of the musical score and the print as the performance. It was essential that the negative recorded all the information in the scene and that the printing process rendered this information in the print. The Zone System provided the necessary information to select appropriate exposure and processing for each scene's dynamic range.

The final stage was to control the amount of exposure for each local part of the image (dodging and burning) to render all the desired information from a high dynamic range scene into a low-dynamic

Figure 6.5   Ansel Adams's print of the negative with dodging and burning. © The Ansel Adams Publishing Rights Trust.

Figure 6.6   John Sexton's print of the Adams' negative without dodging and burning. © The Ansel Adams Publishing Rights Trust.

Film Response

Decreased
Actual dynamic range

+   =

Uniform pre-exposure

Increased
Apparent dynamic range

Figure 6.7   Pre-exposing film increases the apparent range of the print. © The Ansel Adams Publishing Rights Trust.

range print. Not only can these techniques preserve detail in high and low exposures, they can be used to assign a desired tone value to any scene element. For Adams the final print was a record of his visualization of his desired image, not a simple record of the radiances from the scene. In fact, Ansel's Zone System process was the 1940s equivalent of an all-chemical Photoshop®.

There are a great many parallels between the Adams Zone System and HDR photography. They share the same goals and use many parallel techniques. Controlling the capture of useful information and optimizing the storage of the scene information is the first essential part of the process. Recording all the useful information is possible with good single exposure and quantization practice. The rendition of the final image has two components. The stored information needs to be adjusted to fit in the range of the print paper or display device. Meaningful dynamic range compression requires spatial image processing to avoid trading off improvements in one part of the tone scale for degradation in another. Dodging/burning, or spatial image processing, reconstruct the image content to preserve local detail while fitting the HDR world into the LDR rendition. Successful HDR imaging synthesizes a new image from the usable light capture to meet the aesthetic rendering intent. This is exactly what Ansel continues to teach us in his technical writings, and what Ansel and John Sexton show us in their photographs (Adams,1963; 1979; 1983b; Sexton, 1990;1994).

6.10   References

Adams A (1956) *Artificial Light Photography: Basic Photo Five*: New York Graphic Society, Boston.
Adams A (1957) *Natural Light Photography: Basic Photo Four*, New York Graphic Society, Boston.
Adams A (1963) *The Eloquent Light*, Sierra Club, San Francisco.
Adams A (1967) *The Camera: Basic Photo One*, Boston: New York Graphic Society, Boston.
Adams A (1978) *Polaroid Land Photography*, Boston: New York Graphic Society, Boston.
Adams A (1979) *Yosemite and the Range of Light*, Little Brown & Company, Boston.
Adams A (1981) *The Negative Exposure and Development: Book Two*, New York Graphic Society, Boston.
Adams A (1983a) *The Print: Book Three*, Boston: New York Graphic Society, Boston.
Adams A (1983b) *Examples: The Making of 40 Photographs*, Little Brown & Company, Boston.
Sexton J (1990) *Quiet Light*, Little Brown, Boston.
Sexton J (1994) *Listen to the Trees*, Little Brown, Boston.

7

Electronic HDR Image Processing: Analog and Digital

7.1 Topics

Electronic High-Dynamic-Range image capture and display began in the 1960s. At first, it used cumbersome analog processing, but quickly moved to digital image processing. The earliest ideas applied the same principles found in human spatial vision to picture making. This chapter reviews research on spatial vision at that time. It summarizes early electronic algorithms and reviews later pixel-based electronic processes.

7.2 Introduction

Advances in electronic digital imaging removed the fundamental restriction of silver halide photography. Images were no longer restricted to the quanta catch at a pixel. With digital arrays of entire images it became possible for distant pixels to interact with each other in image processing. With this major limitation of chemical photography removed, Prometheus was unchained. Spatial imaging became possible. The electronic equivalent of dodging and burning meant that digital photography was free of tone scale compromises. Electronic spatial processing could handle HDR scenes the way artists had done for centuries.

Artists successfully render high-dynamic-range scenes in media with less range. This is also true of human vision. Although rod and cone retinal receptors have a dynamic range of more than 10 log units, optic nerves that transmit the retinal response to the cortex have a range of only 2 log units. Vision makes a low-dynamic-range representation of HDR scenes.

7.3 Human Spatial Vision

Over the past century psychophysical and physiological experiments have provided overwhelming evidence that vision is a result of spatial processing of receptor information. Selig Hecht and others showed that the threshold detection mechanism uses pools of retinal receptors (Pirenne, 1962). Rod

The Art and Science of HDR Imaging, First Edition. John J. McCann, Alessandro Rizzi.
© 2012 John Wiley & Sons, Ltd. Published 2012 by John Wiley & Sons, Ltd.

and cone receptors respond to light over a dynamic range of over 10 billion:1. That is the range of radiances from snow on a mountaintop to the half-dozen photons needed for a dark-adapted observer to say he saw the light. Kuffler (1953) and Barlow (1953) showed that the signal traveling down the optic nerve has spatial-opponent signal processing. In one example, the center of the cell's field of view is excited by light (more spikes per second). The receptors in the surround of the cell's field of view are inhibited by light (fewer spikes per second). The net result is the cell does not respond to uniform light across its field of view, but is highly stimulated by edges, or scene content. It has the greatest response to a white spot in a black surround. Hartline and Ratliff (1957) found spatial processing in the compound eye of Limulus Polyphemus. Dowling (1987) showed pre- and post-synaptic behavior of the retina establishing post-receptor spatial interactions in many species.

Edwin Land (1964) proposed Retinex theory, asserting that information from cone types acts to form independent color channels, in which the response was determined by their spatial interactions within a channel. The phenomenon of color constancy is best explained by independent long-, middle-, and short-wave spatial interactions. Zeki (1980, 1993) found color constant cells in V4 with predicted spatial properties. Hubel and Wiesel (2005) reviewed their studies of the organization of the primary visual cortex's response to stimuli projected on a screen in front of the animal. In each small region of the cortex they found a three-dimensional array of different representations of the visual field. Each segment of the visual field has columns of cortical cells that report on the left-eye image next to a column for the right-eye image. The cells perpendicular to the left/right eye columns respond to bars of different orientations. The third cortical dimension has cells with different retinal size segments of the field of view. Campbell and colleagues showed that there are independent spatial-frequency channels corresponding to bar detectors of different visual angle (Blakemore & Campbell, 1969). The noted Cornell psychologist, J J Gibson (1968) described the importance of bottom-up spatial image processing.

The late 1950s and early 1960s provided a decade of new evidence that human vision had mechanisms using spatial comparisons. These physiological and psychophysical experiments provided a background for making algorithms that mimic human vision.

7.4   Electronic HDR Image Processing

The earliest HDR rendering mimicked human vision in that it applied models of spatial image processing. Much later HDR models reverted to approaches used in film: pixel processing, multiple exposures, and tone-scale look up tables (Tumblin & Rushmeier, 1993; Larson et al., 1997; Reinhard et al., 2006).

7.4.1   The Black and White Mondrian

Gelb had shown that the same luminance from the same paper could look white or black, changing other objects in the scene (Gelb, 1929). Land extended this observation by demonstrating this fact in the same scene, at the same time. Land's OSA Ives Medal Address included the Black and White Mondrian. This experiment showed that two areas with identical luminances generated the white and black lightnesses in the same scene, at the same time (Land & McCann, 1971a).

The Black and White Mondrian was a large array of black, white and gray rectangular papers. They had different sizes and shapes, so as to suppress afterimages (Daw, 1962). The illumination was highly directional. A footlight on the floor was very close to a low-reflectance paper at the bottom of the Mondrian (Figure 7.1). The experimenter used a telephotometer to measure the luminance coming from that paper to the observer's eye. The experimenter then used the telephotometer to find a high-reflectance paper near the top of the Mondrian that had the same luminance. The experimenter asked the observer to describe the apparent lightness of the two areas. The high-reflectance paper (in dim light) appeared near white, and the low-reflectance paper (in bright light) appeared near black, despite the fact that they had identical luminances. Photographs of these areas have identical optical densities,

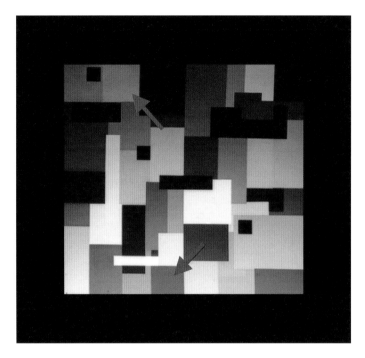

Figure 7.1   Black and White Mondrian. Red arrows identify equal luminances.

or digit values. Tone-scale manipulation of films, or digital images, cannot change the equal responses to these very different papers. Only dodging and burning, and other spatial processes, can render the high-reflectance paper nearer to white, and the low-reflectance paper nearer to black.

In this Black and White Mondrian experiment (Figure 7.1), the same scene had white and black appearances from exactly the same luminances. In other words, the entire range of sensations between white and black were generated by identical colorimetric stimuli, or pixel values (Land & McCann, 1971a).

How can we calculate apparent lightness from luminance? Hans Wallach (1948) had shown that constant differences in sensation were associated with constant ratios of luminance in simple targets. This property is a necessary, but not a sufficient model of lightness. Appearance of one scene area can be influenced by all other parts of the scene. Experiments that quantify the effects of spatial scene content are described in Sections C and D.

A model of lightness needs to use the ratios of luminances for local comparisons. In addition, it needs spatial interactions to introduce long-distance influence. That operation in Retinex models is achieved by replacing the ratio with the ratio-product. The calculation starts with the current value at one pixel and multiplies that value by the ratio of the luminance of the comparison pixel divided by the luminance at the starting pixel. This *New Product* is the value of the output. Models that iterate this ratio-product operation successfully predict the lightness sensations reported by observers (Chapter 32).

7.4.2   Analog Electronic Spatial Rendering

In 1967, Land demonstrated the first electronic (analog) HDR rendering in his Ives Medal Address to the Optical Society of America (Land, 1967; Land & McCann, 1971a; McCann, 2004). The device is

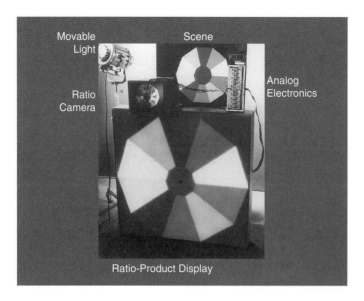

Figure 7.2   Retinex analog image-processing using spatial comparisons.

shown in Figure 7.2. Here, the intent was to render HDR images using spatial comparisons that mimic human vision. This paper suggested that lightness correlated with spatial ratios and expanded it beyond the restraints of uniform illumination. The idea was that what we see was synthesized from the ratio at an edge multiplied by the ratios at all other edges. This process synthesized an image based on the relationship of all edges in the scene, independent of the luminances of each. (Chapter 32) The history of the development of this idea is found in Land's Friday Evening Discourse to the Royal Institution, London (Land, 1974).

In Figure 7.2 the scene consisted of 10 pie-shaped papers illuminated with a movable spotlight. The *Ratio Camera* sat on top of the large display in the foreground. Ten pairs of light sensors measured the ratios of light across edges in the camera's image of the scene. The ratios were multiplied to form the product, reset to the maximum value, and used as the signal to drive the light coming from the *Ratio-Product-Reset Display* (Chapter 32).

The pair of Land and McCann (1971b) and Land et al. (1972) patents described analog embodiments of calculating matches from arrays of luminances. The second patent introduced the idea of non-linear reset to the maxima that is critical in distinguishing Retinex processing from subsequent spatial-frequency filtering. Human vision normalizes to local maxima, rendering them as white or near white (McCann, 2007, Section C). This property of vision is modeled by the reset to maximum in Retinex algorithms (Chapter 32).

Tom Stockham saw one of Land's frequent lectures demonstrating the Black and White Mondrian experiment at MIT. He became interested in the application of Fergus Campbell's and Arthur Ginsburg's spatial frequency approach to imaging. Stockham (1972) wrote a paper on rendering high-dynamic-range scenes using a low-spatial frequency filter to compress the image. This, along with Wilson and Bergen's (1979) spatial-frequency models, and Marr (1974) and Horn's (1974) work on gradients, became the foundation of substantial interest in spatial-frequency techniques of rendering images. All of these models designed specific low-spatial frequency filters, and then applied them to all images (Chapter 31).

Human vision is different. Human vision uses the equivalent of scene-dependent spatial-frequency filtering. The reset in Retinex introduces this special kind of spatial response to the scene. Human vision and the reset-Retinex algorithm with fixed model parameters generates scene-dependent rendering (Chapter 32).

7.4.3 Digital Electronic Spatial Rendering

The practical embodiment of Land and McCann needed two technologies: first, the digital image processing hardware, and second, an efficient algorithmic concept that reduced the enormous number of pixel to pixel comparisons to a practical few, enabling rapid image synthesis. The hardware became commercially available in the mid 1970s for the display of digital satellite and medical images. The efficient image processing began with the Frankle and McCann's (1983) patent using International Imaging Systems (I^2S) image processing hardware with multi-resolution software. The explanation of this work and its relation to other multi-resolution and pyramid processing is found in the literature (McCann, 2004; Chapter 32).

Figure 7.3 shows the rendition of "John at Yosemite" an example of a very efficient digital, multi-resolution HDR algorithm, using spatial-comparisons. It was first shown in the Annual Meeting of the Society of Photographic Scientists and Engineers in 1984.

At the original scene for this photograph, spot photometer readings showed that the illumination in the sunlit foreground is 32 times brighter (5 stops) than that in the shade under the tree. Spot meter

Figure 7.3   HDR scene processed with spatial comparisons using the Frankle and McCann process. The illumination on the white card in the shadow is $1/32^{th}$ that on the black square in the sun. Both the white card in shade and black square in sun have the same luminance. The spatial processing converted equal input digits into very different output digits, thus rendering the HDR scene into the small range of the reflective print.

readings showed that the sunlit black square has the same scene luminance as the white card in the shade. Prints cannot reproduce 32:1 in sun, plus 32:1 in shade, (dynamic range 32^2) because the entire print range is only 32:1 in ambient light. Using the spatial comparison algorithms described by Frankle and McCann (1983), it is possible to synthesize a new 32:1 image that is a close estimate of appearance and render it to film. This was the first practical example of capturing scene radiances, calculating color sensations, and printing them on film (Chapter 32).

7.4.4   Electronic HDR Pixel Processing

At Siggraph courses in 1984 and 1985 McCann described HDR image capture using low-slope slide duplication film, a graphic-arts scanner and Retinex digital image processing (McCann, 1988). Although these conferences were early in the Siggraph series, they were very well attended by enthusiastic computer scientists eager to develop Computer Graphics (Chapter 8).

McCann HDR scene capture data agreed with that of Jones and Condit (1941). Typical sun and shade images had a range of roughly 3.0 log units. New in this study was the effect of changes in spectral composition of illumination. Tungsten light, without shadows showed a 3.0 log unit range because the amount of long-wave light was significantly greater than the short-wave light. Conversely, in skylight the amount of short-wave light was significantly greater than long-wave light. Both showed a 3.0 log unit range, but this was due to spectral shifts in illumination composition, not due to sun and shade (McCann, 1988). As we will see later, this 3.0 log unit range is all that can be accurately recorded for typical real scenes (Chapter 11).

Ochi and Yamanaka (1985) of Sony patented a multiple-exposure CCD system using one imaging lens, a beam splitter, and two CCD image regions. An object of the invention was to "produce a still image of excellent quality and which prevents deterioration of the image quality which results from smearing and blooming."

Alston et al. (1987) of Polaroid patented an electronic imaging camera for substantially expanded dynamic exposure range by combining two successive exposures with different durations.

S. Mann (1993) of MIT investigated a series of different digital image fusion techniques. One of these techniques was to merge different exposures to capture a wider dynamic range. This work was expanded in a second paper (Mann & Picard, 1995). They called the collection of multiple exposures a "*Wyckoff set*" in honor of Charlie. Their concept was to use multiple digital frames to make "*undigital*" images with floating point precision. They wrote:

"Double precision (64 bit) floating point number is close to analog in spirit and intent."

For them the problem was the loss of image quality from each successive image-processing step, leading to gaps in histograms.

Debevec and Malik (1997) used multiple exposures and least-square fits to solve for the camera response function and the luminance of each pixel in the scene. Although some people have described its use of multiple exposures as revolutionary, in fact this paper is most significant because it asserted that camera digit can be used to calculate scene luminances. Many papers are based on the Debevec and Malik attempt to record scene luminances (Chapters 8, 11, and 31). The work of Paul Debevec has had a major influence on Computer Graphics (Chapter 8).

7.5   Summary

Electronic image processing, first with analog, then with digital image processing and digital photography addressed the rendition of HDR scenes. Early systems set out to mimic human vision's example of capturing HDR information and rendering it in LDR media. These systems use HDR spatial information from the entire scene to calculate appearances and write them on prints and displays. Ironically, later HDR algorithms mimicked film by reverting to multiple exposures, and universal tone scales with pixel-based image processing. Today, however, there is considerable interest in using spatial-image processing to make the best renditions of HDR scenes (Section F).

7.6   References

Alston L, Levinstone D, Plummer W (1987) Exposure Control System for an Electronic Imaging Camera Having Increased Dynamic Range, US Patent 4,647,975, filing date 10/30/1985, issue date 3/3/1987.

Barlow H (1953) Summation and Inhibition in the Frog's Retina, *J Physiol*, **119**, 69–88.

Blakemore C & Campbell F (1969) On the Existence of Neurons in the Human Visual System Selectively Sensitive to the Orientation and Size of Retinal Images, *J Physiol*, **203**, 237–60.

Daw N (1962) Why After-Images Are Not Seen In Normal Circumstances, *Nature*, **196**, 1143–45.

Debevec P & Malik J (1997) Recovering High Dynamic Range Radiance Maps from Photographs, *ACM SIGGRAPH'97*, 369–78.

Dowling J (1987) *The Retina: An Approachable Part of the Brain*, Belknap Press, Cambridge.

Frankle J & McCann J (1983) Method and Apparatus of Lightness Imaging, U.S. Patent, 4384336, issued 5/17/1983.

Gelb A, (1929) Die "Farbenkonstanz" den Sehdinge. Handbuch der normalen und patholgischen, *Physiologie*, **12**, Springer, Berlin.

Gibson J (1968) *The Senses Considered as Perceptual Systems*, Allen & Unwin, London.

Hartline H & Ratliff F (1957) Inhibitory Interaction of Receptor Units of the Limulus Eye, *J Gen Physiol*, **40**, 357.

Horn B (1974) Determining Lightness from an Image, *Comp Gr Img Proc*, **3**, 277–99.

Hubel D & Wiesel T (2005) *Brain and Visual Perception*, Oxford University Press, New York.

Jones L & Condit H (1941) The Brightness Scale of Exterior Scenes and the Computation of Correct Photographic Exposure, *J Opt Soc Am*, **31**, 651–78.

Kuffler S (1953) Discharge Patterns and Functional Organization of the Mammalian Retina, *J Neurophysiol*, **16**, 37–68.

Land E (1964) The Retinex, *Am Scientist*, **52**, 247–64.

Land E (1967) Lightness and Retinex Theory, *J Opt Soc Am*, **57**, 1428A; For citation see (1968) *J Opt Soc Am*, **58**, 567.

Land E (1974) The Retinex Theory of Colour Vision, *Proc Roy Institution Gr Britain*, **47**, 23–58.

Land E, Ferrari L, Kagen S & McCann J (1972) Image Processing System which Detects Subject by Sensing Intensity Ratios, US Patent 3,651,252, issued 3/21/1972.

Land E & McCann J (1971a) Lightness and Retinex Theory, *J Opt Soc Am*, **61**, 1.

Land E & McCann J (1971b) Method and System for Reproduction Based on Significant Visual Boundaries of Original Subject, US Patent 3,553,360, issued 6/5/1971.

Larson G, Rushmeier H & Piatko C (1997) A visibility Matching Tone Reproduction Operator for High Dynamic Range Scenes, *IEEE Trans Vis & Comp Graphics*, **3**(4), 291–6.

Mann S (1993) Compositing Multiple Pictures of the Same Scene, *Proc. IS&T Annual Meeting*, **46**, 50–2.

Mann S & Picard R (1995) On Being 'Undigital' with Digital Cameras: Extending Dynamic Range by Combining Different Exposed Pictures, *Proc. IS&T Annual Meeting*, **48**, 442–48.

Marr D (1974) The Computation of Lightness by the Primate Retina, *Vision Res*, **14**, 1377–88.

McCann J (1988) Calculated Color Sensations applied to Color Image Reproduction, Image Processing Analysis Measurement and Quality, Bellingham WA: *Proc. SPIE*, **901**, 205–14.

McCann J (2004) Capturing a Black Cat in Shade: Past and Present of Retinex Color Appearance Models, *J Electronic Imaging*, **13**, 36–47.

McCann J (2007) Aperture and Object Mode Appearances in Images, Human Vision and Electronic Imaging XII, eds. Rogowitz, Pappas, Daly, Bellingham, WA: *Proc. SPIE*, 6292-26.

Ochi S & Yamanaka S (1985) US Patent, 4,541,016, filed 12/29/1982, issued 9/10/1985.

Pirenne MH (1962) *The Eye: 2 The Visual Process*, Davson H ed. Academic Press, New York.

Reinhard E, Ward G, Pattanaik S, Debevec P (2006) *High Dynamic Range Imaging Acquisition, Display& Image-Based Lighting*, Amsterdam: Elsevier, Morgan Kaufmann.

Stockham T Jr. (1972) Image Processing in the Context of a Visual Model. *Proc. IEEE*, **60**(7), 828–42.

Tumblin J & Rushmeier H (1993) Tone Reproduction for Realistic Images, *IEEE Computer Graphics and Applications*, **13**(6), 42–8.

Wallach H (1948) Brightness Constancy and the Nature of Achromatic Colors, *J Exptl Psychol*, **38**, 310–24.

Wilson H & Bergen G (1979) A Four Mechanism Models for Threshold Spatial Vision, *Vision Res*, **26**, 19–32.

Zeki S (1980) The Representation of the Cerebral Cortex, *Nature*, **284**, 412–18.

Zeki S (1993) *Vision of the Brain*, Williston, Vermont, Blackwell Science Inc.

8

HDR and the World of Computer Graphics

8.1 Topics

This chapter briefly summarizes the extensive history of Computer Graphics that has become a dominant force in imaging today, and how HDR imaging is influencing its future.

8.2 Introduction

Not all images are still; some are computer generated, or animated images. This is the territory of *Computer Graphics (CG)*. CG is a pervasive and important field in the computer industry and entertainment world. In fact, CG not only deals with the creation of beautiful virtual images, but it also deals with all graphics hardware and software necessary to interact with present day computer systems and electronic appliances. Since the appearance of the first video games on personal computers, gaming has been one of the most relevant driving forces of CG development, bringing ever increasing efficiency and power to graphics boards and to specific game consoles. The second most important driving force has been the movie industry, whereas the first computer generated movies were TV logos. While animation was a relatively small part of early movie making, usually reserved for children's cartoons, the fusion of animation and scientific computation created a whole new world of imaging.

Since the 1970s the most important annual event, for aggregating scientists, movie makers, and the computer industry, is the ACM Siggraph Conference. Since 1997 this community has developed an interest in HDR imaging. For this reason and for its importance for the field of digital imaging we recall some milestones of CG history and its relationship with HDR.

A brief history of CG can be described from many viewpoints: the evolution of graphics hardware, the growth of the market of graphics systems and hardware components, and the continuous birth of new ways of using CG for improving human-computer interactions. However, we prefer to focus on the evolution towards new solutions to the constant problem of progressing from simple line drawings on a plotter to the creation of realistic images and video. CG history is strictly linked to the advancements in image processing; indeed we can consider digital imaging as the middle ground

The Art and Science of HDR Imaging, First Edition. John J. McCann, Alessandro Rizzi.

between synthesis and analysis of images. This strict link, in conjunction with the profusion of digital cameras, has caused research and development to close the gap between photography and digital imaging. Last, but not least, an important aspect is the strict link between CG and human computer interactions: we can also consider the very first graphics programs to be the first interactive computer applications.

8.3   Early Years: the 60s

In 1963 Ivan E. Sutherland (1963) created a line drawing program called Sketchpad: using a light pen, Sketchpad allowed its user to, not only draw interactively line drawings on a computer controlled cathode ray tube, but also to save and recall drawings later. The solution by Sutherland also allowed the user to create drawings and select, move, and delete single elements. Sketchpad was a turning point, computers started to work as easy tools for graphics. The first video game, *Spacewar*, was conceived in 1961 and implemented on a DEC PDP-1 in 1962 by Steve Russell and others. It was an instant success to the point that the engineers at DEC started to use it as a diagnostic program on every new PDP-1 before shipping it. In a short time, the major corporations started taking an interest in computer graphics and by the mid 60s, IBM released the IBM 2250 graphics terminal, the first commercially available graphics computer. Behind such devices many basic graphics problems were solved, like scan conversion of lines by JE Bresenham (1965). The hidden line problem to represent 3-D models was solved by different algorithms such as Appel (1967), and more efficiently by ME Newell et al. (1972). During these years the major focus was on designing efficient and low cost displays. It is relevant to recall that Sutherland also designed and implemented the first head-mounted display in 1968 (Sutherland, 1968).

8.4   Early Digital Image Synthesis: the 70s

The availability of low cost displays and the interest in creating nice pictures of geometric 3-D objects attracted research aimed at introducing new solutions to the hidden surface problem, and coloring objects with realism. H Gouraud (1971) proposed the first algorithm to smooth gray shaded polyhedral shapes and BT Phong (1975) introduced a new development by simulating reflections and highlights in color. Color displays were already available, but expensive, based on the principle of frame buffer, designed by Kajiya, Sutherland and Cheadle (1975). While research on human-computer interaction continued by creating more powerful interactive graphics programs (Newman, 1968; Foley & Wallace, 1974), a new line of research emerged, that was later called the "quest for realism"; the objective being to produce pictures that were more and more similar to photographs of a real scene. To this end the first improvement was to apply textures, taken from photographs, on shapes (Blinn & Newell, 1976). The study of light-material interaction was pioneered by J Blinn (1977). Ideas from image processing and signal theory solved the problem of jagged lines using antialiasing (Crow, 1977). Other important problems like Z-Buffer and rendering curved surfaces began to be addressed by E Catmull (1972) in his Ph.D. thesis in computer science. He ended up in the Lucasfilm CG lab working on movie production.

The 70s saw also the introduction of computer graphics in the world of television. The Computer Image Corporation (CIC) developed complex hardware and software systems and Bell Telephone and CBS Sports were among the many that made use of them.

In 1971, Nolan Kay Bushnell along with a friend formed Atari. He would go on to create an arcade video game called Pong in 1972 and start one of today's big video game industries. In 1979 Atari launched Asteroids: a game with very simplified graphics, composed entirely of lines drawn on a black and white vector monitor and which enjoyed tremendous success. These have been just two of the many successes in the video game industry that is nowadays as important as the Hollywood majors in terms of income.

In 1973 the Association of Computing Machinery's (ACM) Special Interest Group on Computer Graphics (SIGGRAPH) held its first conference. In 1976, ACM allowed for the first time, exhibitors into the annual SIGGRAPH conference. This attracted 10 companies who exhibited their products. By 1993 this would grow to 275 companies with over 30 000 attendees. From that period on, CG has become more and more pervasive in all forms of entertainment, its improvements in quality were dramatic, and the popularity of the discipline spread considerably.

8.5   The Turning Point: the 80s

Digital imaging accelerated the spread of CG with its application to special effects in the movies, to TV station break animations, and to the visual arts. Depending on the degree of evolution of CG, some movies represent a milestone. Here is a brief sample of the movies that had a big impact on wide audiences, making CG a celebrated discipline:

> *Bedknobs and Broomsticks* in 1971 was a pioneer of CG in which animated cartoon melded with classic footage.
>
> *The Star Wars Saga*, in 1977, 1980, 1983, is still a standard for its amazing special effects and is still a reference for today sci-fi movies.
>
> *Tron* in 1982 was a celebration of the growing field of computer games together with *Final Fantasy*, in 1987, the videogame, later the movie.

Another milestone for computer graphics was the founding of Silicon Graphics Inc. (SGI) by Jim Clark (1982). He designed the Geometry Engine, a special processor dedicated to the computation of fundamental operations necessary for color rendering. SGI created the highest performance graphics computers available at the time with built-in 3D graphics capabilities, high speed RISC (Reduced Instruction Set Chip) processors and symmetrical multiple processor architectures.

Tom Brigham, surprised the audience at the 1982 SIGGRAPH conference with a video sequence which showed a woman distorting and transforming herself into the shape of a lynx. This was the birth of the "Morphing" technique. This technique was further improved by the work by Beier and Neely (1992).

In 1984, AT&T formed the Electronic Photography and Imaging Center (EPIC) to create PC-based videographic products. In the following year they released the TARGA video adapter for personal computers. This allowed PC users for the first time to display and work with 32-bit color images on the screen.

The advancements in technology were accompanied by advancements in rendering methods designed to create even more realistic pictures. Whitted (1980) proposed the ray tracing method to simulate transparency and specular reflections, while in 1984 C Goral, K Torrance and D Greenberg at Cornell University (Goral et al., 1984) published a new method called Radiosity based on the computation of the radiation exchange in a closed environment, thus addressing the problem of computing ambient lighting; it made complex and mixed lighting configurations much faster to compute. 1984 was also the year of the first Macintosh computer release.

The computer graphics division of Industrial Light and Magic (ILM), the George Lucas company for CG production and research, split off to become Pixar in 1986. The Pixar Animation Group had its first success on March 29, 1989 by winning an Oscar at the Academy Awards for the short animation, "Tin Toy". It was created entirely with 3-D computer graphics, and was directed by John Lasseter (1987), who had already animated Luxo, inspired by the creativity of Disney cartoons.

Other fundamental innovations for the movie industry and for high quality CG came in the form of the invention of numerical methods to compute complex events, like explosions or fires with particle systems (Reeves, 1983). The modeling of natural scenes was greatly improved by the application of

Mandelbrot's Fractal theory (Mandelbrot, 1983) by Fournieret et al. (1982). Reynolds (1987) improved the animation of large groups of flocks and herds based on behavioral methods.

From this period, CG started to be used widely in military applications, weather patterns, flight simulation, and surgery.

8.6   Computational Photorealism: from the 90s

From then on, the quest for photorealism accelerated, facing problems of resolution of details, computational costs, textures, cast shadows, fog, antialiasing, etc. Modeling that built scene images from only abstract objects and points was too time consuming, and was not realistic enough in some cases. Thus, research started to use two-dimensional images of a scene to generate a three-dimensional model to render. This process is called image-based modeling and rendering (IBMR). This approach has multiplied the connections of computer graphics with the image processing discipline, mostly focused on the opposite process of detecting, grouping, and extracting features (edges, faces, etc.), and then trying to interpret them as three-dimensional clues.

A milestone in IBMR was QuickTime VR by Apple, born from research by Chen and Williams (1993) on image interpolation: multiple views of an object or a scene interpolated and assembled to create a 3-D panorama or a 3-D view of an object, that can be explored and navigated like a traditional 3-D scene. Plenoptic modeling, by McMillan and Bishop (1995), was the second step: 2-D mapping on a cylindrical surface or a sphere of interpolated pictures provided the simplest kind of plenoptic function, i.e. the intensity of light for any wavelength and any angle.

The important work of Paul Debevec used the plenoptic function to develop techniques to acquire from a real scene, the geometry and the intensity of the light field, in order to reproduce it in the virtual model. (Reinhard et al., 2006) This is called *Image Based Lighting (IBL)* and extended the lighting models to more complex and realistic ones. One of the methods is to take a picture, or a series of pictures, with a reflecting sphere inserted in the scene. This allows one to detect the light sources and their mutual intensity. To increase the effectiveness of the method and detect the light source intensities with more precision HDR scene capture is necessary. Debevec and Malik (1997) reformulated multiple exposure techniques for a different purpose, following upon work by Mann and Picard (1995). They wanted to record accurate scene radiances over a wider dynamic range. If possible, they could, in principle, use these values and image processing to control the desired rendition. This approach has been actively adopted by the computer graphic community and then expanded to include image and video processing. Paul Debevec describes his techniques in detail in his book (Reinhard, et al., 2006).

The recent research on HDR, IMBR and IBL has revamped the interest in digital photography, opening up new research: Computational Photography. Here, the full panoply of techniques and methods from computer graphics, image processing and digital imaging provides new applications for digital photography.

8.7   Summary

In this chapter we have presented, with the help of Daniele Marini, an overview of the history of a rich and complex field of research. A more complete history can be found by reading Crow (1987) and Wolfe (1998) and the following web sites:

www.siggraph.org/

http://cs.fit.edu/~wds/classes/graphics/History/history/history.html

http://design.osu.edu/carlson/history/lessons.html

http://hem.passagen.se/des/hocg/hocg_1960.htm

8.8 References

Appel A (1967) The Notion of Quantitative Invisibility and the Machine Renderings of Solids, *Proc ACM National Conference*, 387–93.

Beier T & Neely S (1992) Feature Based Image Metamorphosis, *Computer Graphics (SIGGRAPH'92)*, **26**(2), 35–42.

Blinn JF & Newell M (1976) Texture and Reflection in Computer Generated Images, *Comm ACM*, **19**(10), 542–6.

Blinn JF (1977) Models of Light Reflection for Computer Synthesized Pictures, *Computer Graphics (SIGGRAPH '77 Proc)*, **11**(2), 192-198.

Bresenham JE (1965) Algorithm for Computer Control of a Digital Plotter, *IBM Sys J*, **4**(1), 25–30.

Catmull E (1972) A System for Computer Generated Movies, *Proc. ACM Nat Conference*, 422–31.

Chen SE & Williams L (1993) View Interpolation for Image Synthesis, *Computer Graphics (SIGGRAPH '93)*, 279–88.

Clark JH (1982) The Geometry Engine: A VLSI Geometry System for Graphics, *Computer Graphics (SIGGRAPH '82)*, **16**(3), 127–33.

Crow F. (1977) The Antialiasing Problem in Computer Generated Shaded Images, *Comm ACM* **20**(11), 799–805.

Crow F (1987) The origin of the teapot, *IEEE Computer Graphics & App*, **7**(1), 8–10.

Debevec P & Malik J (1997) Recovering High Dynamic Range Radiance Maps from Photographs, *Proceedings of SIGGRAPH 97, Computer Graphics Proceedings, Annual Conference Series*, 369–78.

Foley JD & Wallace VL (1974) The Art of Natural Graphic Man-Machine Conversation, *Proc IEEE*, **62**(4), 462–71.

Fournier A, Fussel D, Carpenter L & Hanrahan P (1982) Computer Rendering of Stochastic Models, *Comm ACM*, **15**(6), 371–84.

Goral CM, Torrance KE, Greenberg DP and Battaile B (1984) Modeling the interaction of light between diffuse surfaces, *Computer Graphics (SIGGRAPH '84)*, **18**(3), 213–22.

Gouraud H (1971) Continuous shading of curved surfaces, *IEEE Trans on Computers*, **2**(6), 3–629.

Kajiya JT, Sutherland IE & Cheadle E (1975) A Random Access Video Frame Buffer, *Proc. IEEE Computer Graphics, Pattern Recognition and Data Structures*, 1–6.

Lasseter J (1987) Principles of Animation as APplied to 3D Character Animation, *Computer Graphics (SIGGRAPH 87)*, **21**(4) 35–44.

Mandelbrot B (1983) *The Fractal Geometry of Nature*, Freeman, New York.

Mann S & Picard RW (1995) Extending dynamic range by combining different exposed pictures, *Proc. IS&T Ann Conf.*, 442–8.

McMillan L & Bishop G (1995) Plenoptic Modeling: An Image-based Rendering System, *Computer Graphics (SIGGRAPH '95)*, 39–46.

Newell ME, Newell RG & Sancha TL (1972) A Solution to the Hidden Surface Problem. *Proc ACM National Conference*, 443–50.

Newman WM (1968) A System for Interactive Graphical Programming, *Proc AFIPS Spring Joint Conf*, Washington, 47–54.

Phong Bui-Tuong (1975) Illumination for Computer Generated Pictures, *Comm ACM*, **18**(6), 311–17.

Reeves WT (1983) Particle Systems: A Technique for Modeling a Class of Fuzzy Objects, *Computer Graphics (SIGGRAPH '83)*, **17**(3), 359–76.

Reinhard E, Ward G, Pattanaik S & Debevec P (2006) *High Dynamic Range Imaging Acquisition, Display and Image-Based Lighting*, Elsevier, Morgan Kaufmann, Amsterdam, ch 9.

Reynolds CW (1987) Flocks, Herds and Schools: a Distributed Behavior Model, *Computer Graphics (SIGGRAPH '87)*, **21**(4), 25–34.

Sutherland IE (1963) Sketchpad: A Man-Machine Graphical Communication System. *Proc AFIPS Spring Joint Computer Conference*, Washington, 329–46.

Sutherland I.E. (1968) A Head-Mounted Three-Dimensional Display, *Proc AFIPS Fall Joint Computer Conference*, Washington, 57–764.

Whitted T (1980) An improved illumination model for shaded display, *Comm of the ACM*, **23**(6), 343–9.

Wolfe R. ed (1998) Seminal Graphics, Pioneering Efforts that Shaped the Field, *ACM SIGGGRAPH*.

9

Review of HDR History

9.1 Topics

The many disciplines used in reproducing scenes with a high range of light have a five-century history that includes painting, photography, electronic imaging, computer graphics and image processing. HDR images are superior to conventional images. This chapter summarizes the work of painters, photographers, scientists and engineers working with HDR scenes. It summarizes the differing image rendition goals and techniques.

9.2 Summary of Disciplines

This text, so far, has described a long list of HDR imaging over five centuries. The best way to categorize these examples is to organize them by rendering intent. Painters from da Vinci to modern photorealists rendered HDR images so that the scene's illumination was as important as its people and objects. The rendering was a combination of both aesthetic and range compression intent. Robinson's early examples of multiple exposure silver-halide photography had the same intent as painters: Render a scene of extended luminance range to a limited one with aesthetic design.

The Mees (1920) example of multiple exposures was not so much an artistic technique, as it was a demonstration of an improvement in image quality. Mees, as director of Research at Kodak for half a century, led the development of negative films that can capture the entire range of light possible using glare limited lenses. These films captured all the possible range in single exposures (McCann & Rizzi, 2007). Such film designs were the result of extensive photographic research, with Jones and Condit (1941), as an example. This work led to a single tone-scale reproduction function used in all manufacturers' color films for the second half of the twentieth century. Innumerable experiments in measuring users' print preferences led the massive amateur color print market to use a single tone-scale-system response. Even digital camera/printer systems mimicked this function (McCann, 1998). It is important to note that this tone-scale function is not slope 1.0. It does not accurately reproduce the scene. It compresses the differences in luminances in both whites and blacks, enhances the mid-tones (increasing color saturation) and renders only light skin tones, and near-blacks accurately.

The Art and Science of HDR Imaging, First Edition. John J. McCann, Alessandro Rizzi.
© 2012 John Wiley & Sons, Ltd. Published 2012 by John Wiley & Sons, Ltd.

The Ansel Adams Zone System combined the chemical achievements of capturing wide ranges of luminances in the negative with dodging and burning to synthesize his aesthetic intent. Controlling exposure and development captures all the desired scene information. Spatial manipulations (dodging and burning) fit the captured range into the limited print range.

Land and McCann's (1971a) Retinex, starting with analog electronics and quickly expanding to digital imagery, used a new approach. It incorporated the initial stage of the Mees and Adams HDR scene capture as its first step. Instead of using the Adams aesthetic rendering, it adopted the goal that image processing should mimic human visual processing. The Retinex process writes calculated visual sensations onto print film, rather than recording scene luminances (McCann, 1988a; b). To this aim, Retinex renders scenes using spatial computations that take into account ratios among areas. These renditions required extremely efficient algorithms (Frankle & McCann, 1983). In computing these spatial relationships the reset step is essential to mimicking vision. It is a powerful non-linear operator that applies the equivalent of a scene-dependent spatial frequency filter (McCann, 2006).

Rizzi and colleagues introduced a new family of advanced Automatic Color Equalization (ACE) that used information from the entire image so that the output values depended on scene content (Chapter 33).

Stockham's spatial filtering of low-spatial frequency image content intended to combine Mondrian experiments with Fergus Campbell's multi-channel spatial frequency model of vision. This concept was the basis of a great many image processing experiments and algorithms. It differs from the original Retinex algorithm because it lacks the non-linear reset, which locally normalizes images to maxima (Chapter 32).

Subsequent digital processes (Ochi and Alston) provided methods for increased digital camera sensors range to approach that which is possible in negative films. Mann's Wyckoff Set of multiple exposures had the rendering intent of better digital segmentation between max and min.

Debevec and Malik (1997) and related papers had a new and different rendering intent for the very old multiple-exposure technique. Using the simple, but efficient, single pixel processes they attempt to make accurate records of scene luminances. This process assumed that one can calculate scene luminance from digital camera values. This led to proposals for digital image files covering extended dynamic ranges up to 76 log units (Reinhard et al., 2006, Chapter 3). It also led to the development of HDR display technology (Seetzen et al., 2004) with a modulated DLP projector illuminating an LCD display.

This approach uses a simple tautology: namely, a display that accurately reproduces all scene radiances, must look like that scene. The weakness in the argument is that it is impractical and nearly impossible to reproduce every single pixel in the entire color space accurately. The measurements of scenes and captured scenes in Section B demonstrate the difficulties of accurate reproduction of every scene pixel.

Figure 9.1 summarizes the dates and rendering intents for six important HDR examples.

9.3   Review

Section A has developed a central theme of this text, namely, the distinction between *pixel-based* tone-scale approaches and spatial synthesis of what we see.

> The tone-scale approach is indifferent to the content of the image. Every pixel with an input luminance capture value of I has the same final output value O.

> The *spatial synthesis* approach is indifferent to the input capture value. Every pixel with a capture value I can have any output value.

Recalling the Black and White Mondrian experiment (Land & McCann, 1971a) is very helpful here. The display was an array of rectangular achromatic papers with a single bright light on the floor. More light fell on the papers at the bottom. Using a spot photometer, the experimental procedure was to find a black paper at the bottom with the same luminance as a white paper at the top. This was easy to do

Image	Year	Author	Rendering Intent
	1851	Edouard Baldus 10 negatives	Scene Rendition
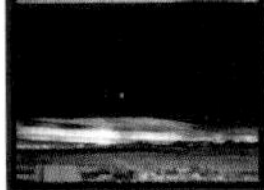	1939	Adams Adams Zone System Dodge & Burn (spatial)	Make the Visualized Image
	1968	Land & McCann Analog Product Spatial Comparisons	Display Calculated Appearances
	1978	Frankle & McCann Digital Multi-resolution Processing	Efficient Spatial Computation
	1984	McCann Scan HDR Positive Film Single Exposure	Scan HDR Scene Capture on Film
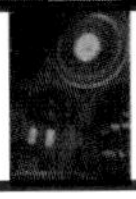	1997	Debevec & Malik Multiple Exposures	Capture Scene Radiances

Figure 9.1   Important experiments in photographic and electronic HDR imaging.

by adjusting the distances between the light and the white and black papers. The experimenter asked observers about the appearances of the equal luminances. They reported that one was white, and the other was black. The fact that they have equal luminances is inconsequential to the experiment that measures what we see.

The Black and White Mondrian experiment is central to understanding how HDR images can improve the reproduction of HDR scenes.

First, it disposes of the idea that a single tone-scale function can be helpful in rendering all HDR images. Pixel-based global tone scale functions cannot improve the rendition of both the black and the white areas in the Mondrian with equal luminances. Tone-scale adjustments designed to improve the rendering of the black do so at the expense of the white in less illumination. As well, improvements to white make the blacks in the bright light worse. When two Mondrian areas have the same luminance, tone-scale manipulation cannot improve the rendering of both white and black.

Second, Land and McCann, (1971a) made the case that spatial algorithms can perform spatial rendering automatically, doing what dodging and burning did to compress HDR scenes into the limited range of prints.

In order to mimic human vision, output response must be independent of the sensor's quanta catch. Early HDR algorithms (Land & McCann, 1971b; Land et al., 1972; Frankle & McCann, 1983; McCann, 2004) never attempted to determine actual scene radiance, since radiance is almost irrelevant to what we see (Sections C, D, and E).

Instead, these spatial algorithms mimicked vision by synthesizing LDR visual renditions of HDR scenes using spatial comparisons. The intent of Land and McCann's electronic HDR imaging was to

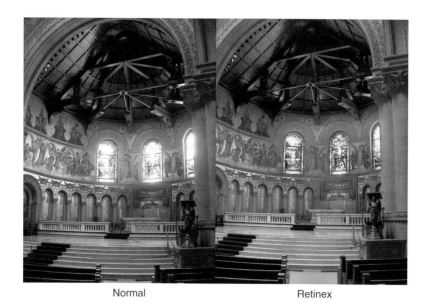

Normal   Retinex

Figure 9.2   Images, with and without spatial comparisons, both taken with a commercial HP 945 camera to make hand-held single-exposure images: (left) uses normal processing; (right) uses the Retinex based Adaptive Lighting/Digital Flash option. The spatial processing removes the over-exposure of the windows, while lightening red rug, white marble altar and pews.

render high-dynamic range scenes in a manner similar to human vision. They made the case for spatial comparisons as the basis of HDR rendering in the B&W Mondrian experiments.

The advent of electronic imaging made possible spatial manipulation of entire images. Such extensive spatial processing is not possible in silver halide chemical photography. Quanta catch at a pixel determines the system response, namely density of the image. Digital imaging processing, or its equivalent, had to be developed in order for each pixel to be able to influence every other pixel. Digital image processing unchained imaging from being bound to universally responsive pixels. Spatial interactions became technologically possible. Ironically, recent HDR tone-scale processes impose pixel-value-dependent global restrictions on digital systems. Global tone-scale functions re-chain Prometheus unchained.

Lloyd Jones's work in the 1920s showed that the accuracy of the luminance record is unimportant. The spatial relationships of objects in shadows are preserved in multiple exposures. Spatial-comparison image processing has been shown to generate successful rendering of HDR scenes. Such processes make use of the improved differentiation of the scene information. Since 1968, there have been many different examples of spatial algorithms used to synthesize improved images from captured image plane luminances. Efficient digital spatial algorithms, such as Frankle and McCann (1983), have been used to display high-range scenes with low-range media. Section F of this text provides an extensive discussion of HDR image processing.

HDR imaging is successful because it preserves local spatial details. HDR has shown considerable success in experimental algorithms, and in commercial products. Figure 9.2 shows the results of spatial image processing from a single exposure using automatic firmware in an amateur camera.

9.4 Summary

We have seen four different rendering intents used extensively in HDR. They are:

1. An artist's aesthetic intent.
2. Improved image quality.
3. Calculated sensations written to film and display.
4. Making accurate records and reproductions of scene radiances at every pixel.

Calculating aesthetics is beyond the scope of this text, but calculating an optimal HDR reproduction is not. The improvement in HDR images, as compared to conventional photography, does not correlate with accurate luminance capture and display. Accurate reproduction has never been the goal of amateur photography. The improvement in HDR images is due to better digital quantization, and the preservation of relative spatial information. Successful HDR image processing algorithms mimic processes developed by human vision, chiaroscuro painters, and early photographers. They render HDR scenes in low-range outputs accessible to human vision.

9.5 References

Debevec P & Malik J (1997) Recovering High Dynamic Range Radiance Maps from Photographs, *ACM SIGGRAPH'97*, 369–78.

Frankle J & McCann JJ (1983) Method and Apparatus of Lightness Imaging, U.S. Patent, 4,384,336, filed 7/29/1980; issued 5/17/1983.

Jones LA & Condit H (1941) The Brightness Scale of Exterior Scenes and the Computation of Correct Photographic Exposure, *J Opt Soc Am*, **31**, 651–78.

Land EH & McCann JJ (1971a) Lightness and Retinex Theory, *J Opt Soc Am*, **61**, 1.

Land EH & McCann JJ (1971b) Method and System for Reproduction Based on Significant Visual Boundaries of Original Subject, U.S. Patent 3,553,360, 6/5/1971.

Land EH, Ferrari L, Kagen S & McCann JJ (1972) Image Processing System which Detects Subject by Sensing Intensity Ratios, US Patent 3,651,252, 3/21/1972.

McCann JJ (1988a) The Application of Color Vision Models to Color and Tone Reproductions, *Proc. Japan Hardcopy'88*, 196–99.

McCann JJ (1988b) Calculated Color Sensations applied to Color Image Reproduction, *Proc. SPIE 901*, 205–14.

McCann JJ (1998) Color imaging systems and color theory: past, present, and future, *Proc. SPIE*, **3299**, 62–82.

McCann JJ (2004) Capturing a Black Cat in Shade: Past and Present of Retinex Color Appearance Models, *J Electronic Imaging*, **13**, 36–47.

McCann JJ (2006) High-dynamic-range-scene Compression in Humans, Human Vision and Electronic Imaging XI, SPIE, San Jose, CA, USA, *Proc. SPIE*, **6057**, 605707-11.

McCann JJ & Rizzi A (2007) Camera and Visual Veiling Glare in HDR Images, *J Soc Info Display*, **15**(9), 721–30.

Mees CEK (1920) *The Fundamentals of Photography*, Eastman Kodak, Rochester.

Reinhard E, Ward G, Pattanaik S & Debevec P (2006) *High Dynamic Range Imaging Acquisition, Display & Image-Based Lighting*, Elsevier, Morgan Kaufmann, Amsterdam.

Seetzen H, Heidrich H, Stuerzlinger W, Ward G, Whitehead L, Trentacoste M, Ghosh A, & Vorozcovs A (2004) High Dynamic Range Display Systems, *ACM Trans Graphics*, **23**(3), 760–68.

Section B

Measured Dynamic Ranges

10

Actual Dynamic Ranges

10.1 Topics

This chapter introduces Section B on the measurement of the actual ranges of light in images. It discusses the paradox that successful HDR reproductions do not accurately reproduce scenes. It reviews sensors' impressively large range of light responses, and the displays that combine light emission and light transmission technologies. The section introduces the issue that pixels in capture and display systems influence each other. As well, it describes the measurement of the range of light in images in cameras and on retinas.

10.2 Introduction

There are two fundamental scientific issues in the practice of HDR image capture and reproduction. First, we need to record as wide a range of light information as possible. Second, we need to display that information in the best way for humans.

One obvious approach to accurately reproducing high-dynamic range scenes is to improve the imaging technology. Namely, if we can improve the range of the light sensors, increase the number of digital quantization levels, and improve the range of light emissions of displays, we could make accurate reproductions over a greater range of light.

Figure 10.1 (top) illustrates common impressions of conventional digital imaging. We have silicon sensors that capture scene ranges of 2 to 3 log units. We encode digital files that have three (RGB) 8 bit quantization (0–255 levels). Displays and print make reproductions with 2 to 3 log unit ranges.

Figure 10.1 (bottom) shows three desirable improvements:

1. If silicon sensors had a wider range of responses to light they could increase the captured dynamic range.
2. If the analog to digital converters had an increased number of quantization levels, then pixels would have more gray levels.
3. If the displays had a greater range of light output, then such an improved system should be able to present a much more accurate reproduction of the original scene.

The Art and Science of HDR Imaging, First Edition. John J. McCann, Alessandro Rizzi.
© 2012 John Wiley & Sons, Ltd. Published 2012 by John Wiley & Sons, Ltd.

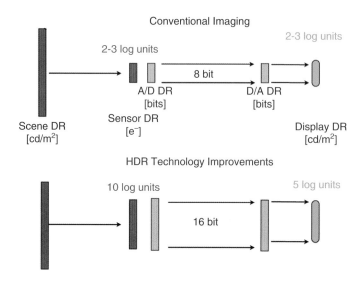

Figure 10.1 Three technological changes that might improve conventional imaging: wider range sensors; more bits per pixel; and a greater range of display luminances.

The interesting paradox is that the traditional reproductions do not attempt to reproduce scenes accurately. We have seen in *Section* A that the earliest technology was not up to the task of capturing the range and reproducing it. Nevertheless, in the 1850s Baldus and Robinson were successful in making HDR images. Then, we saw the work of Jones and Mees that defined a different combination of objective and subjective goals. Employing the spatial techniques of painters, Baldus, Robinson, and Adams synthesized the appearances of the HDR scene in LDR media.

In order to really understand this paradox, we need to resolve this question of whether it is a good idea to try to have an accurate capture and reproduction of the light coming from the scene.

This boils down to three specific objective, physics questions:

1. How large a dynamic range can sensors measure?
2. How large a range of light can displays emit?
3. What are the interactions of having millions of pixels together in an image?

As well there is a subjective question:

   Do people prefer accurate reproductions?

10.3   Dynamic Range of Light Sensors

It is possible to build electronic sensors with a dynamic range of 10, or more log units. Traditional CCD and CMOS technology works by counting the electrons in the pixel's well. The dynamic range for a sensor is the ratio of the full well capacity to the noise floor (dark noise level). Manufacturers can increase the well capacity with bigger, more costly pixels, or by reducing noise with cooling, and other techniques. Either, or both, can increase the sensor's range.

The development of Active Pixel Sensors (APS) (Matsumoto et al., 1985) and the integration of smart functions into imagers on a single chip (Yadid-Pecht et al., 1991; Mendis et al., 1997) led to considerable expansion of digital cameras. The desire for more and more pixels, and the scaling trends in CMOS technology facilitate higher resolution imagers. Smaller pixels have an adverse effect on performance because of limited quantum efficiency, increased leakage current, and decreased dynamic range. More recent sensors can respond actively during exposure in order to increase range. Active pixel processing has led to techniques that can increase the range of measurement previously limited by well capacity (Yadid-Pecht, 1999; Kavusi & Gamal, 2004; Spivak et al., 2009, Fowler, 2011).

Spivak et al. provides quantitative assessments of sensor performance within different categories, as well as overall qualitative comparisons between the different categories. Their analysis discusses the advantages and drawbacks of each of seven sensor categories:

1. Companding sensors that compress their response to light due to their logarithmic transfer function.
2. Multimode sensors that have a linear and a logarithmic response at dark and bright illumination levels, (switches between linear and logarithmic modes of operation).
3. Clipping sensors, a capacity well adjustment method;
4. Frequency-based sensors, sensor output is converted into pulse frequency.
5. Time-to-saturation [(TTS); time-to-first spike] sensors, signal is the time to saturated pixel.
6. Sensors with global control over the integration time.
7. Sensors with autonomous control over the integration time, where each pixel has control over its own exposure.

Spivak's analysis describes the size of pixels, power consumption, intrapixel capacitance, noise limits, maximal SNR and sensor sensitivity, in addition to the sensor's dynamic range. These theoretical results are truly impressive with many examples of more than 100 dB, or 10 log unit, dynamic ranges. Spivak did not emphasize the measurements of actual working devices.

Stoppa et al. (2007) designed and characterized an active pixel, implementing a mixed Active Pixel Sensory APS and TTS architecture, using a 140x140-pixel CMOS array. They report a sensor with over 120 dB range using simple monotonic or two piece waveforms. They showed image of a shadowed figure and lamp filament which they reported to have 120 dB range.

The sensor studies show the ability to read many log units of light falling on the sensor. The least successful active pixel sensors have five to six log unit ranges. It is practical to design sensors that have 10 log units of response to light at the sensor.

The question becomes: "How much of the 10 log unit range of the sensor can we use in camera imaging?

10.4   Bits per Pixel

The most common standard for digital imaging is 8 bits in each of three color channels, 24 bits in all. The number of bits defines the theoretical maximum number of gray, or lightness, levels in a color channel, 256 for 8 bit and 65 536 for 16 bit. The performance of the sensor, and of the display, determines whether all possible quantization levels are used succesfully. Usually, near maximum and minimum, the limits of technology, such as sensor blooming, and sensor noise floor, affect the shape of the sensor's tone-scale curve (Janesick, 2001). Because of these limitations, the quantization levels represent different increments in luminance near black, middle-gray and white. We cannot estimate the range of the sensor by raising 2 to the power of the number of bits. That would be backwards. We have to measure the range of the sensor between saturation from too much light, and noise-level threshold near to too little light.

Increasing the number of gray levels from 256 to 65 536 has no effect on the range of the system. Range is the measured max divided by the measured min. It simply increases the number of subdivisions of that fixed dynamic range. Using the *Italian salame* example, the size of the salame is fixed, and the number of *salame* slices changes. The danger in having too few gray levels is that the observer will see artificial quantization bands such as seen in Figure 31.1. If the range of light that falls into a single quantization level is too big then smooth gradients will appear as visible steps.

It is the observer that chooses the number of gray levels necessary. The most efficient system is one that uses just enough gray levels so that artifacts do not appear in any gradients. More gray levels just increase the cost of the system without benefit. The problem becomes more complex when we consider future processing, such as additional tone-scale maps, further along the processing chain. Stretching local regions of the tone map can make invisible artifacts very visible. In discussions of the optimum number of bits per pixel, we find the term *headroom,* that describes additional bits per pixel to allow for future alterations without artifact problems.

10.5 Dynamic Range of Display Devices

The dominant display technology today is Liquid Crystal Display (LCD). It works by using an electric field to rotate the plane of polarization of nematic liquid crystals between crossed polarizers. The crossed polarizers make minimum light transmission. With a variable electric field, the liquid crystal rotates the plane of polarization and increases the light transmission. LCD technology is fabricated using technology similar to silicon chips for CMOS sensors and memory chips. LCD has the capacity for very high resolution. LCDs vary between 2 and 3 log units range. One way to significantly increase the dynamic range of a display is to modulate the local illuminance on the backlight of the LCD screen. Seetzen et al. (2004) used a photographic unsharp mask technique to illuminate a high resolution LCD display with an out of focus array of LED illuminators. The Dolby embodiment, called Dolby Locally Modulated LED claims to have a 5 log unit range (Dolby, 2010). Since both LED and LCD signals are digital, HDR input image can be decomposed into low- and high-spatial frequency and redistributed to the separate LED and LCD display (Ledda et al., 2007).

In addition, the Dolby display has at least 1500 cd/m2 screen luminance, compared to 450 to 650 cd/ m^2 for standard displays, an increase of a factor of three. Dolby's citations of black level are: −0.001, compared to −.0.25 standard displays, a factor of 250 for large areas. The combined range increase is reported to be 750:1 or 2.9 log units.

There are significant problems in error-free rendering of the information in computer memory on a display device. Extensive calibration of all image areas throughout the full 3-D color space are needed to avoid hardware limitations. The hardware have many operations that affecting the light emission. The pixel value in memory is continuously sent, via a graphics card, to the display pixel (refreshed at the rate specified by the hardware). The physical characteristics of the display (spectral emission, number and size of pixels); the time budget (refresh rate and response times); the color image processing in the graphics card; and the circuitry in the display all influence the display's light output at each pixel. The amounts of light output does not always correspond to image digits in computer memory. A good example is that the EMF of the display signals in the screen wiring introduces image-dependent color shifts. (Feng & Daly, 2005) Hardware systems introduce image-dependent transformations of the input signals that on average improve the display's appearance. (Feng & Yoshida, 2008). HDR display with two active light modulators introduce even more complexity with high-resolution LCDs and low-resolution LEDs that are integrated with complex, proprietary, spatial filtering of the image data. It is not a simple matter to verify the accuracy of a display over its entire light-emitting surface, for all light levels, at full resolution, for its entire 3-D color space.

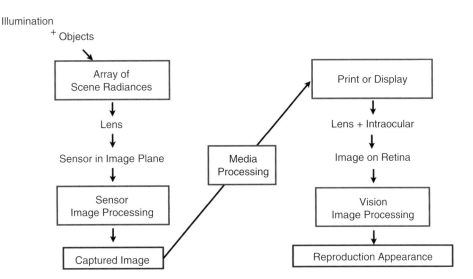

Figure 10.2   A schematic of the processes in making a reproduction.

10.6   Interactions of Pixels in Images

The schematic framework of our analysis starts with the illumination falling on the objects in a scene, and continues to the appearance of those objects in the reproduction (Figure 10.2).

10.6.1   Capture to Reproduction

The original scene is made up of one or more *Objects* in an unknown *Illumination*. A lens collects light from the scene. The light arriving at the lens is a complex integration of properties of the light, the surface of the objects and the spatial positions of the illumination, the objects and the lens. For example, a shiny black object can have very high radiance, if the illumination and object's surface form a specular reflection. Alternatively, that same black object can have a very low radiance in another position, or orientation, in the illumination. At the viewpoint of the lens there is a measurable *Array of Scene Radiances* that characterize that scene from that particular point of view.

The reproduction process uses a camera to attempt to record those scene radiances. The optical properties of the *Lens* introduce a transformation of the scene radiance involving focus, depth of field, aberrations and veiling glare. The lens creates an image on the camera's *Sensor in Image Plane* that represents the scene.

The *Sensor in Image Plane* records the quanta catch and the *Sensor Image Processing* transforms it into the *Captured Image*. In silver halide photography, the negative film is developed and fixed. In digital photography, the analog quanta catch is digitized, signal processed to transform it into a two-dimensional array of values with a specific encoding. In this Section B we will pay particular attention to the effects that high-dynamic-range scenes have on image recording. Ideally, each camera pixel captures all the light from a small area in the array of scene radiances.

Between the *Scene Radiances* and the *Captured Image* on the left side of Figure 10.2 there are two items that introduce practical limits to ideal performance:

With real lenses there are limits introduced by *veiling glare*. The sensor in the image plane records the pixel's desired light from its portion of the array of scene radiances, plus the unwanted scattered light from all the other pixels. Chapter 11 measures the limits of dynamic range on the sensor image plane caused by veiling glare in the lens.

All electronic cameras have complex digitization firmware that reads the signal captured by the silicon, and transfers that signal to digital memory. These processes affect the quanta catch signal, particularly for near the maximum and minimum values. The digital partially-processed image stored in memory is sometimes called the *Raw Image*.

Between the left and right sides of Figure 10.2 there is a *Media Processing* box that processes the *Raw Image* to de-mosaic the colored filters, apply desired tone-scale and color gamut adjustments, anticipate the rest of the reproduction process and compress it. (Fraser & Schewe, 2009; Green, 2010, Morovic, 2008; Chapters 32, 35) The output of the Media Processing is sent to printers and display devices.

10.6.2   Reproduction to Perception

The right side of Figure 10.2 describes the human response to the reproduction. Just as glass and plastic lenses have *veiling glare*, the tissues in the *cornea*, *lens* and *intraocular media* limit the dynamic range of light making the *Image on Retina*. Chapter 12 measures the limits of dynamic range of the image on the retina caused by veiling glare in the lens and intraocular media.

The *Vision Image Processing* box identifies the complex transformation that makes the study of vision so interesting. Sections C, D and E, that is much of the text, describe experiments measuring human response to spatial content in images.

10.7   Summary

Before we can turn our attention to the ideal signal processing of HDR systems, we need to measure the influence of the image on the individual pixel. The content of the image controls veiling glare, the veiling glare controls the range of the image.

Section B describes measurements of how much veiling glare limits HDR imaging in image capture. Glare is the scene- and camera-dependent scattered light falling on image sensors. First, glare limits the range of luminances that can be measured accurately by a camera, despite multiple exposure techniques. We used 4.3 log unit dynamic range test targets and a variety of digital and film cameras. In each case, the camera's response to constant luminances varied considerably with changes in the surrounding pixels. HDR image capture cannot record *accurately* the luminances in these targets.

As we will see in the following chapters, the improvement in HDR images, compared to conventional photography, does not correlate with accurate luminance capture and display. Accurate capture in a camera is not possible, and accurate rendition is not essential. The improvement in HDR images is due to better preservation of relative spatial information that comes from improved digital quantization. Spatial differences in highlights and shadows are not lost. Reproductions of scenes can look more like they actually appear. Spatial HDR image processing algorithms can mimic processes developed by human vision, by chiaroscuro painters, and by early photographers that render HDR scenes in low-range outputs.

10.8   References

Dolby (2010) Dolby's High-Dynamic-Range (HDR) Technologies: Breakthrough TV Viewing http://www.dolby.com/ uploadedFiles/zz-_Shared_Assets/English_PDFs/Professional/dolby-hdr-video-technical-overview.pdf.
Feng X & Daly SC (2005) Improving Color Characteristics of LCD. *Proc. SPIE*, **5667**, 328–35.

Feng X & Yoshida Y (2008) Improving the Gray Tracking Performance of LCD, *Proc.* CGIV 08/IS&T, *Terrassa*, **4**, 327–30.

Fowler B (2011) High dynamic range image sensor architectures, Proc SPIE 7876, 7876-01.

Fraser B & Schewe J (2009) *Real World Camera Raw*, Peachpit Press, Berkeley.

Green P ed (2010) *Color Management Understanding and Using ICC Profiles*, Wiley/IS&T, Chichester.

Ledda P, Chalmers A, Troscianko T & Seetzen H (2007) Evaluation of Tone Mapping Operators Using High Dynamic Range Display, *Proc ACM SIGGRAPH/EUROGRAPHICS Symp*, 24.

Janesick JR (2001) *Scientific Charge-Coupled Devices*, SPIE Press, Bellingham.

Kavusi S & Gamal A (2004) Quantitative Study of High Dynamic Range Image Sensor Architectures, *Proc. SPIE*, 5301, 264–75.

Matsumoto K, Nakamura T, Yusa A & Nagai S (1985) A New MOS Phototransistor Operating in a Non-Destructive Readout Mode, *Jpn J Appl Phys*, **24**, L323.

Mendis S, Kemeny S, Gee R, Pain B, Staller C, Kim Q, & Fossum E (1997) CMOS Active Pixel Image Sensors for Highly Integrated Imaging Systems, *IEEE J. Solid-State Circuits*, **32**(2) 187–97.

Morovic J (2008) *Color Gamut Mapping*, Wiley IS&T, Chichester.

Seetzen H, Heidrich W, Stuerzlinger W, Ward G, Whitehead L, Trentacoste M, Ghosh A, & Vorozcovs A (2004) High Dynamic Range Display Systems, ACM SIGGRAPH.

Spivak A, Belenky A, Fish A & Yadid-Pecht O (2009) Wide Dynamic-range CMOS Image Sensors: A Comparative Performance Analysis, *IEEE Trans on Electron Devices*, **56**, 2446-61.

Stoppa D, Vatteroni M, Covi D, Baschirotto A, Sartori A & Simoni A (2007) A 120-db Dynamic Range CMOS Image Sensor with Programmable Power Responsivity, *IEEE Journal of Solid-State Circuits*, **42**, 1555–63.

Yadid-Pecht O (1999) Wide Dynamic Range Sensors, *Opt Eng*, **38**, 1650–60.

Yadid-Pecht O, Ginosar R, & Shacham-Diamand Y (1991) Random Access Photodiode Array for Intelligent Image Capture, *IEEE Trans Electron Devices*, **38**(8), 1772–80.

11

Limits of HDR Scene Capture

11.1 Topics

High-dynamic-range images are superior to conventional images. The experiments in this chapter measure camera responses to calibrated HDR test targets. We made a 4.3 log unit test target, with minimal and maximal glare from a changeable surround. Glare is an uncontrolled spread of an image-dependent fraction of scene luminance in cameras. We use this standard test target to measure the range of luminances that can be captured on a camera's image plane. We discuss why HDR is better than conventional imaging, despite the fact that reproduction of luminance is inaccurate.

11.2 Introduction

Recently, multiple exposure techniques (Debevec & Malik, 1997) have been combined with LED/LCD displays that attempt to reproduce scene luminances accurately (Seetzen et al., 2004). However, veiling glare is a physical limit to HDR image acquisition and display. We performed camera calibration experiments using a single test target with 40 luminance patches covering a luminance range of 18 619:1 (4.3 log units). Veiling glare is a scene-dependent physical limit of the camera and lens (McCann & Rizzi, 2006a; 2006b; 2007a; 2007b; 2007c). Multiple exposures cannot reconstruct scene luminances accurately beyond the veiling glare limit.

11.3 HDR Test Targets

While the ISO Standard for glare (9358:1994) provides a standard to compare different lenses and apertures; we wanted to measure the effects of veiling glare on HDR imaging. We used a single calibrated test target with 40 test luminance sectors (dynamic range = 18 619:1). Nearly 80 % of the total target area was an adjustable surround; 20 % of the area was luminance test patches. Using opaque masks to cover the surrounding portions of the scene, we photographed three sets of HDR test scenes with different amounts of glare. The experiment compares camera digits with measured scene luminance

The Art and Science of HDR Imaging, First Edition. John J. McCann, Alessandro Rizzi.
© 2012 John Wiley & Sons, Ltd. Published 2012 by John Wiley & Sons, Ltd.

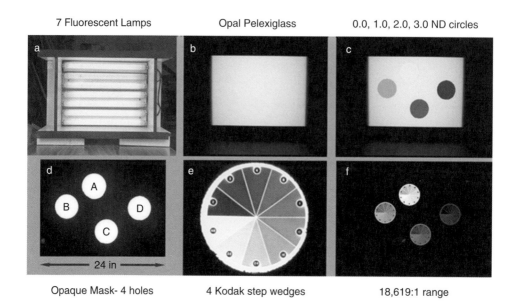

Figure 11.1 The apparatus. Panel a shows the light source with seven fluorescent tubes (20 W); Panel b shows an opal-plexiglas diffuser placed 15 cm in front of the lamps; Panel c shows the addition of three circular neutral density filters attached to the Plexiglas with densities of 1.0, 2.0, and 3.0; Panel d shows an opaque mask that covered the entire lightbox except for four circular holes registered with the N.D. filters; Panel e shows an enlarged view of a single Kodak Projection Print Scale; Panel f shows the assembled 4scaleBlack target with a dynamic range of 18,619:1 (2049 to 0.11 cd/m^2).

over a very wide range of luminances and exposure times. This experiment measured the extent that veiling glare distorts camera response in situations common to HDR practice.

In 1939, Kodak patented the Projection Screen Print Scale for making test prints (Gillon, 1940). It is a circular step wedge with 10 pie-shaped wedges, each with a different transmission. (Figure 11.1e) The range of transmissions was 20:1. After focusing a negative in an enlarger on the print film plane, darkroom technicians would place this scale on top of the unexposed print film in the dark. The measured transmissions of the wedges were 82%, 61%, 46%, 33%, 25%, 17%, 14%, 9%, 8% and 4% of incident light, so as to make a quick and accurate test print to select the optimal print exposure. We used these Scales to make 10 different test luminance sectors.

The components of our test display are shown in Figure 11.1. The display is made of transparent films attached to a high-luminance light-box. There are four Kodak Print Scale transparencies mounted on top of 0.0 (*Scale A*), 1.0 (*B*), 2.0 (*C*), and 3.0 (*D*) neutral density filters. The 40 test sectors are constant in all test targets.

For low glare, we covered all parts of the display except for the pie-shaped gray scales with an opaque black mask (*4scaleBlack*) (see Figure 11.2 middle). To measure the luminances of the sectors we used additional opaque black masks to cover all other sectors. We measured the luminance of each sector of *Scales A, B, C* and *D* using a spot luminance meter (Konica-Minolta LS-100C); one wedge at a time in a completely dark room. The highest transmission area in *Scale A* measured 2049 cd/m^2 (maximum); the lowest transmission area in *Scale D* measured 0.11 cd/m^2 (minimum). The paper masks prevented veiling glare from high-luminance test areas having an adverse effect on the accuracy of the spot meter. The dynamic range of the test target was 18 619 to 1.

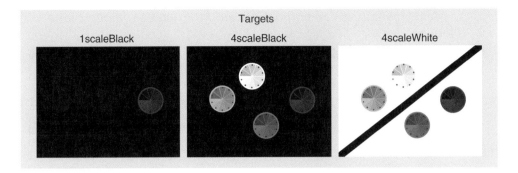

Figure 11.2   A diagram of very-low-glare, low-glare, and high-glare test targets.

For maximal glare, we removed the opaque black mask so that the zero-glare surround was replaced with a maximal glare surround (*4scaleWhite*). The diagonal line in *4scaleWhite* is an opaque strip in front of the display.

To measure the cameras' response to light, we further reduced glare; we covered the background and scales A, B and C, leaving only the light coming from *Scale D* (*1scaleBlack*) with a 20:1 range. Figure 11.2-left shows one very-low-glare target (range = 20:1); 11.2-middle the low-glare and 11.2 right high-glare targets (range = 18 619:1).

11.4   Camera Veiling Glare Limits

We made separate sets of measurements, first with a digital camera, then with a 35 mm film camera using both slope 1.0 slide duplication and conventional color negative films. We also used a lensless pinhole camera. We used all three HDR calibrated targets to measure the camera response. With the *1scaleBlack* target we measured the camera response using only the lowest luminances with a 20:1 range. With the *4scaleBlack* target we measured the camera response using a wider display range of 18 619:1 with minimal glare. With the *4scaleWhite* target we measured the camera response using the same display range with maximal glare (McCann & Rizzi, 2007c).

11.4.1   Digital Camera Response

Cameras respond to the number of photons captured by their sensor. The integrated amount of light (flux) on a camera's image plane is the product of the scene luminance (cd/m^2) and the time of exposure (sec). The idea is simply to assume that a pixel's scene flux [(cd/m^2) * sec] generates a unique camera digit. If we divide the pixel's flux value by the known exposure time we get the scene luminance of the pixel. Longer exposures collect more scene photons, and can move a dark portion of the scene up onto a higher region of the camera response function. This higher exposure means that the darker portions of the scene have better digital segmentation, more digits for better quantization of luminances. Digital HDR multiple-exposure techniques (Ochi & Yamanaka, 1985; Alston et al., 1987; Mann, 1993; Mann & Picard, 1995, Debevec & Malik, 1997) claim to extend the camera's range by calibrating flux vs. camera digit. Debevec and Malik make the specific argument that calculations using multiple-exposure data measure high-dynamic-range scene luminance. For reference, this technique will be described below as *Multiple Exposure to Scene Luminance (ME2SL)*. This is to distinguish this use of multiple exposures from the traditional use in photography.

11.4.2   Measurements of Luminous Flux on the Camera's Image Plane (1scaleBlack)

We measured the veiling glare's influence with 16 photographs taken with variable exposure times and the same f-7.3 aperture. We selected a Nikon Coolpix 990 camera because it has manual controls for both aperture and time of exposure. The *1scaleBlack* photographs have the lowest veiling glare and provide an accurate measure of the camera sensor response function. The only sources of glare are the test patches themselves (range 20:1). The camera response is the average digital value (from 491 pixels) calculated in Photoshop® for a circular area falling inside the pie-shaped luminance sectors (Figure 11.2, left).

Figure 11.3 shows the desired coincidence of camera digits and flux. The curve provides us with an accurate camera response function. The sensor digit values saturate at 247 with a flux of 78.4 sec*cd/m2; at digit value 1 the flux is 0.107 sec*cd/m2. The camera dynamic range is 731:1, or 2.9 log units. The black + symbols plot lookup table data is derived from the average measured digital data. This lookup table assigns an average luminance value to each digital value from 0 to 255 and is the measured camera response function.

The very high degree of overlap of multiple exposure responses in Figure 11.3 is possible because of the accuracy of the camera's exposure-time mechanism and the very low level of veiling glare found in the 20:1 test targets. The Multiple Exposure to Scene Luminance *(ME2SL)* technique works well in these conditions. The data in Figure 11.3 is plotted as flux (luminance * time) because that is related to the total number of photons falling on the camera's CCD sensor. We cannot derive actual photon counts without detailed knowledge of the camera's spectral responses, and anti-blooming, noise reduction and tone scale circuits. Nevertheless, we can plot (measured scene luminance*exposure time) to define the scene flux before interaction with the camera. The results in Figure 11.3 provide a consistent measure of camera response and a Look-Up Table (LUT) that allows us to convert camera digit to estimated scene luminance. This estimate is accurate as long as the *ME2SL* technique is error free, as shown in Figure 11.3.

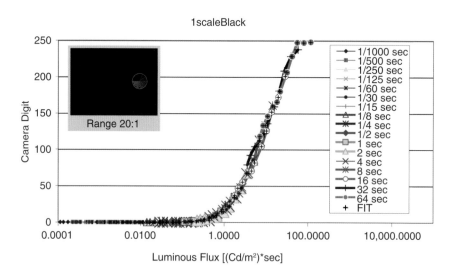

Figure 11.3   Plot of 1scaleBlack (range 20:1) flux vs.camera digits.

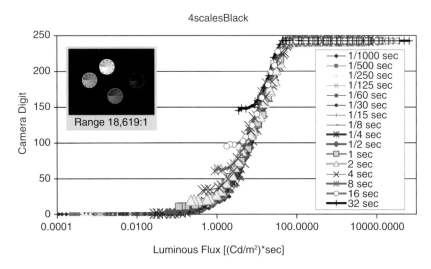

Figure 11.4   Plots of 4scaleBlack (range 18,619:1) flux vs. camera digits.

11.4.3   Measurements of Luminous Flux on the Camera's Image Plane (4scaleBlack)

In *4scaleBlack* the camera's responses attempt to capture a scene dynamic range of 18619:1 (Figure 11.4). This target measures the minimum glare for a scene with this range, because it has an opaque black surround. The only sources of glare are the 40 test patches that vary from 2094 to 0.11 cd/m².

Figure 11.4 shows the effects of glare for this range and configuration using a black surround between test scales. For an optimum exposure (1/2 sec) the sensor digits saturate at 242 with a flux of 119 sec*cd/m²; at digit 11 (departure from camera response curve) with a flux of 0.84 sec*cd/m². The glare limited dynamic range is 141:1, or 2.2 log units for this exposure. The effects of glare are seen as departures from a single camera function at the bottom of each exposure plot.

The data from Figure 11.4 shows that camera digit does not predict scene flux because the data for the darkest circle (*Scale D*) does not to fall on the single camera response function measured in 1scale-Black measurements. The *4scaleBlack* measurements show that the same scene luminance generates different (exposure-dependent) digits. This is important because this display was intended to measure the minimal glare for an 18619:1 scene. Despite the fact that 80 % of the target area does not contribute to glare, we measure that there is a problem with the *ME2SL* technique. The problem of glare can be seen easily in an exposure that does well in recording the light from the darkest circle. Figure 11.5 is the 16 second exposure. Scale D, (middle right) is a good record of the target. Scale C is overexposed with a large scene reflection on top of it. Glare from Scale B makes its saturated image size larger than the scene's image, while more glare makes the image of Scale A much larger. There is a larger non-inverted reflection on top of Scale A. The corners of the image are much darker than the center, even though the scene's black mask is uniformly opaque. Glare limits the range of light falling on the image plane of the camera.

Sowerby (1956), in his *Dictionary of Photography* discusses the reflection of light in lenses as the "diversion of an appreciable portion of the incident light from its intended path". The small percentage of light reflected from each air-glass surface is called a parasitic image. Parasitic images that are completely out of focus give rise to a general fog that limits the dynamic range of the image falling on the

Figure 11.5   Photograph of 4scaleBlack with 16 sec exposure for optimal recording of the luminances in Scale D.

image plane. The actual *4scaleBlack* image (Figure 11.5) shows a magnified inverted in-focus parasitic image, as well as the out-of-focus fog. Multiple exposures improve the digital quantization, and thus the camera's performance. Nevertheless, multiple exposures have no effect on the dynamic range of the image falling on the sensor. Veiling glare is a constant fraction of scene luminance that varies with scene content.

The formula to count parasitic images is $2n^2-n$, where n is the number of lenses. The digital camera used here (Nikon Coolpix 990) has nine elements and 153 parasitic images. The film camera used below (NikonFM with a Nikkor 50 mm 1:2 lens) has seven elements and 91 parasitic images (McCann & Rizzi, 2007c). Kingslake (1992) describes many examples of parasitic images that contribute to veiling glare.

11.4.4   Measurements of Luminous Flux on the Camera's Image Plane (4scaleWhite)

When we remove the black mask covering the lightbox in the background, we go to the situation with maximal veiling glare (*4scaleWhite*). Nearly 80 % of the pixels are making the highest possible contribution to veiling glare (Figure 11.6).

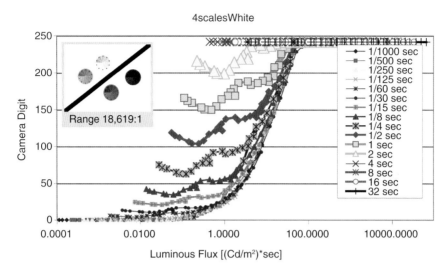

Figure 11.6 Plots of flux vs. camera digits from 16 different exposure times for 4scaleWhite, the high-glare target.

Figure 11.6 shows that the influence of glare is dramatic. For *Scales C* and *D* camera digits are controlled as much by glare as by luminance. The many large departures from a single line are due to scene-dependent glare. All departures from the camera response function are errors in the *ME2SL* technique.

The data from the three sets of photographs are different. Data from *1scaleBlack* (Figure 11.3) provides a single camera sensor response function. Camera digit correlates with scene flux. Data from *4scaleBlack* show a lack of correlation for low luminances at some exposures. Data from *4scaleWhite* show that glare corrupts camera digit correlation with scene flux. This is a major problem for the Multiple Exposure to Scene Luminance (*ME2SL*) technique. It works well only when veiling glare is low. Veiling glare is both scene and camera dependent.

11.4.5   Errors in Estimated Scene Luminance

We took the data of Scale D from *1scaleBlack* to generate a lookup table that describes camera digit as a function of flux and its inverse (See FIT in Figure 11.3). We then used this camera response lookup table to convert the camera digits from *4scalesBlack* and *4scalesWhite* to get the camera estimate of flux. Figure 11.7 plots camera estimated flux for each exposure for both low- and high-glare targets. When the ME2SL technique is working we expect to see the camera response plots shifted in a series of parallel lines. Since the scene luminances are identical, the plots for *4scaleBlack* and *4scaleWhite* should be identical. Any departures from parallel plots indicate errors in the ME2SL process.

To measure the size of the errors created by veiling glare we took the ratio of camera-estimated flux to actual measured flux. This ratio is a measure of *ME2SL* error of each target sector. If the camera digit predicted scene flux accurately, then this ratio equals 1.0. Ratios greater than 1.0 measure the magnitude of errors introduced by glare. Figure 11.8 plots these error ratio values vs. scene luminance. If camera digits measure scene luminance (*ME2SL*) accurately, then all the data must fall on a horizontal line with a value of 1.0. For the *4scalesBlack* the results are close to 1.0 for log luminance above 0.0.

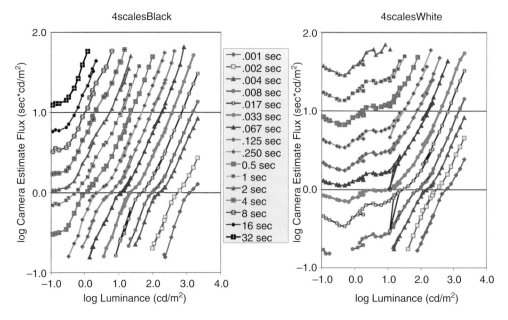

Figure 11.7 The log camera-estimated flux for 4scaleBlack (left) and 4scaleWhite (right) vs. log scene luminance of the test target. If camera digit measures scene luminance (ME2SF) accurately, then all the data must fall on a set of parallel lines separated by exposure time. Both sets of curves should be identical. Note that four curves (4 to 32 sec exposures) have disappeared because all pixels were at saturation. The ME2SF process has only limited success for the highest exposures.

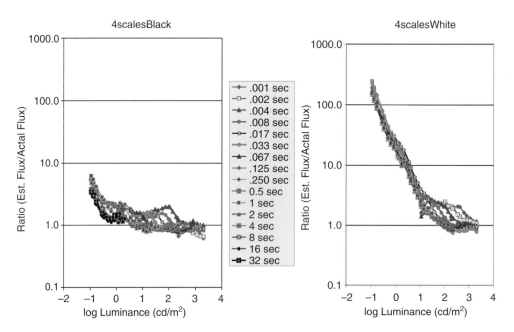

Figure 11.8 (left) Ratio of (Camera-Estimated Flux/Actual Flux) for 4scaleBlack; (right) 4scaleWhite vs. log scene luminance of the test target. Error-free estimates of scene luminance fall on the horizontal line of Ratio = 1.0.

The maximum error (estimated camera flux/actual flux) was 6.15, or 615% for the lowest luminance test area.

For the *4scalesWhite* there are no results close to 1.0 for log luminance below 1.5 log luminance. The maximum error (estimated camera flux/actual flux) was 243.83, or 24 383%, again for the lowest luminance test area.

The test target has a dynamic range of 18 619:1, or 4.3 log units. We can use the Multiple Exposures to Scene Flux (*ME2SF*) technique for approximating luminance over a range of about 3 log units with the low-glare *4scaleBlack* target. With maximal white surround glare that range is reduced to 1.5 log units. If we hypothesize a variety of different surrounds to substitute for the all white, or all black surround, all possible luminance backgrounds, around *Scales A, B, C,* and *D,* will fall in between the white and the black data sets. The usable range of this camera is between 1.5 and 3.0 log units depending on the scene content.

Camera digits from multiple exposures cannot provide a reliable means of measuring HDR scene luminance. Camera digits cannot capture HDR scene luminance accurately because of glare.

11.5   Glare in Film Cameras

As discussed earlier in Chapter 5, Jones and Condit (1941) described techniques to measure camera scenes and scene dependent glare. In this section we performed the same experiments using two types of photographic films. First, we used an industrial film for copying transparencies (Ektachrome EDUPE). It has the desirable properties of having a large dynamic range and a linear response to light. This film does not have the S-shaped response found in ordinary films (Figure 5.6).

11.5.1   Duplication Film-Camera Response

We made another set of photographs with a typical high-quality 35 mm film camera (NikonFM with a Nikkor 50 mm 1:2 lens). This follows the single exposure HDR capture technique described by McCann (1988) in tutorials at Siggraph conferences in 1984 and 1985. Slide duplication film has slope 1.0 on a log exposure vs. log luminance plot. In other words, output luminance is equal to input luminance. Since it is a color film it can be scanned for color and does not require calibration to remove the color masks found in color negative film. Here we use multiple exposures to capture both 18 619:1 displays (*4scaleBlack* and *4scaleWhite*). The exposed film was developed with the standard E6 process. All exposures were mounted in a single 35 mm film holder so that all images were scanned at the same time with the same scanner settings. The scanner was an Epson Perfection Photo with calibration for E6 films. Figures 11.9 and 11.10 plot scanned positive film digit vs. log flux.

Figure 11.9 data show that this particular camera-film-scanner system has less veiling glare than the digital camera in Section 11.4.

Figure 11.10 plots the scanned digits from the set of exposures of the *4scaleWhite* target.

Although there may be small differences between this data and the digital camera's response, the same scene-dependent glare dominates both results.

11.5.2   Negative Film-Camera Response

We made another set of photographs with the same NikonFM camera using Kodak Max 200 negative film. The exposed film was developed with the standard C41 process. Again, all exposures were mounted for a single scan at the same time with the same scanner settings. The scanner was an Imacon Flextight Precision with calibration for C41 films.

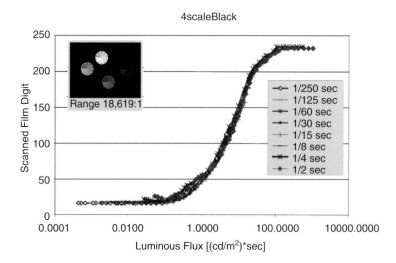

Figure 11.9 Plots of scanned film digits from Ektachrome EDUPE photographs vs. luminous flux for eight exposures. The data superimpose to form a single function, except for the very lowest luminance sectors.

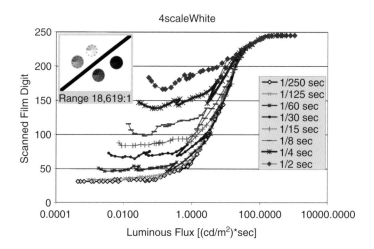

Figure 11.10 Scanned digits from Ektachrome EDUPE photographs of the 4scaleWhite data. Here the white surround adds veiling glare to generate eight different response functions.

First, we used seven different exposures to measure the camera-film-scanner process using the very low glare 20:1 single scale (*1scaleBlack*). The red squares in Figure 11.11 plot the combined response of film, camera, development process, and film scanner. The negative process can record fluxes accurately from 2639 to 0.238 sec*cd/m^2 (dynamic range of 11 100:1, or 4.05 log units). This is much greater than the 2.9 log unit range of the digital camera in Section 11.4.

We then photographed *4scaleBlack* and *4scaleWhite*, using the same single exposures to capture the 18 619:1 displays. In the case of the white surround, the gray scales had high digital values, covering less than 50 digits. We made an additional single exposure called *4scaleWhiteb* negative with a

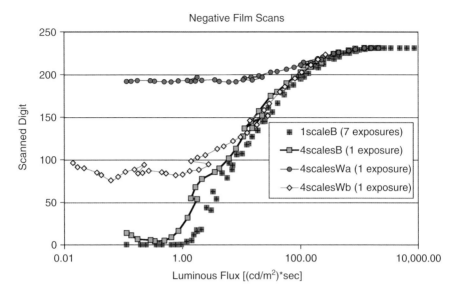

Figure 11.11   Plots of scanned single negative digits vs. log luminous flux for three targets.

shorter exposure time. This scanned film had a digit range of 150, allowing better quantization of the scene.

The green squares in Figure 11.11 plot the single exposure curve from *4scaleBlack*. It saturates at 1181 sec*cd/m² and inverts at 0.34 sec*cd/m². The inversion is caused by glare that limits usable range. The accurate dynamic range is 6345:1, or 3.54 log units. The single exposure data from *4scaleBlack* show a small effect of glare from the addition of 30 higher luminance test pie-shaped areas. This glare reduces the dynamic range of the image in the camera to 3.5 log units.

There are two different single exposures for the *4scaleWhite* target; one is eight times longer than the other. The data from *4scaleWhite (a; blue circles)* saturates at 1181 cd/m² and inverts at 6.17 cd/m². The curve from *4scaleWhite (b: yellow diamonds)* has a maximum digit at 2094 cd/m² and inverts at 6.17 cd/m². The glare from the white surround reduces the range to 2.3 *(a)* and 2.5 *(b)* log units.

The fact that the dynamic ranges for the two exposures of the *4scaleWhite* target are almost the same is important. Their response curves in Figure 11.11 are very different. The *4scaleWhite* scanned digits have a max of 231, and a min of 191. The *4scaleWhite (b)* scanned digits have a max of 223, and a min of 94. The range of digits representing the scene is only of secondary importance. The range of digits describes the number of quantized levels used to represent the image. It controls discrimination, but does not control the dynamic range of the image.

There is an interesting characteristic of scanned negatives. The lowest density film (highest transmission) response measures the lowest scene luminance levels. These areas send the most light to the scanner and have the best signal-to-noise scanner response. As shown in Figure 11.11, there is a long "toe" region that stretches from 100 to 10000 sec*cd/m2. Over that 2 log unit scene range the scanned digits are all above 200. Although the digital quantization is poor, the sampling (average digital value) discrimination is excellent because of the strong signal read by the scanner.

Conventional negative film can capture a greater range of luminances than that which falls on the camera image plane from these targets. The dynamic range of a single exposure negative-film-scanner process exceeds the glare limited *4scaleBlack* scene by 0.5 log units and glare limited *4scaleWhite* scene by 1.6 log units. Multiple exposures with negative films serve no purpose. The glare-limited range of the camera for these HDR scenes is smaller than the film system range.

Jones and Condit (1941) measured the luminance range in 128 typical outdoor photographic scenes. (Chapter 5.8.3) They reported a minimum range of 27:1; the maximum was 750:1; and the average was 160:1. One can increase the scene range by including the light source and specular reflections. The data above suggest that in high, and in average glare scenes, the glare-limited image dynamic range on the film/CCD image plane will be less than 3.0 log units. Only in special cases, very low-glare scenes, will the image plane's dynamic range exceed 3.0. The data here shows that the designers of negative films did well in optimizing the process. They selected the size distribution of silver halide grains to make the negative have a specific dynamic range, around 4.0 log units. Thus with low glare scenes, single-exposure negatives capture the entire range possible in cameras. For most scenes, this image-capture range provides a substantial exposure latitude, or margin of exposure error. Color negative film captures all the light possible in cameras for the vast majority of scenes. After reading the papers of Mees, Jones, and Condit, it is easy to believe that this fact is not a coincidence (Jones & Condit, 1941; Mees, 1937; Mees & James, 1966).

11.5.3   Pinhole-Camera Response

Jim Larimer suggested that we measure the dynamic range of images made with a lensless pinhole camera. We made a pinhole out of soft black plastic, counter bored with a 1 cm drill and pierced with a needle. The pinhole was slightly elliptical. The average diameter was 376 microns (392 by 362). It was placed 50 mm from the film plane. Each point source in the scene is diffracted by the pinhole. The Airy pattern formed by diffraction is the distribution of light on the image plane from a single point source of light (Padley, 2005). The major lobe (peak to first minimum), called Airy's disc, is 83.7% of the light falling on the image plane (Ditchburn, 1991). The diameter of that lobe in 550 nm light in this camera is 0.178 mm. The remaining 16.3% of the light is diffracted outside the disc to form a diffracted fog that limits dynamic range.

The pinhole images are less sharp, and the *4scaleWhite* has a smaller luminance range than the digital images made with a lens. Figure 11.12 plots the scanned digits from negatives taken in the

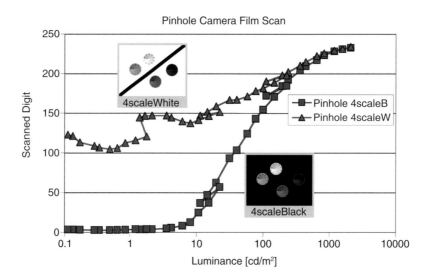

Figure 11.12   Plots of pinhole camera digits scanned from negative film images.

pinhole camera with 180 sec exposures of the *4scaleWhite* and *4scaleBlack* targets. The diffraction fog from the white surround increases the responses of the gray sectors compared to the same luminances in the black surround. As before with lenses, we were unable to record scene luminance accurately using a pinhole camera.

11.6   Review

Summarizing, regardless of the type of camera, film and lens, HDR images have strong optical limits. The range of light falling on the sensors is limited by veiling glare from parasitic images in glass lenses and diffracted fog in pinhole images. Although the glare is formed by reflections in one case, and diffraction in the other, they both show limited dynamic range. That range depends on both the camera and the scene. Scene dependence is a substantial problem for any *ME2SL* HDR algorithm attempting to measure scene luminance.

Veiling glare limits HDR imaging because camera glare limits the luminance range that can be measured accurately. Multiple exposures improve the quantization of digital records, but fail to record scene luminance accurately.

Accurate capture of scene luminances from camera images is impossible to achieve. We were unable to make accurate camera estimates of scene luminance for the 4.3 log dynamic range scenes studied here. The comparison of white and black surrounds shows dramatic scene dependence. It may be tempting to look for some type of average-flux curve that represents data with smaller errors, but that idea is in conflict with the fundamental aim of the process, namely recording accurate scene luminance. Multiple-exposure HDR (*ME2SL*) is limited by veiling glare that is scene-, lens-, aperture-, and camera-body-dependent. The accuracy of scene-luminance estimates varies with all these parameters.

Some HDR algorithms attempt to correct for glare (Xiao et al., 2002; Reinhard et al., 2006; Bitlis et al., 2007). Given the characteristics of the camera, they calculate the luminances in the scene. The glare spread functions of commercial lenses fall off very rapidly with distance to a very small value. We might think that such small glare values cannot affect distant pixels. However, there are millions of pixels that contribute glare to all other pixels. The light at each pixel is the sum of scene luminance plus scattered light from all other pixels. The sum of a very large number of small contributions is a large number. Sorting out these millions of scene-dependent contributions would be required to correct precisely for glare. The ISO 9358:1994 Standard (1994) states unequivocally that: "the reverse (deriving luminance from camera response) calculation is not possible."

Claims are made that recent multiple-exposure HDR algorithms capture wider scene luminances, or colors than previously possible (Reinhard et al., 2006; Qui et al., 2007). These claims are severely limited by scene and camera veiling glare. As shown above, the designers of negative films selected a 4.1 log response range. That range exceeds the camera glare limit for almost all cameras and scenes.

11.7   Summary

This chapter describes measurements of how much veiling glare limits HDR imaging in image capture. Glare is the scene- and camera-dependent scattered light falling on image sensors. First, glare limits the range of luminances that can be measured accurately by a camera, despite multiple exposure techniques. We used 4.3 log unit dynamic range test targets and a variety of digital and film cameras. In each case, the camera response to constant luminances varied considerably with changes in the surrounding pixels. We were unable to make accurate camera estimates of scene luminance for the scenes studied here using HDR multiple exposure image capture.

The ranges measured here are scene dependent, and cannot be used to make conclusions about other scenes. These ranges are also camera and lens specific. Ironically, the only data that is free of veiling glare is from the 20:1 very low-dynamic-range test target.

11.8   References

Alston L, Levinstone D & Plummer W (1987). Exposure Control System for an Electronic Imaging Camera Having Increased Dynamic Range, US Patent 4,647,975, Filed date 10/30/85, Issued 3/3/1987.

Bitlis B, Jansson PA, & Allebach JP (2007) Parametric point spread function modeling and reduction of stray light effects, *Proc. SPIE*, Bellingham, 6498–27.

Debevec P & Malik J (1997) Recovering High Dynamic Range Radiance Maps from Photographs, *ACM SIGGRAPH'97*.

Ditchburn R (1991) *Light*, Dover, New York, 165.

Gillon J (1940) Exposure Tablet, US Patent 2,226,167 filed, 11/30/1939, issued 12/4/1940.

ISO 9358: 1994 Standard (1994) Optics and Optical Instrument: Veiling Glare of Image Forming Systems. Definitions and Methods of Measurement. International Organization for Standardization, Geneva.

Jones L & Condit H (1941) The Brightness Scale of Exterior Scenes and the Computation of Correct Photographic Exposure, *J Opt Soc Am*, **31**, 651–78.

Kingslake R (1992) *Optics in Photography*, SPIE Press, Bellingham.

McCann JJ (1988) Calculated Color Sensations applied to Color Image Reproduction, *Proc. SPIE*, **901**, 205–14.

McCann JJ & Rizzi A (2006a) Optical Veiling Glare Limitations to In-Camera Scene Radiance Measurements, *ECVP 2006 Abstracts, Perception*, **35** Supplement, 51.

McCann JJ & Rizzi A (2006b) Spatial Comparisons: The Antidote to Veiling Glare Limitations in HDR Images, *Proc ADEAC/SID&VESA*, Atlanta, 155–58.

McCann JJ & Rizzi A (2007a) Veiling Glare: The Dynamic Range Limit of HDR images, *Proc. SPIE, Bellingham*, 6492–41.

McCann JJ & Rizzi A (2007b) Spatial Comparisons: The Antidote to Veiling Glare Limitations in Image Capture and Display, *Proc, IMQA 2007*, Chiba, E-1.

McCann JJ & Rizzi A (2007c) Camera and Visual Veiling Glare in HDR images, *J Soc Info Display*, **15/9**, 721–30.

Mann S (1993) Compositing Multiple Pictures of the Same Scene, *Proc. IS&T Annual Meeting*, **46**, 50–2.

Mann S & Picard R (1995) On Being 'Undigital' with Digital Cameras: Extending Dynamic Range by Combining Different Exposed Pictures. *Proc. IS&T Annual Meeting*, **48**, 442–8.

Mees CEK (1937) *Photography*, The MacMillan Company, New York.

Mees CEK & James T (1966) *The Theory of Photographic Process*, 3rd edn, MacMillan Company, New York.

Ochi S & Yamanaka S (1985) US Patent 4,541,016, filed 12/29/1982, issued 9/10/1985.

Padley P (2005) Diffraction From a Circular Aperture http://cnx.org/content/m13097/latest/.

Reinhard E, Ward G, Pattanaik S & Debevec P (2006) *High Dynamic Range Imaging Acquisition, Display and Image-Based Lighting*, Elsevier, Morgan Kaufmann, Amsterdam.

Seetzen H, Heidrich H, Stuerzlinger W, Ward G, Whitehead L, Trentacoste M, Ghosh A, & Vorozcovs A (2004) HighDynamicRange Display Systems, *ACM Trans Graph*, **23**(3), 760–8.

Sowerby A (1956) *Dictionary of Photography*, 18th ed. Philosophical Library, New York, p 568.

Qiu G, Reinhard E & Finlayson E (2007) Guest Editorial: Special issue on high dynamic range imaging, *J Vis Commun Image R*, **18** 357–8.

Xiao F, DiCarlo JM, Catrysse PB & Wandell BA (2002) High Dynamic Range Imaging of Natural Scenes, *Proc. IS&T/SID Color Imaging*, **10**, 337–42.

12

Limits of HDR in Humans

12.1 Topics

We measured the appearance of the test luminance patches used to calibrate camera glare. If displays perfectly reproduced the flux from a scene would human vision be able to use that information?

12.2 Introduction

This chapter describes measurements on humans of the effects of veiling glare on the appearance of the same test targets used in Chapter 11. We saw that camera glare limits the range of luminances falling on the camera sensor plane. Human glare is caused by Tyndall scattering by macromolecules in the intraocular media, reflections from the whitish blind spot, as well as the layers of neurons between the lens and the sensors. Scatter limits the eye's dynamic range more than glass lenses limit cameras. Here we will describe the range of discrimination and the corresponding range of retinal luminances. We will measure observed appearance for both *4scalesBlack* and *4scalesWhite* test targets.

12.3 Visual Appearance of HDR Displays

We asked observers to evaluate the appearance of the *4scaleBlack* and *4scaleWhite* displays using *magnitude estimation*. Observers sat 1.9 m from the 61 cm wide display. The radius of each sector was 5.1 cm; subtending 2.4 degrees. Three observers were asked to assign 100 to the "whitest" area in the field of view, and 1 to the "blackest". They were given a diagrammatic map of the display and asked to write the values 100 and 1 in the sectors they selected. We then instructed them to find a sector that appeared middle gray and assign it the estimate 50, and identify it on the map. We then asked them to find and record sectors having 25 and 75 estimates. Using this as a framework the observers continued to assign estimates to all 40 sectors. The data from each observer was analyzed separately. No significant difference between observers was found.

The average magnitude estimate results (Figure 12.1) show very different responses to luminance. The *4scalesBlack* estimates (blue diamonds) show that observers can discriminate between all the

The Art and Science of HDR Imaging, First Edition. John J. McCann, Alessandro Rizzi.
© 2012 John Wiley & Sons, Ltd. Published 2012 by John Wiley & Sons, Ltd.

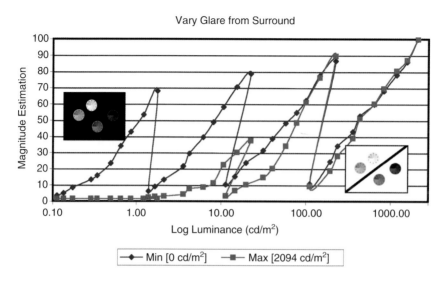

Figure 12.1 Plots of magnitude estimation of appearance vs. measured luminance for the 40 sectors in 4scalesBlack and 4scalesWhite test targets.

segments in A, B, C, and D. With minimal glare from black surrounds, observers can discriminate between the entire 4.3 log range. Observers report four response curves with 10 different lightness estimates for each test circle.

The local maxima (highest transmission in A, B, C, D) in *4scalesBlack* have estimates ranging over 100 to 69. They vary in appearance between white and light gray. The appearances below these maxima decrease rapidly with luminance. The four segments A, B, C, and D appear almost the same, despite large changes in luminances. Discriminable edge contrast builds significant visual differences from small luminance differences on the retina.

The effects of adding veiling glare are seen in the *4scalesWhite* appearances. The *4scalesWhite* estimates (Figure 12.1, squares) show that observers cannot discriminate between the pie sectors in D, and that those in C are very similar to each other. We can hypothesize that intraocular scatter has added stray light across the retinal image of the target. That could explain why observers are unable to see differences below 2 cd/m². We will see later that this assumption is correct.

Although the scene luminances are exactly equal the appearances are not. With a black surround observers can discriminate between all 10 sectors in all four displays. With a white surround observers cannot discriminate below 2 cd/m2. Veiling glare makes the gray test sectors brighter, but they look darker.

The *4scalesBlack* and *4scalesWhite* appearance estimates overlap for only the top six luminances in Scale A. Below that, equal scene luminances appear darker in the white surround. The plots of appearance vs log luminance for Scale B have a higher slope in the white surround, than in the black one. Equal scene luminances look darker, despite the added glare. The white surround makes the local maxima in *Scales C* and *D* darker than in the zero-luminance surround. Scattered light from the white surround severely limits all discrimination below 2 cd/m².

In *4scalesBlack*, the pie-shaped sectors with the highest luminance in each scale all appear light (Estimates: A = 100, B = 90, C = 80, D = 69). As shown in other experiments, the local maxima generate appearances that change slowly with luminance. (McCann, 2005; 2007). The local maxima in the scene change very slowly with their luminance values. The other segments near these local maxima, change more quickly with luminance. This is due to *neural contrast,* vision's spatial image processing (Section C). The different rates of change of appearance with change in luminance is key to understanding HDR imaging. The appearance of maxima change slowly, and areas less than local maxima change rapidly with different luminances. The spatial pattern of luminances controls the appearance of these darker areas.

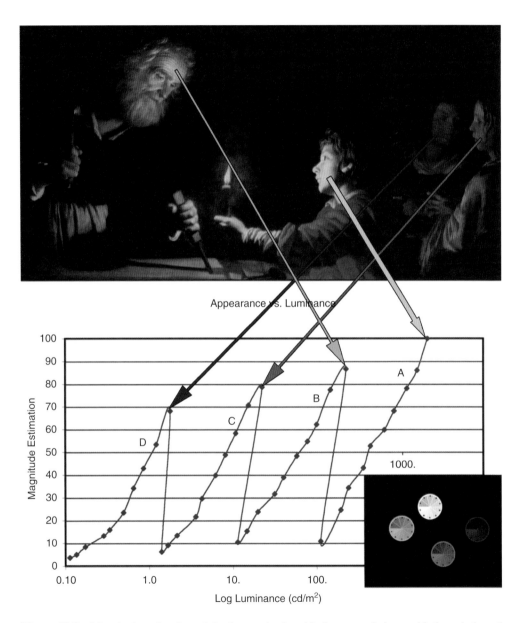

Figure 12.2 Magnitude estimation of the four scales in a black surround along with the painting of "The Childhood of Christ".

12.4   von Honthorst's Painting and the 4scaleBlack HDR Target

The experimental data for the black surround (diamonds in Figure 12.1) show great similarity to Gerrit van Honthorst's faces in the painting "Childhood of Christ" (Chapter 4.6). Each Scale is the analog for the four figures in the painting. As the distance between the candle and the faces grew, the tones rendering the faces got slightly darker. Each person is rendered slightly darker, but the spatial contrasts for each are very similar. Note the correlation with observers' assignment of almost the same tone-scale magnitude estimates to *Scales A, B, C,* and *D* in which each started and ended a few percent lower in magnitude estimates, despite the substantial decreases in luminance. Figure 12.2 assigns numbers to 16th century observations. Chiaroscuro painters did not render luminances; rather they rendered what they saw.

12.5   HDR Displays and Black and White Mondrian

One can think of the *4scalesBlack* experiment as a 4-scale version of Land's "Black and White Mondrian" (Chapter 7; Land & McCann, 1971). In that experiment the illumination was adjusted so that a high-reflectance paper in dim light had the same luminance as a white paper in bright light. Despite the fact that the two areas sent the same light to the eye, one appeared white and the other black. In the *4scaleBlack* experiment we have, side by side, four identical transparencies in four different illumination intensities. When the targets are isolated in the opaque surround, we have three different examples of white and black appearances generated by the same luminances (brown arrows in Figure 12.3).

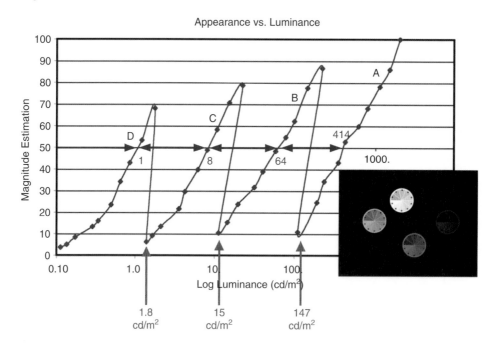

Figure 12.3   Three different luminances that have both near white and near black appearances (brown arrows). As well, the red arrows identify the four luminances (1, 8, 64, 414 cd/m²) that appear as the same gray, equal to magnitude estimate 50.

For luminance $147 \, \text{cd/m}^2$ there is a pair of magnitude estimates: 17 in A and 78 in B. For $15 \, \text{cd/m}^2$, the magnitude estimates are: 7 in B, and 71 in C. For $1.8 \, \text{cd/m}^2$ the magnitude estimates are: 10 in C, and 68 in D.

We can also analyze this data by looking at the luminances that created a constant appearance. Figure 12.3 used red arrows along magnitude estimates of 50. We have four different luminances (1.06, 8.4, 63.5 and 414.0) that generated the same appearance (ME = 50). The same holds for all magnitude estimates except for the lightest and darkest values. As argued in the *Black and White Mondrian* experiment (Land & McCann, 1971), there often is no correlation of a pixel's luminance and its appearance. Models of appearance require a strong spatial component.

12.6   HDR and Tone Scale Maps

Recently, there has been a lot of interest in digital tone scale maps that convert captured scene luminance (Debevec & Malik, 1997) to ideal renditions for humans. (Tumblin & Rushmeier, 1993; Larson et al., 1997; Reinhard et al., 2006) The observer data in Figure 12.3 show dramatically that tone scales for HDR images are of little use when discussing the general solution of rendering all scenes. Statistically the best tone scale is a straight line that starts at magnitude estimate 100 and goes to 1, because it has the lowest average error. One can use Look-up Tables to enhance one region of the *4scalesBlack,* or a region of the *4scalesWhite*. A single Global Look-up Table is unable to improve both *4scalesBlack,* and *4scalesWhite* at the same time. Just as Land pointed out with his *Black and White Mondrian* experiment, pixel-based- processes (silver halide film, LUTs, and Global Tone Scales) cannot render white and black values succesfully from identical inputs.

12.7   HDR Displays and Contrast

Our targets have exactly the same scene luminance in each of the 40 segments. The *4scalesWhite* target is created by removing the opaque black mask from the *4scalesBlack* target. The transparent pie-shaped segments send the same light to the eye. The only physical difference at the target is the surround. Figure 12.1 illustrated that the scene radiances are altered by veiling glare to create the image on the retina. The *4scalesBlack* has minimal, and the *4scalesWhite* has maximal glare. That means that the amount of light on the retina for each of the 40 segments is greater in the white surround, than in the black surround.

In the white surround the sector's appearances are darker in *Scales B, C,* and *D* than in the black surround, despite the fact that they have much more light on the retina.

The important paradox here is that we have two different visual mechanisms working against each other. They are both spatial mechanisms that involve all the pixels in the scene. One is veiling glare that increases the amount of light on the retina. The other is neural contrast from spatial interactions that makes *"more light"* look darker.

12.8   Summary

Here, human observer experiments that measure the appearances of identical gray areas show two independent and opposing visual mechanisms. Intraocular veiling glare reduces the luminance range on the retina, while neural contrast increases the apparent differences in lightness.

12.9   References

Debevec P & Malik J (1997) Recovering High Dynamic Range Radiance Maps from Photographs, *ACM SIGGRAPH'97*, 369–78.

Land EH & McCann JJ (1971) Lightness and Retinex Theory, *J Opt Soc Am*, **61**, 1–11.

Larson G, Rushmeier H & Piatko C (1997) A Visibility Matching Tone Reproduction Operator for High Dynamic Range Scenes, *IEEE Trans Vis & Comp Graphics*, **3**(4), 291–6.

McCann JJ (2005) Rendering High-Dynamic Range Images: Algorithms that Mimic Human Vision, Maui: *Proc. AMOS Technical Conference*, 19–28.

McCann, JJ (2007) Aperture and Object Mode Appearances, *Proc. SPIE*, 6292–26.

Reinhard E, Ward G, Pattanaik S & Debevec P (2006) *High Dynamic Range Imaging Acquisition, Display and Image-Based Lighting*, Elsevier, Morgan Kaufmann, Amsterdam.

Tumblin J & Rushmeier H (1993) Tone Reproduction for Realistic Images, *IEEE Computer Graphics and Applications*, **13**(6), 42–48.

13

Why Does HDR Improve Images?

13.1 Topics

HDR photographs render a wider range of luminances in the highlights and shadows than conventional ones. They improve the colors of objects in the scene. Careful measurements in previous chapters showed that cameras have a limited, scene-dependent, range of accurate luminance measurement. Even if cameras could capture accurate scene information, and displays could reproduce it, humans' intraocular veiling glare modifies the display information depending on the scene content. This chapter summarizes Section B and describes the reasons for improved HDR images.

13.2 Introduction

There must be reasons, other than accurate luminance rendition, that explain the improvement in HDR images.

Veiling glare limits HDR imaging in two distinct ways. First, camera glare limits the luminance range that can be measured accurately (Chapter 11). Multiple exposures improve the quantization of digital records, but fail to record scene luminance accurately. Second, observers report variable appearances from the same luminance. Just like cameras, they show scene dependent limits to dynamic range. Unlike cameras, they show increased apparent contrast with decreased contrast on the retina. (Chapter 12)

Accurate portrayal of scene luminances from camera images is both not possible to achieve, and not essential to the visual process.

Veiling glare for human vision is worse than for cameras. Nevertheless, human vision has a much greater apparent dynamic range than camera systems. Humans can see details in highlights and shadows much better than conventional cameras. Although the rods and cones in the retina respond to more than a 10 log unit range, the ganglion cells that transmit the retinal response to the brain have only a 2 log unit range. There is no simple correlation between retinal quanta catch at a pixel and appearance (Chapters 7 and 12).

The interplay between glare and neural contrast is very complex. They act in opposition to each other, with *neural contrast* tending towards canceling glare (Section C). When a negative photographic

The Art and Science of HDR Imaging, First Edition. John J. McCann, Alessandro Rizzi.
© 2012 John Wiley & Sons, Ltd. Published 2012 by John Wiley & Sons, Ltd.

image is combined with a positive with the same H&D slope, they make a uniform middle-gray image. Since *glare* and *neural contrast* mechanisms are so different, there is no exact image-wise cancellation. Intraocular glare, because it is the sum of contributions from all other pixels, adds a scene-dependent low-spatial-frequency mask to the scene.

This counteraction of glare by neural contrast is fundamental to how we see. If the visual environment were only stars at night, then these low-glare scenes are easy to interpret because appearance tracks luminance (Zissell, 1988; McCann, 2005; 2007; Bodmann et al., 1979; Takahashi et al., 1999). As scenes change from the starry night to grays-in-a-white-surround, glare decreases the edge ratios on the retina, and neural contrast increases the relative differences in appearance.

13.3 Why are HDR Images Better?

If not accurate records of luminance, what do multiple exposures accomplish?

The luminous flux falling on the camera's image plane is the sum of the flux from the scene and the flux from glare. If we begin with an underexposed scene capture, we find scene information in the highlights, because the short exposure limits the absolute amount of glare. Here, there is insufficient flux to differentiate details in the shadows. Increasing the exposure to the best-average response does a good job of differentiating the mid-tone values. Increasing the exposure further, differentiates the details in the shadows while detail separation in the highlights is lost.

Multiple exposures preserve spatial information over a wider range of luminances. It does not matter that we cannot unscramble the scene luminance from the camera's image plane flux. Multiple exposures change the number of quantization levels assigned to pixels across an edge between two objects. Different exposures improve edge ratios. Multiple exposures improve local information – pixel A is darker than pixel B. That information is essential for synthesizing visual appearances of HDR scenes. The improved contrast allows displays to present better spatial information to humans. When human vision looks at high-dynamic range displays, it processes scenes using spatial comparisons.

As we saw in the discussion of film's tone scale curves (Figure 5.7), the S-shaped response expands the luminance ratios in the high-slope regions, and compresses them in the low-slope regions. By combining the best parts of multiple-exposure images we can remove the compressed and saturated responses leaving just the high contrast renditions. We then shape the combined image into the color space that the display device expects. Recall the analogy of HDR imaging as the task of moving the contents of a castle into a cottage. The multiple exposures provide all of the spatial information about each room's contents. The task of the lookup table and local manipulations is to make the best rendering in the display's limited output space. The user gets to reproduce some rooms accurately, some rooms with expanded space, and the rest with much less space. The user makes this selection based on what looks best for him. This process works well for individual scenes, but is not helpful in our search for the general solution of finding the best HDR process for all scenes. This general solution is found in human vision, so we know that it exists.

13.4 Are Multiple Exposures Necessary?

Improved digital quantization, which allows discrimination of adjacent objects, can be used in spatial comparison algorithms. Unlike analog silver-halide film density responses, digital imaging, particularly in the 1960s and 70s, was limited in the number of bits available in electronic imaging devices. Although appropriately spaced 24 bits of data per pixel (three channels; 8 bits each) is close to being able to represent the entire range detectable to human vision, it lacks the digital resolution to handle computations, such as demosaicing, image processing, printing and displaying satisfactory pictures. The number of bits, and more important, the luminance spacing between each bit is critical to having an artifact-free image. If digital cameras had photon well sizes and pixel digitizer circuits with a 4.1 log range comparable to silver halide negative film, then multiple exposures would not be necessary. Such devices are technologically possible (Chapter 10.3).

As discussed above, the significant role of intraocular glare has always prevented retinal receptors from seeing actual scene luminances. Neural mechanisms, often described as Simultaneous Contrast, work to reduce the adverse effects of glare. It follows that computational approaches to render HDR scenes for humans should use spatial comparisons as the essential tool in synthesizing the optimal display (McCann, 2004; Section E). The best approach to HDR is to follow the lead of artists' spatial rendering techniques, but with computational mechanisms. In Section C we take a closer look at the interplay of glare and neural processing to learn how to apply what we know of human vision to the general solution of HDR imaging.

13.5   Summary

The improvement in HDR images, compared to conventional photography, does not correlate with accurate luminance capture and display. Accurate capture in a camera is not possible, and accurate rendition is not essential. The improvement in HDR images is due to better preservation of relative spatial information that comes from improved digital quantization. Spatial differences in highlights and shadows are not lost. Spatial HDR image processing algorithms can be used to mimic processes developed by human vision (Section C), by chiaroscuro painters, and by early photographers that render HDR scenes in low-range outputs. (Section A).

13.6   References

Bodmann H, Haubner P, Marsden A (1979) A Unified Relationship between Brightness and Luminance, *CIE 50, CIE Proc. 19th Session* (Kyoto), 99–102.

McCann J, ed. (2004) Retinex at Forty, *J Electronic Img*, **13**, 1–145.

McCann J (2005) Rendering High-Dynamic Range Images: Algorithms that Mimic Human Vision, *Proc. AMOS Technical Conference*, Maui, 19–28.

McCann J (2007) Aperture and object mode appearances in images, *SPIE Proc.* 6492, 64920Q-64912.

Takahashi H, Yaguchi H, Shiori S (1999) Estimation of Brightness and Lightness in All Adaptation Levels, *J Light & Vis Env*, **23**, 38–48.

Zissell R (1988) Evolution of the "Real" Visual Magnitude System, *JAAVSO 26*, 151.

Section C

Separating Glare and Contrast

14

Two Counteracting Mechanisms: Glare and Contrast

14.1 Topics

What we see in High Dynamic Range scenes is the result of two spatial mechanisms that counteract each other. Appearance is controlled by intraocular glare (optical) and neural contrast (human image processing). This section describes experiments that measure the range of appearances with variable dynamic range. They show that intraocular scatter limits the possible range of light on the retina. As well, the calculated retinal image correlates with observer estimates of appearance. Glare changes images made up of uniform patches into a complex array of edges and gradients. Neural contrast makes high-glare scenes appear to have high apparent contrast.

14.2 Introduction

In general, the lightness of a pixel cannot be predicted by its scene luminance. There are many examples that include:

Natural scenes with reflected light, shadows, and multiple illuminants.

Mondrian displays in non-uniform illumination.

Gray squares on large white and black surrounds (Simultaneous Contrast).

Mach bands.

Gradients of luminance.

Changes in the overall intensity of illumination.

Lightness cannot be expressed as a unique function of luminance at a point. Lightness appearance is a function of the luminance at that point, and the luminance at all other points in the scene.

Only in special cases do we find that lightness is a simple function of luminance. Those cases are in the psychophysical laboratory, where we use isolated uniform illumination falling on uniform

The Art and Science of HDR Imaging, First Edition. John J. McCann, Alessandro Rizzi.
© 2012 John Wiley & Sons, Ltd. Published 2012 by John Wiley & Sons, Ltd.

reflectance test samples. Nevertheless, these special case experiments have been of great interest since the birth of psychophysics. The work of Wundt, James, Fechner, Boring, and Stevens applied the principles of physics to understanding human vision (Boring, 1942). The form of the function that describes appearance as a function of the amount of light is a cornerstone of the approach. Fechner asked a painter friend to mix a gray paint that was half-way between white and black. That middle gray became the 18% reflectance standard gray card in photography, Munsell's value 5/, and L*=50 in CIELAB.

Wyszecki and Stiles (1982) review many experiments that measured change in lightness with luminance. Since they all agree quite well, CIE L*, the cube root of luminance, has become the most used standard function. But, why should appearance be the cube root of luminance, when the retinal receptors have a logarithmic response to light?

14.3   Two Spatial Mechanisms

Human observer experiments show two independent and opposing visual mechanisms. Intraocular veiling glare reduces the luminance range on the retina while neural contrast increases apparent differences (McCann 2007a; McCann & Rizzi, 2007). To make sense of HDR imaging we need to isolate these competing mechanisms.

Figure 14.1 shows the classic Simultaneous Contrast experiment. If we consider the gray patch surrounded by white, it will have much higher glare, due to the white surround (Stiehl, et al., 1983). If retinal luminance predicted appearance, then it follows that the gray patch in white should appear lighter than the other identical patch in the black surround. However, the visual response to Simultaneous Contrast makes the gray-in-white look darker.

14.4   Calculated Retinal Image

Numerous experiments have studied the relationship between luminance and sensation. White, gray, and black sensations, called achromatic lightnesses, or gray scales, have been studied by Munsell et al.

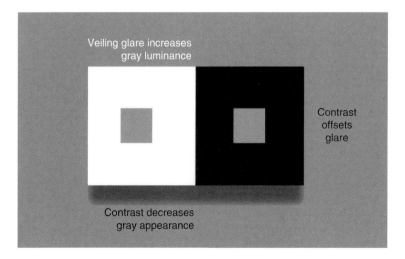

Figure 14.1   Simultaneous contrast shows two opposing visual mechanisms that contribute to the final appearance of the gray patches.

(1933), Ladd and Pinney, 1955; Glasser et al. (1958), Jameson and Hurvich (1961), Stevens and Stevens, 1963, Semmelroth (1970), and Bodmann (1979), and were reviewed by Wyszecki and Stiles (1982) and McCann (2007b). In these studies lightnesses, the sensations generated by the visual system, are compared with relative luminances, measures of the amount of light coming from objects to the eye. We describe such measurements as *luminance at the eyepoint*.

A more-relevant physical quantity for models of the psychophysics of human vision is the amount of light arriving at the retina. By accounting for the physical effects of the scattering of light passing through the ocular media, Stiehl, et al. (1983) calculated the relative stimulus at the retina. They convolved luminance at the eyepoint with the point-spread function (PSF) from Vos et al. (1976) to determine the *calculated retinal radiance*.

14.4.1   Making a Standard Lightness Scale

Figure 14.2 is a diagram of a transparent target that we called a *standard lightness display*. We use this display in lightness-matching experiments. A similar target and its use have been described by McCann, Land and Tatnall (1970). The standard lightness display was made using a bisection procedure similar to Munsell's procedure. Observers selected equal lightness steps by choosing from a series of neutral filters of various densities on a light box with a luminance of 3426 cd/m2. First, they chose a filter that appeared midway in sensation between white and black (assigned lightnesses of 9.0 and 1.0, respectively) and labeled this middle gray area with a lightness value of 5.0. Then the observers bisected the sensations of white (9.0) and middle gray (5.0); this lightness was labeled 7.0. Then they bisected the sensations of 5.0 and 1.0 to get 3.0, 9.0 and 7.0 to get 8.0, etc. The averages of the energies chosen by several observers determined the optical densities of the seven areas with integral values of lightness in the *standard lightness display*.

The luminances of the remaining shaded areas in Figure 14.2 were interpolated along a smooth curve used to fit the nine integral values. The standard lightness display was constructed with these extra steps to facilitate the observer's judgment in lightness-matching experiments. The 25 squares are 2.5° on a side in a uniform white field that subtends 30° by 25°. A small black square (1.25° on a side) is adjacent to each of the 25 large squares. The luminances of the 9.0 area and the background are 3426 cd./m2; the area labeled 1.0 and all the black squares are opaque.

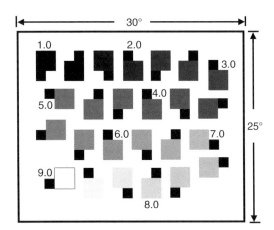

Figure 14.2   Diagram of standard lightness display.

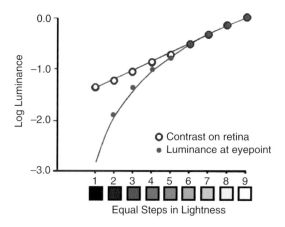

Figure 14.3 Luminances of equally spaced increments in lightness: solid red circles at eyepoint; solid curve is the Glasser's lightness function. The open blue circles are the nine calculated values of retinal radiance. Since they fall on the straight line, lightness is proportional to the logarithm of the retinal stimulus.

The luminances of the areas in Figure 14.2 that are labeled 2.0–9.0 are plotted as filled red circles in Figure 14.3 (the opaque area, labeled 1.0, cannot be represented on this logarithmic scale). These areas are equally spaced in lightness. The values of luminance are normalized to the luminance of the white area (lightness of 9.0). The solid curve fits a power law that conforms to the Glasser et al. (1958) lightness scale, and to the CIEL*a*b* cube root function.

Stiehl et al. (1983) convolved the entire test standard lightness display (two-dimensional spatial distribution of luminance from the 30-by-25-degree target) with the *Point Spread Function* (PSF of the eye to determine the pattern of contrast at the retina. They looked for correlation between the calculated image on the retina and the lightness of the patches in the scene.

14.4.2   Scatter Calculation

Stiehl used the *Point Spread Function* (PSF) summarized by Vos et al. (1976). The PSF is the foveal image of a point source; this image is degraded because of optical aberrations and intraocular light scattering. Stiehl used the PSF for the 2.0-mm pupil for these calculations.

Stiehl's convolution was performed with an angular resolution of 1′, the separation of retinal receptors in the fovea. There are 2 700 000 points at 1′ separation in a 30-by-25 degree field, and the calculation of the amounts of scattered energy from so many points onto any single receptor required about 34.5 min on Stiehl's 1976 PDP 11/60. The calculation of scatter-corrected luminance at all points at full resolution would have required about 177 years. Stiehl used a significant economy in calculation with no loss of accuracy for his purpose. Beyond about 10′ from the peak, the magnitude of the scatter function begins to diminish sufficiently gradually that sampling the scatter function and the target at 1′ intervals is not necessary. They reduced the number of samples needed for the calculation of scattered energy at a point from 2 700 000 to as few as 1304, thus eliminating over 99.95 % of the execution time. Also, they did not need to determine the scatter-corrected values of luminance at all points in the retinal image, but at only a few selected ones, thereby further reducing the total execution time.

The convolution was performed in the spatial domain rather than in the Fourier-transform, spatial-frequency domain. The combination of the large display size and the high-angular resolution needed

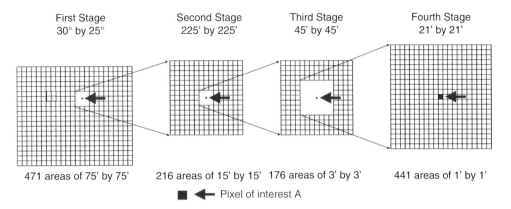

Figure 14.4 Diagram of the four stages of resolution for the calculation of scatter-corrected luminance. Stiehl calculated the total luminance scattered into the 1′ square area (A) in four stages. The first stage calculated the scatter from distant pixels with coarse angular resolution (75′ by 75′). In each successive stage they calculated the scatter from areas closer to A with successively finer resolution. The technique reduced the time of calculation drastically without significant loss of accuracy.

to sample the inherently narrow PSF accurately made the Fourier-transform technique unwieldy. Because they were interested in calculating the scatter at only a few points, the spatial convolution is more efficient.

Stiehl divided the source into four regions that are resolved into square areas of different sizes, with high angular resolution near the center of calculation, and much coarser resolution away from it. This manner of sampling the scatter function is valid because the sizes of the different areas at each stage of resolution are chosen to be sufficiently small so that the scatter function is nearly flat over each area and can be sampled accurately by the value at the center of each area.

Figure 14.4 depicts the different resolution sizes that Stiehl used to calculate the scatter corrected luminance at an arbitrary point A on the retina. The upper diagram shows 471 square areas, each 75′ on a side. They excluded the 225′ region about A because the scatter function must be sampled more finely over this region. The program samples the luminances at the centers of the 471 areas and computes the amounts of energy scattered from each of these areas to A.

In the second stage of processing, they subdivided this 225′ region into 225 square areas, each 15′ on a side. They excluded the central 45′ region around A and computed the energy scattered onto A from each of the remaining 216 areas. In the third stage, they further subdivided the 45′ region into 225 square areas, each 3′ on a side. They excluded a 21′ region around A and computed the energy scattered from the remaining 176′ areas. In the fourth, and final, stage of processing, they resolved the remaining 21′ region into 441 areas, each 1′ on a side, and computed the energy scattered from the pixels nearest A.

14.4.3 Results of Scatter Calculations

Figure 14.5 shows the luminance profiles, before and after scatter, along a diagonal through the middle of the black area with a lightness of 1.0 on our standard lightness display. The curves are normalized to the peak luminance of the target. On this plot, 100 %, or the white of the lightbox, has optical density 0.0. The contrast between this opaque area and the white background was the greatest of the entire display, and, consequently, the effect of intraocular scatter was greatest. The sharp edges in the source

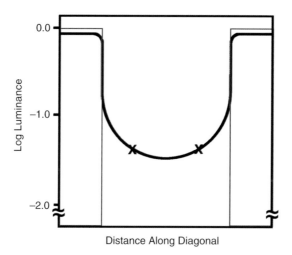

Figure 14.5 Calculated retinal luminance along diagonals through the opaque area in the standard lightness display (lightness of 1.0). The thin blue lines plot scene luminance.

(the blue lines in Figure 14.5) were replaced by steep gradients, and thus there was no single value of luminance to be ascribed to this entire test patch area. They used the average of the luminances at the four points on the two principal diagonals midway between the center and the four corners.

The average was calculated in this manner for each of the nine areas in the field that are labeled in Figure 14.2, and these scatter-corrected amounts of light are plotted as open blue circles in Figure 14.3.

The effect of scatter on these nine areas was to increase the luminance in the areas of low lightness, with proportionally less increase for areas of lightness approaching that of the background. The curve plotted through these data corrected for intraocular scatter had a different form: a straight line when the logarithm of scatter-corrected luminance was plotted against lightness. Consequently, whereas lightness is a power function of target luminance, lightness was also a logarithmic function of scatter-corrected retinal luminance.

14.4.4   Retinal Contrast with Different Backgrounds

Stiehl et al. (1983) created four additional targets that were designed to measure the effects of differing backgrounds. They measured bisection lightness scales, and calculated their retinal images. The backgrounds were chosen to provide differing distributions of scattered light within the eye. Target A, with a uniform white background, had large amounts of scattered light. The background in target D had far less scattered light. It was dark with a remote, thin, high-luminance perimeter. Targets B and C were designed to study the effects of intermediate amounts of scatter. In B there were thin (0.6°) high-luminance strips adjacent to the test areas, while in C the width of the strips was increased to 2.5 deg and had a separation of 1.88 deg. Both B & C targets were designed so that the same amount of light scatters from the surround into the central test area.

Observers performed the bisection experiments on these four targets. The results used the average bisection data for all five observers, for each of the four different backgrounds. Lightness data varied with the surround. The bisection data from target D with a predominantly dark surround are reasonably well fitted by a straight line in a plot of lightness vs. log scene luminance. Plots of observed lightness data from targets A, B, and C were not linear.

As we saw in Figure 14.3, calculated retinal contrast changed the shape of the plots dramatically for bright backgrounds. All four equally spaced lightness plots were fit by three linear log retinal luminance functions. The white (A) surround covered the range of white to black in 1.1 log units of retinal contrast. The dark (D) surround needed 2.3 log units of retinal contrast for the same white to black range. The intermediate (B, C) background data overlapped, and required 2.0 log units. In summary, equally spaced lightnesses have four logarithic retinal luminance functions with three different slopes.

The purpose of the study of scatter was to isolate neural contrast mechanisms affecting targets with different surrounds. By neural mechanism we mean spatial interactions that occur after the visual receptors. The intraocular media scatter light from bright areas into darker areas, thereby reducing the real contrast of the retinal image. Nevertheless, neural mechanism generates the darkest sensations in the cases of greatest amounts of scatter.

14.4.5   *Stiehl et al.'s Conclusions*

Stiehl et al. reported two important findings:

> First, the sensations of lightness are proportional to the logarithm of scatter-corrected retinal stimuli. Different backgrounds have different slopes. The reason that previous experiments have shown lightness to be a power function, or cube root of luminance at the eyepoint is that the scattering of incident light within the ocular media had not been considered.

> Second, the fact that different backgrounds yield lightness scales with different slopes shows that lightness is a function of both luminance at a point, and spatial interactions with other points in the field of view. In other words, correcting for intraocular scatter does not diminish, but rather increases, the need for understanding spatial interactions that generate lightness.

14.5   Measuring the Range of HDR Appearances

In this Section C, Chapter 15 describes a set of magnitude estimation experiments for a series of pairs of test targets. In the first pair, the surround around the 40 different gray patches were a half-white/half-black surround. Observers measured the appearances of *Single-Density* and *Double-Density* targets. Their dynamic ranges were markedly different, but the half-white/half-black surround had minimal effect on glare, and did not introduce changes due to Simultaneous Contrast. The measured range of useable light information for 50% white surround target was 2.3 log units. In addition, other pairs of displays varied the surround and found different ranges of useable information.

14.6   Calculating the Retinal Image

Chapter 16 describes a new technique for calculating the retinal image from the measured scene luminances. Instead of calculating individual pixels (Stiehl, 1983) this technique calculates the retinal radiance of the entire scene. The close agreement between the observed estimate of appearance and retinal radiance shows that the limit of useable information in a display correlates with the veiling glare in the retinal image. The useable range of light is scene dependent. With the possible exception of stars in a very dark sky, most scenes have limits of less than 3.0 log units.

14.7   Visualizing the Retinal Image

Chapter 17 describes pseudocolor visualizations of the HDR scenes and retinal images. They show that veiling glare reduces a 250 000:1 scene to be very similar to a 500:1 scene. For this pair with constant glare, but very different dynamic ranges, we find nearly the same image on the retina. Almost all of the increase in scene dynamic range was removed by glare.

This study of the retinal image also shows, as one would expect, that glare increased the values of the black pixels. It also shows that small white areas have lower values because of a decrease of scattered light from nearby white pixels. Uniform patches in the scene become complex gradients on the retina. This discussion points out the need for different size-dependent spatial contrast mechanisms.

14.8   HDR and Uniform Color Space

Chapter 18 describes a study of whether HDR imaging requires that we re-evaluate Uniform Color Spaces (UCS).

14.9   Summary

Glare distorts the luminances of the scene in the image on the retina. It lowers retinal contrast. Neural contrast works to counteract glare. The lower actual-contrast image appears to have higher apparent contrast.

Both glare and neural contrast are spatial processes. The changes they generate depend on the contents of the entire scene. In order to measure the glare limits of vision, we needed to design a new experiment that isolates the effects of glare from those of neural contrast. We needed to measure the effect of pre-retinal glare independent of post-retinal processing.

This section measures the appearance of the same scene made with different ranges of light. Substantially increasing the range of light beyond 3 log units makes almost no change in appearances. The experiments document that veiling glare limits the range of light that is useful in a display. That range varies with scene content.

14.10   References

Bodmann H, Haubner P & Marsden A (1979) A Unified Relationship between Brightness and Luminance, *CIE Proc. 19th Session* (Kyoto) CIE 50–1979, 99–102.

Boring EG (1942) *Sensation and Perception in the History of Experimental Psychology*, Appleton-Century-Crofts, New York.

Glasser G, McKinney A, Reilley C, & Schnelle P (1958) Cube-Root Color Coordinate System, *J Opt Soc Am*, **48**, PPR-246, 736–40.

Jameson D & Hurvich L (1961) Complexities of Perceived Brightness, *Science*, **133**,174–9.

Ladd H & Pinney J (1955) Empirical Relationships with the Munsell Value Scale, *Proc. Inst Radio Eng*, **43**, 1137.

McCann JJ (2007a) Art, Science, and Appearance in HDR images, *J Soc Info Display*, **15**(9), 709–19.

McCann JJ (2007b) Aperture and Object Mode Appearances in Images, in Human Vision and Electronic Imaging XII, S. Daly, *Proc. SPIE*, Bellingham, WA, 6292–26.

McCann JJ & Rizzi A (2007) Camera and Visual Veiling Glare in HDR images, *J Soc Info Display*, **15**(9), 721–30.

McCann JJ, Land EH, & Tatnall S (1970) A Technique for Comparing Human Visual Responses with a Mathematical Model for Lightness, *Am J Optom*, **47**, 845–55.

Munsell E, Sloan L, & Godlove I (1933) Neutral Value Scales. I. Munsell neutral value scale, *J Opt Soc Am*, **23**, 394–411.

Semmelroth C (1970) Prediction of Lightness and Brightness on Different Backgrounds, *J Opt Soc Am*, **60**, 1685–89.

Stevens JC & Stevens SS (1963) Brightness Function: Effects of Adaptation, *J Opt Soc Am*, **53**, 375–85.

Stiehl W, McCann J, & Savoy R (1983) Influence of Intraocular Scattered Light on Lightness Scaling Experiments, *J Opt Soc Am*, **73**, 1143.

Vos J, Walraven J & Van Meeteren A (1976) Light Profiles of the Foveal Image of a Point Source, *Vision Res*, **16**, 215–19.

Wyszecki G & Stiles W (1982) *Color Science*, 2nd edn,: John Wiley & Sons, Inc, New York.

15

Measuring the Range of HDR Appearances

15.1 Topics

In Section B we increased the amount of glare by increasing the fraction of white in the surround. That changed both the amount of glare and the neural contrast response function. Here, we studied targets that increased the range while holding glare nearly constant. To do this we made a pair of targets that had half-white and half-black surround areas. The white areas contributed nearly all the glare, while much darker blacks had little effect despite an increase in dynamic range of 500 times. We measured lightness vs. log luminance of the pair with LDR and HDR dynamic ranges. We measured the change in lightness, and hence the influence of neural contrast with increase in dynamic range. Changing dynamic range had little effect on appearance. In parallel experiments, changing the amount of white in the surround changed the amount of glare, the limits of range on the retina, the plot of lightness vs. log luminance, and the neural contrast response functions. We found different usable dynamic ranges that varied with scene content.

15.2 Introduction

What is the advantage of increasing the range of luminances in a display? To measure this one has to increase the range of the display, while holding constant both the influence of glare, and the influence of neural contrast. The goal of this chapter is to measure the usable dynamic range of luminance using targets with a fixed amount of glare and without changes caused by neural contrast.

We need to have very high-dynamic-range images with very precise control of light. We need to vary the dynamic range, but not change the veiling glare. Rizzi, Pezzetti, and McCann (2007) made multiple copies of computer generated silver-halide film transparencies. When the observer viewed one transparency the range was 2.7 log units. Superimposing two identical transparencies doubles the densities to make a range of 5.4 log units. We had 40 gray test squares in a half-white and half-black surround. By doubling the photographic densities we left the white luminance nearly constant and squared the percentage transmissions of the blacks. This pair of targets radically changes range, while leaving glare nearly constant. In the single density experiment, the blacks in the surround, with optical density 2.7, make

The Art and Science of HDR Imaging, First Edition. John J. McCann, Alessandro Rizzi.
© 2012 John Wiley & Sons, Ltd. Published 2012 by John Wiley & Sons, Ltd.

almost no contribution to glare. The black areas are too dark. Doubling the black densities does not reduce glare significantly. Glare comes from the white parts of the surround covering 50 % of the area.

15.3   Design of Appearance Scale Target

We studied patches of light that are uniform in the target. We use a surround that is, on average, equal to the middle of the dynamic range (50 % max and 50 % min luminances). Further, these surround elements are made up of different size min and max blocks, unevenly distributed spatially, so that the image has energy over a wide range of spatial frequencies. It avoids the problem that Simultaneous Contrast depends on the relative size of the white areas and of the test patch (McCann, 1978; McCann & Rizzi, 2003). Plots of the radial spatial-frequency distribution vs. frequency approximate the 1/f distribution (Tolhurst et al., 1992) found in natural images.

Figures 15.1 and 15.2 show the layout of our 50 % surround test target. The display subtended 15.5 by 19.1 degrees. It was divided into 20 squares, 3.4 degrees on a side. Two 0.8 degree gray patches are within each square along with various sizes of max and min blocks. The two gray square length subtends an angle approximately the diameter of the fovea. The smallest block (surrounding the gray patches) subtends 1.6 minutes of arc and is clearly visible to observers. Additional blocks 2×, 4×, 8×, 16×, 32×, 64× are used in the surround for each gray pair.

15.3.1   Single- and Double-Density Targets

The observers made *Magnitude Estimates* (MagEst) of the appearance of patches in *Single-* and *Double-Density transparencies*. The *Double-Density* target is the aligned superposition of two identical 4 by 5 inch photographic (*Single Density*) films. Two transparencies double the optical densities. The whites in each transparency have an optical density (O.D.) of 0.19; the blacks have an O.D. of 2.89. The *Double-Density* images have a min of 0.38 and a max of 5.78 O.D. (see Table 15.1). Both transparency configurations are backlit by four diffused neon bulbs. The experiments were done in a dark room. The only source of light was the target. The lightbox had an average luminance of 1056 cd/m2 (chromaticities x = 0.45, y = 0.43).

Veiling glare is a property of the luminance of each image pixel and the *glare spread function (GSF)* of the human optical system. Surrounds made up of half-max and half-min luminances have very interesting glare properties for *Single-* and *Double-Density* test targets. The average luminance of the *Single-Density* target is 50.10 % of the maximum luminance, from a display with a range of ~ 500:1. The average luminance of the *Double-Density* target is 50.00 % of its maximum luminance, from a display with a range of ~ 250 000:1. The effect of glare on the luminances of the gray test areas will be very nearly the same, despite the fact that the dynamic range has changed from 500:1 to 250 000:1. In other words, the black (min luminances) in both *Single-* and *Double-Density* targets are so low they make only trivial contributions to glare. The white (max luminances) in both targets are almost equal and generate virtually all the glare. The layouts of both targets are constant, keeping simultaneous contrast

Table 15.1   Luminances and optical densities of the min and max areas in Single and Double Density displays.

Target	Max cd/m^2	Min cd/m^2	Range:1	%Average cd/m^2	O.D. Min	O.D. Max	O.D. Range
Single Contrast	684	1.36	501	50.01%	0.19	2.89	2.70
Double Contrast	441	0.0018	251,189	50.00%	38	5.78	3.40

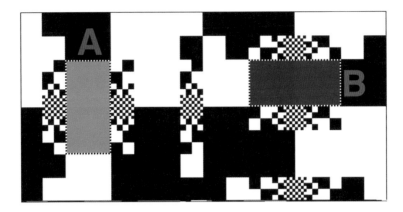

Figure 15.1 Magnified view of two of the 20 gray pairs of luminance patches. The left half (A) has the same layout as the right (B) rotated 90° counterclockwise. The gray squares have slightly different luminances: A (top & bottom); B (left & right). Each size of surround blocks has equal numbers of min and max squares.

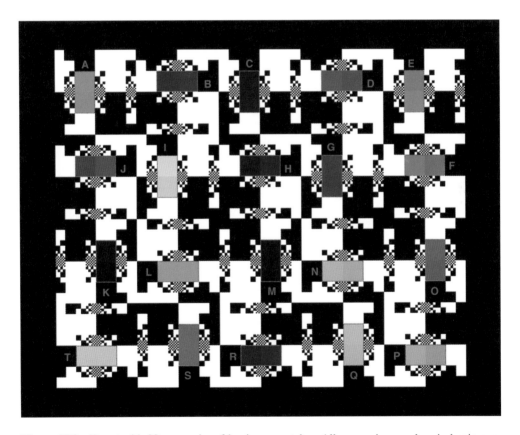

Figure 15.2 Target with 20 gray pairs of luminance patches. All gray pairs are close in luminance, but some edge ratios are larger than others.

stable. By comparing the magnitude estimates of appearance of these *Single-* and *Double-Density* targets, we can measure the effects of constant glare on very different dynamic-range displays.

If we can make use of the *Double-Density* image (range 250 000:1), then we expect to see a greater range of appearances in this image. If the veiling glare limit has been reached in the *Single-Density* image, then adding 500 times more range will have little, or no, effect on appearances.

15.4   Magnitude Estimation Experiments

Five observers made magnitude estimates of the appearance of each test patch between white and black. The observers were university students and workers between 18 and 23 years of age, with 20/20, or corrected 20/20 acuity. The five observers were asked to assign 100 to the "whitest" area in the field of view, and 1 to the "blackest" appearance. We then instructed them to find a sector that appeared middle gray and assign it the estimate 50. We then asked them to find gray squares having 25 and 75 estimates. Using this as a framework the observers assigned estimates to all sectors (A-T in Figure 15.2). Each of five observers repeated the experiment five times, not consecutively. They gave estimates for each half of the gray areas. We repeated the experiment with the same observers with *Single-* and *Double-Density* displays.

15.4.1   Average luminance = 50% max luminance

The first experiment measured the target shown in Figure 15.2. The average results are shown in Figure 15.3. The plots for Single and Double density nearly superimpose. In the *Single-Density* image the

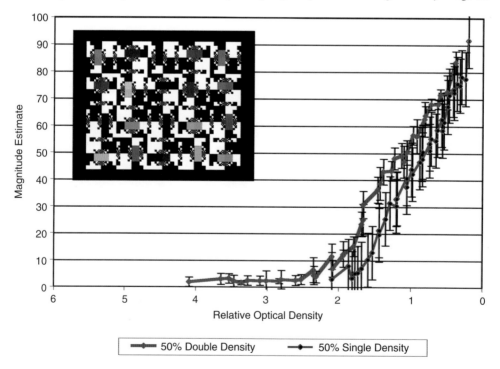

Figure 15.3   Appearances of Single- (SD) and Double-Density (DD) displays with 50% average luminance surround. Observers estimate the same range limit of 2.3 log units.

highest luminance gray (Area I) has a relative optical density of 0.19, and an appearance estimate of 92. In that target the lowest luminance gray (Area K) has an optical density of 2.1 and an appearance estimate of 3.0. In the *Double-Density* image Area K has an optical density of 4.1 and an appearance estimate of 1.8. The average of all observers on both targets show the same asymptote to black at density 2.3.

These *Single-* and *Double-Density* images have nearly the same veiling glare and Simultaneous Contrast pattern. The curves overlap over most of their range. The range of appearances from white to black is seen over 2.3 O.D. units.

The results are consistent with veiling glare determining the visible ranges. The effect of increasing the stimulus range has little or no effect because the *Single-Density* image is at, or near the maximum range possible on the retina for this scene. The plots in Figure 15.3 are the optimal tone scale function for these 50% White Surround scenes.

15.4.2 *Average luminance = 8% max luminance*

The second experiment kept the dynamic ranges constant and changed the Simultaneous Contrast characteristics of the displays by changing the fraction of white in the surround. It studied another pair of *Single-* and *Double-Density* targets. We reduced the area of the white to 8% of the background, leaving the black to cover 92% of the surround. The effect of reducing the white area was to decrease the amount of veiling glare, while holding the dynamic range constant. The results are shown in Figure 15.4.

Although different than the 50% white results, these curves are also similar over most of their range. The curves show the same asymptote at white and black. The range of appearances from white to black

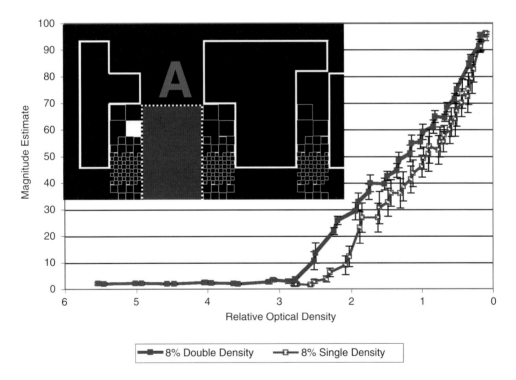

Figure 15.4 Appearances of Single-(SD) and Double-Density (DD) displays with 8% average luminance surround. Insert shows a magnified view of a portion of the target. Usable dynamic range is 2.9 log units.

is seen over 2.9 OD units. The results are consistent with veiling glare determining the visible ranges. The effect of increasing the stimulus range has only a small effect because the *Single-Density* image is at or near the maximum usable range on the retina for this scene.

15.4.3   Control Surrounds – White and Black

As control experiments, we measured gray patches with a completely white, and a completely black background in *Single-* and *Double-Density*.

For 100% White Surround (Figure 15.5), the *Single-Density* target, the optical density 1.8 has a MagEst of 2. In the *Double-Density* target, the optical density 3.4 has the darkest MagEst of 1. The two curves overlap above a density of 2.0. Both curves are shifted to the right of the *50% White* curves. The *100% White Surround* has the lowest usable dynamic range.

For the black (0% White) surround displays, we find that doubling the dynamic range shows changes in appearance. In the *Single-Density* target the optical density 2.7 has a MagEst of 5.2. In the *Double-Density* target the optical density 5.0 has the darkest MagEst of 1 (Figure 15.6). In complete darkness observers can use densities between 2.7 and 5.0 to discriminate between different levels of black (MagEst of 5.2 and MagEst of 1). The *Single* and *Double Density* curves separate around MagEst 30. That means that adding 3.0 log units of display dynamic range improves the rendition of only the darkest 30% of the range of lightnesses, only in a black surround.

For the three pairs of displays containing white (100%, 50% and 8%), we find that doubling the dynamic range shows only small changes in appearance. The slope of the transition from white to black

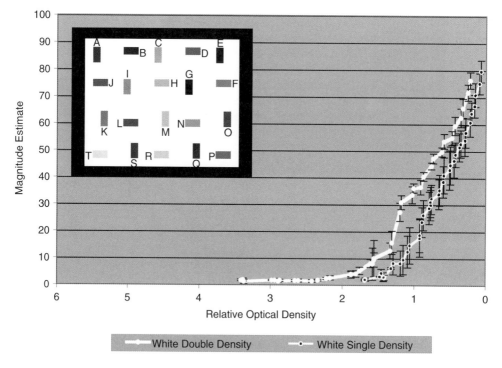

Figure 15.5 Magnitude estimates of SD and DD displays with 100% average luminance surround. Usable dynamic range of 2.0 log units.

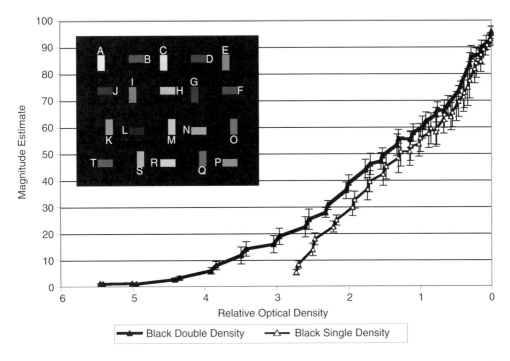

Figure 15.6 Appearances of SD and DD displays with 0% average luminance surround. Usable dynamic range of 5.0 log units.

depends on the amount of white in the background. Glare prevents the appreciation of most of the increase in dynamic-range information provided in the *Double-Density* images.

15.5 Scene Dependent Tone Scale

In previous plots we looked at the limits of usable dynamic range. Now we can analyze the role of percentage White in controlling appearance with neural contrast effect.

The Figure 15.7 shows that for the different backgrounds, there are four different results for changes in lightness vs. scene luminance. The data approximates four different linear log luminance functions. The greater the amount of white the steeper the slope. This data plots the decrements in luminance needed to go from white to black. The decrement is small with the solid white surround. Hence, the range of usable dynamic range is small. As the %White in the surround gets smaller, the slope of lightness vs. luminance is lower, and the usable dynamic range is larger.

In all four surround experiments the *Double-Density* plots superimposed for very light grays, but fell to the left of the *Single-Density* curve for dark grays by a small amount. That is consistent with the results in Figure 15.7. Smaller %White targets fall to the left of those with more light areas. In fact, there are more light-gray test squares in areas in *Single-Density* targets. We have discussed the effects of doubling the film densities on white and black squares in the surround. However, 20% of the area of these targets is made up of gray test areas. Doubling their densities squares their transmissions: 30% becomes 9%; 3% becomes 0.9%. In looking at these results, it is important to treat the targets as a whole by including the influence of the 40 gray test areas, which are darker in *Double-Density* targets. Hence, *Double-Density* falls to the left of *Single-Density* for all backgrounds in Figures 15.3, 15.4, 15.5, and 15.6.

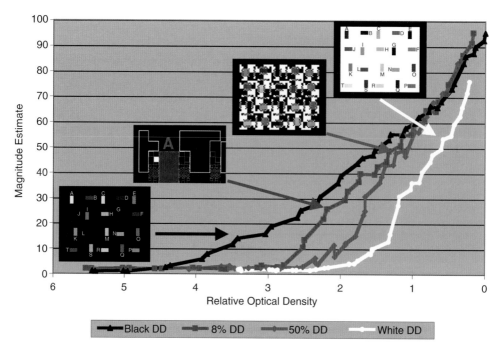

Figure 15.7 Plots of Double-Density displays with 0%, 8%, 50% and 100% average luminance surround. Observers measured different tone-scale slopes and dynamic range limits.

Table 15.2 Comparison of % White Surround area with usable dynamic range of display.

% White	100%	50%	8%	0%
Double Density Targets				
Usable Range (log units)	2.0	2.3	2.9	5.0

15.6 Glare and Contrast

These experiments separate glare from neural contrast. The experiments used four different surrounds with different amounts of white in the surround. We needed four types of surrounds to evaluate the effects of the neural contrast mechanism responsible for Simultaneous Contrast effects. We needed four pairs of *Single & Double-Density* targets to isolate the effects of veiling glare. The pair of 100% White Surrounds targets reached the same asymptote of 2.0 log unit usable dynamic range. That range took lightness from white to black.

The data shown in Table 15.2 shows the maximum usable range of luminance for each target design. Each of the background configurations generates a different amount of glare (stable between *Single-*

and *Double-Density*). Observer estimates show a usable range of 2.0 log units in the highest glare condition. Reducing the amount of white by half increases the usable range to 2.3. We find a usable range of 2.9 with 8 % white background. In the case of the black background observers can discriminate luminances over a 5 log unit range; this can be obtained only with a completely black surround and total darkness in the entire room. These very strict constraints are inconsistent with common scenes and normal viewing situations.

15.7 Summary

We have studied the effect of *Single Density* (0 to 2.7 log units) and *Double-Density* (0 to 5.4 log units) targets with almost no changes in glare. HDR images are limited by scene-dependent intraocular glare. In a white surround, with the maximum glare, observers use an optical density range of 2.0 log units to cover the range of appearances from white to black. By using half-white and half-black surrounds, observers use a range of 2.3 log units for white to black appearances. In a third experiment, we reduced the white to 8 % white, decreasing glare further. Here, observers use a range of 2.9 log units for white to black. Only in a extremely dark surround, in a room with no other light, could observers use 5.0 of the 5.4 log range display.

15.8 References

McCann J (1978) Visibility of gradients and low-spatial frequency sinusoids: Evidence for a distance constancy mechanism, *J Photogr Sci Eng*, **22**, 64–8.

McCann J & Rizzi A (2003) The Spatial Properties of Contrast, *Proc. IS&T/SID Color Imaging Conference*, Scottsdale; **11**, 51–8.

Rizzi A, Pezzetti M & McCann J (2007) Glare-limited Appearances in HDR Images, *IS&T/SID Color Imaging Conference*, **15**, 293–8.

Tolhurst D, Tadmor Y & Chao T (1992) Amplitude Spectra of Natural Images, *Ophthalm Physiol Opt*, **12**, 229–32.

16

Calculating the Retinal Image

16.1 Topics

This chapter continues the study of the HDR test targets. We calculate the intensity of veiling glare in these test scenes using the 1999 CIE Standard for Veiling Glare. We use that glare spread function to calculate the retinal image after intraocular scatter. We compare retinal radiances to the magnitude estimates of appearance reported by observers. By modeling the actual image on the retina we can see how neural contrast and glare determine the range of usable information.

16.2 Introduction

Section C studies two independent and opposing visual spatial mechanisms. Intraocular veiling glare reduces the luminance range on the retina, while neural contrast increases the differences in appearance. Veiling glare is a purely optical phenomenon that reduces the dynamic range of the retinal image. In this chapter we calculate the image on the retina to measure the effects of veiling glare on the three test targets shown in Figure 16.1.

Stiehl et al. (1983) described an algorithm that calculated the contrast of selected pixels in an image on the human retina after intraocular scatter. They calculated the luminance at selected pixels on the retina based on that display pixel's measured luminance and the calculated scattered light from all other scene pixels. The calculation used Vos et al.'s (1976) measurements of the human point spread function. Stiehl measured the actual display luminance and calculated the retinal image (Chapter 14).

Using Stiehl et al.'s algorithm as a guide, we updated the process to use today's more powerful imaging systems. We calculated the luminance on the retina based on the new CIE standard for intraocular glare (Vos & van den Berg, 1999; CIE, 2002; van den Berg et al., 2007). Instead of calculating the retinal value of pixels of interest, we now use the *fast fourier transform (FFT)* to calculate the entire image (Rizzi & McCann, 2009). Our goal is to compute the intensity of light falling onto the retina to separate the effects of glare from those of neural contrast.

The Art and Science of HDR Imaging, First Edition. John J. McCann, Alessandro Rizzi.
© 2012 John Wiley & Sons, Ltd. Published 2012 by John Wiley & Sons, Ltd.

Figure 16.1 Test targets with 100%, 50% and 0% White Surrounds.

We are using the *Double-Density* displays in Chapter 15 as the scene input to the veiling glare model. Its output is retinal radiance. For all points in the image we compute the amount of veiling glare caused by all of the other points, depending on their distance from the pixel of interest. Retinal contrast is the light input to the spatial contrast mechanism of human image processing. We need to know the actual input to the neural contrast mechanisms in order to understand their spatial properties.

16.3   Converting Scene Luminance to Retinal Contrast

We begin with a target made up of film transparencies on a light box. It is a flat, planar display. Our goal is to calculate the changes in that image caused by human veiling glare on the way to the retina. Our "retinal" image is, in fact, the result of a model that accounts for just one of the many transforms performed by the optics of the eye. The Vos and van den Berg model provides us with a glare spread function that we use to calculate a new planar image that includes the veiling glare found in the human eye. The convolution described below calculates the retinal contrast of each scene pixel. It combines each pixel's luminance with the sum of all scattered light from all other pixels. Our calculation does not include many other properties of the actual retinal image. For example, it does not include the spatial transform caused by the curvature of the retina. It does not calculate the absolute photon count on the retina. That calculation would have to include the aperture size of the pupil, and spectral transmission of the ocular media, in addition to optical image transformations. We calculate the relative amount of light compared to the scene target. We will use the term scene luminance to describe the target, because the meter reads luminances. The spherical retinal image alters the absolute calibration, so we will use the term relative retinal radiance to describe the calculated amount of light in the image on the retina.

The following analysis compares the digital array representing the scene with a congruent array of the retinal image.

16.4   Calculating Retinal Radiance

There are three components needed to calculate retinal radiance:

First, measurements of the optical density at each pixel in the target
(*Scene Luminance*)

Second, the glare point spread function of human vision
(*GSF*)

Third, an efficient algorithm that can transform scene luminances to the retinal image. (*Retinal Radiance*)

16.4.1 Scene Contrast – Input Luminance Array

We used a digital editing program to make digital arrays for our targets' designs. We sent the high-resolution digital files to a photofinishing lab that used a film recorder to make 4×5 inch Ektachrome transparencies, two for each target. We measured the transmission of white, black and each gray area using an X-rite 361T densitometer with calibrated standard samples. Since we are concerned with the study of HDR images over a dynamic range of one million to one, we plot our measurements as relative *optical density* (OD = log [1/transmittance]). Using these calibration measurements we made lookup tables (LUTs) to convert target digits to film transmission. When examining the digit vs. optical density (OD) data we found that light scattering in the film recorder system had a small, but consistent, effect on the digit vs. density data. Using these three sets of calibration measurements, we made three different lookup tables (LUTs) to convert target digits to film transmission. We measured the lightbox luminance to be 1056 cd/m2 with a Minolta 100C spot meter. This luminance is associated with OD = 0.

We read the digital array, used to make the targets, and apply the calibration LUT to convert digit to target OD, and then to relative scene luminance. The program reads the 0–255 value in the digital array and replaces it with the floating point luminance value accurate over 6 log units. This accuracy is assured by the densitometer measurements of each separate single transparency.

16.4.2 CIE Veiling Glare Standard

Vos and van den Berg (1999) collected a series of measurements and wrote a recent CIE report in which formulas account for parameters such as the age of the observer and the type of his/her iris pigment.

We used the formula referred to as number eight in the Vos & van den Berg (1999) report. The formula is:

$$
\begin{aligned}
L_{eq}/E_{gl} = &\left[1-0.08*\left(A/70\right)^{4}\right]*\left[\frac{9.2*10^{6}}{\left[1+\left(\theta/0.046\right)^{2}\right]^{1.5}}+\frac{1.5*10^{5}}{\left[1+\left(\theta/0.045\right)^{2}\right]^{1.5}}\right] \\
&+\left[1-1.6*\left(A/70\right)^{4}\right]*\left\{\left[\frac{400}{1+\left(\theta/0.1\right)^{2}}+3*10^{-8}*\theta^{2}\right]\right. \\
&\left.+p\left[\frac{1300}{\left[1+\left(\theta/0.1\right)^{2}\right]^{1.5}}+\frac{0.8}{\left[1+\left(\theta/0.045\right)^{2}\right]^{0.5}}\right]\right\}+2.5*10^{-3}*p
\end{aligned}
\tag{16.1}
$$

where θ is the viewing angle from the point from which the light is spread causing the veiling glare, A is the age of the observer and p is his/her iris pigmentation. This formula measures the equivalent veiling glare in relation to the energy of relative illuminance and is defined as sr^{-1}. Pigmentation types give the origin of correction parameters that range from 0 to 1.21. In the calculation we used an age of 25 and brown Caucasian pigment. We converted this formula into a filter in the frequency domain.

16.4.3 Calculate Retinal Radiances

Ivar Farup and Ale Rizzi wrote code in Matlab® to compute the Glare Spread Function (GSF) from the CIE formula reported above and to convolve it with the calibrated input image converted into floating point luminance values.

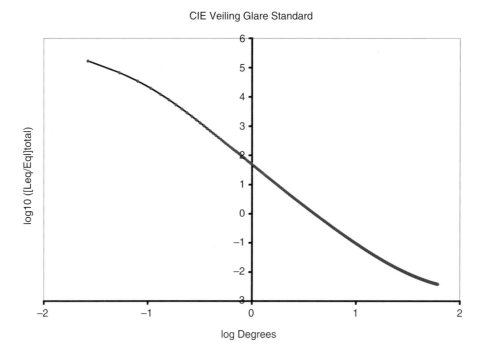

Figure 16.2   Plot of CIE Glare Spread Function filter plotted on log scales.

The steps for the computation are:

- Calculate scene optical density input image using calibration measurements.
- Calculate GSF filter for the input image considering its size and viewing distance (input image is padded with zeros, null luminance). The filter (Figure 16.2) has the shape taken from Vos & van den Berg, 1999, equation 8.
- Each pixel (donor) scatters light into all the other pixels (receivers) depending upon their reciprocal distance. We correct for the amount of light scattered from the center pixel in order to preserve the overall amount of light in the scene.
- Convolve the images with the filter.

Figure 16.2 plots the log of the amount of veiling glare vs. log visual angle between the central point of light and the position of the eccentric retina receiving the glare. To avoid computational problems near the edges of the target, we pad the target image array with null values [luminance = 0.0]. This represents quite well the conditions of the experiment that took place in a dark room in which the target was the only light source, and everything was painted dark in order to avoid reflections. We made a square filter twice the size of the longer target side. This filter calculates the distance-dependent scatter contribution from all other pixels using our equation (equation 16.1). We transformed the filter into the spatial-frequency domain and calculated the glare for each pixel with the *FFT* convolution. Each target pixel is the sum of that pixel's luminance, plus the amount of light scattered into that pixel from all other pixels in the target, minus the light scattered out of that pixel. The CIE standard for veiling glare covers angles from 1/100 to 100 degrees. The dynamic range of veiling glare (vertical axis) in the equation covers 1 000 000 to 1/1000 units of the ratio of equivalent luminance (Rizzi & McCann, 2009).

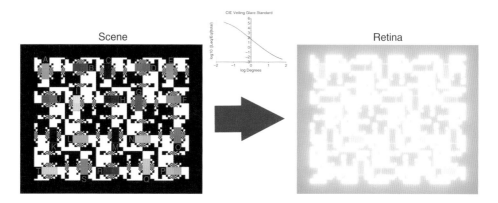

Figure 16.3   Input scene luminance measurements convolved with the CIE Glare filter calculates the retinal image.

One final step in the calculation is to correct the value of the central pixel of interest. The calculation sums all the contributions from all other pixels and adds it to the calibrated value of the source target. That does not account for the loss of light to all other pixels. This means that the sum of all light from the pixel is greater that 1.0. Joris Coppens suggested that we use the departure from 1.0 as the required correction (Franssen & Coppens, 2007).

16.5   Changes in the Retinal Image from Glare

Figure 16.3 illustrates the conversion of scene luminance to retinal contrast. Each target has 20 pairs of gray patches; each gray square patch is adjacent to a very similar gray. We selected a 24 by 24 pixel area inside each of the 40 squares in the calculated retinal contrast images. We measured the mean inside the gray squares for the Double Density targets. In Figure 16.4, we plot each of these 40 calculated retinal radiances vs. scene optical density values.

The plot of the retinal image of the 0 % White Surround (Black) shows that glare does not modify the scene until it is 4. 0 log units darker than white. The 50 % White shows that glare is affecting the retinal image at 1.5 log units; and with maximal glare the retinal image is altered below 1.0 log unit. In areas below middle gray (OD = 0.74), glare reduces the contrast on the retina variable amounts, depending on scene content.

16.6   Appearance and Retinal Image

As we saw in Chapter 11, a conventional negative film is capable of responding to 4.0 log units of light in scenes with very minimal glare. Scenes that add glare reduce the range of light on the film plane and reduce the range of optical densities in the processed film. High-glare scenes reduce the range of response in the negative. Scene content, such as in *4scaleWhite*, reduces the photographic contrast range on the camera's film plane.

Vision is remarkable because increased glare does not reduce apparent contrast. Figure 16.5 plots the magnitude estimates of lightness vs. retinal optical density for these test targets. Increased glare increases the apparent contrast of the scene. The rate of change of lightness vs. optical density is scene dependent.

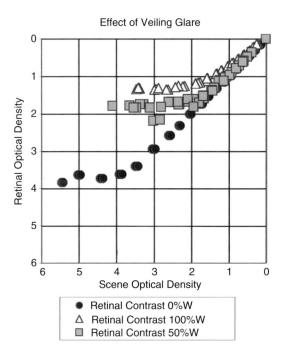

Figure 16.4   Effect of veiling glare.

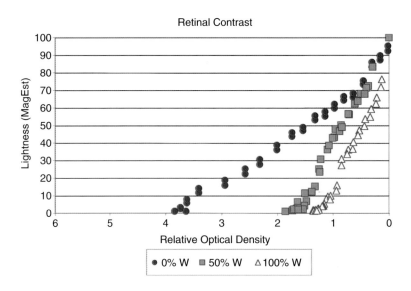

Figure 16.5   The white to black lightness function varies with surround.

In all three backgrounds, we normalized the scene maximum to optical density 0.0. Each has a magnitude estimate of 100, associated with white. Magnitude estimate 1 is black. We plot the calculated retinal contrast as optical density. With a black background (*0% white background*) the appearance black is about 3.7 log units darker than the maximum. With half-white and half-black background, black is about 2.0 log units darker. In the *100% white background*, the appearance of black is associated with patches that are 1.5 log units darker than the maximum. Increased glare causes higher slope lightness functions, or higher apparent contrast.

The results in Figure 16.5 show that:

- Scene content and glare limit the range on the retina.
- Below middle gray, glare reduces the retinal contrast variable amounts, depending on scene content.
- Glare accounts for the cube-root shape of the CIE L* lightness function.
- Lightness is a logarithmic function of retinal radiances (amount of light on the retina).
- Scene content controls the slope of that log retinal contrast function.
- Post-retinal neural processing renders low-retinal-contrast scenes with high apparent contrast.

16.7   Scene Content and Psychometric Functions

In thinking about human vision, and modeling it, we realize that trying to measure human vision using spots of light in a black background is a very bad idea. We see it here in the extreme influence of spatial content in Figure 16.5. The changes in the amount of white in the background take us from needing 1.5 log units of retinal contrast to needing 3.7 units. That is a change in the range of retinal stimulus of 200 times. Instead, if we consider the original scene, that range needed changes from 5 to 2 log units. That is a 1000 times change in the dynamic range needed for the same white-to-black appearances. Applying data from black surround experiments to real scenes will not model vision.

The interplay between scene content and neural spatial mechanisms in vision plays a major role in appearance. Only scenes comprised of just stars have image content on very black backgrounds. What we see in everyday life are scenes with light areas throughout the image. Figure 16.5 shows the remarkable difference between black-background appearances and those found in normal scenes.

No one can dispute that the detailed study of vision in a no-light surround has generated a vast amount of reliable, repeatable data that is very useful in predicting the thresholds of detection. It is also helpful in predicting color matches. The issue is whether these psychometric functions are relevant to the everyday scenes that make up our lives outside of the psychophysical darkroom. Here, we have measured very different responses from complex scenes compared to those from spots of light.

16.8   Summary

We calculated the retinal image of *Double-Density* test targets with dynamic ranges just below 6 log units. We measured the target luminance at each pixel, and we calculated retinal contrast array using the 1999 CIE Standard glare spread function. We calculated intraocular veiling glare that reduced the 6 log unit range of the test target to as little as 1.5 log units on the retina. Intraocular glare reduced the high-dynamic-range target luminances to much smaller ranges. Below middle gray, glare reduces the retinal contrast variable amounts, depending on scene content. However, the interaction of neural mechanisms with scene content tends to counteract glare. Lower actual contrast retinal images have higher apparent contrast. Lightness is a logarithmic function of the amount of light on the retina.

16.9   References

CIE (2002) 146 CIE equations for disability glare, 146/147: CIE Collection on Glare.

Franssen L & Coppens JE (2007) Straylight at the retina: Scattered papers, thesis, Universiteit van Amsterdam, http://dare.uva.nl/record/215172.

Rizzi A & McCann J (2009) Glare-limited appearances in HDR images, *J. Soc. Info. Display*, **17**(1), 3–12.

Stiehl W, McCann J & Savoy R (1983) Influence of Intraocular Scattered Light on Lightness-scaling Experiments, *J Opt Soc Am*, **73**, 1143–48.

van den Berg T, Van Rijn L, Michael R, Heine C, Coeckelbergh T, Nischler C, Wilhelm H, Grabner G, Emesz M & Barraquer R (2007) Straylight Effects with Aging and Lens Extraction, *Am J Ophthalmology*, **144**(3), 358–63.

Vos J & van den Berg T (1999) Disability Glare, CIE Research note 135/1, ISBN 3 900734 97 6.

Vos J, Walraven J & van Meeteren A (1976) Light Profiles of the Foveal Image of a Point Source, *Vision Res*, 16215–219.

17

Visualizing HDR Images

17.1 Topics

This chapter describes a pseudocolor technique used to visualized HDR images. Retinal Radiance ranges for Single- and Double-Density targets are very similar. Glare transforms uniform, equiluminant target patches into different gradients with unequal retinal radiances. Human vision uses complex spatial processing to calculate appearance from retinal arrays.

17.2 Introduction

Veiling glare is a property of the luminance of each image point and the *glare spread function (GSF)* of the human optical system. Surrounds made up of half-max and half-min luminances have nearly the same glare properties for *Single-* and *Double-Density* test targets. The average luminance of the *Single-Density* target is 50.10% of the maximum luminance, from a display with a range of ~500:1. The average luminance of the *Double-Density* target is 50.00% of its maximum luminance, from a display with a range of ~250 000:1. (Chapter 15; Rizzi et al., 2007a, 2007b) The effect of glare on the luminances of the gray test areas will be very nearly the same, despite the fact that the dynamic range has changed from 500:1 to 250 000:1. In other words, the black areas (min luminances) in both *Single-* and *Double-Density* targets are so low, they make only trivial contributions to glare. Nevertheless, the white (max luminances) in both targets are almost equal and generate virtually all the glare. The layouts of both targets are constant. The physical contributions of glare are very nearly constant. By comparing the calculated retinal radiance with the appearance of these *Single-* and *Double-Density* targets, we measured the effects of constant glare on very different dynamic-range displays.

The Art and Science of HDR Imaging, First Edition. John J. McCann, Alessandro Rizzi.
© 2012 John Wiley & Sons, Ltd. Published 2012 by John Wiley & Sons, Ltd.

17.3   Calculated Retinal Image Contrast

This chapter uses pseudocolor lookup tables to compare two types of digital images. The first type represents the high-dynamic range of the targets' scene contrast. This is the visualization of the physical measurements of the scene target. We will call this visualization "Target Contrast". The second type of image is the calculated retinal-image radiance. This is the visualization of the relative amount of light on the retina after intraocular glare. We will call this visualization "Retinal Contrast".

We will evaluate the 50% White target shown in Figure 15.2. The range of target luminances for this target was 2.7 optical density (OD) units (see Table 15.1). It is not possible to reproduce this range on the printed page, or on conventional displays. The *Double-Density* target made the problem much more severe. We use a 64-pseudocolor colormap to render the target luminance range of 5.4 OD units. The maximum luminance was white and the minimum was black. In order of decreasing luminance, the pseudocolors, are white, yellow, green cyan, blue, magenta, red brown, and black. Rendering the range of 5.4 OD target luminances in 64 steps gives a range of 0.084 OD per individual colorbar element. We show the pseudocolor scale in the center of Figures 17.1–17.5. The color map images illustrate the substantially different ranges of luminance measured in the 2.7 OD *Single-Density* and 5.4 *Double-Density* targets.

We used the same colormap to render the calculated retinal radiance of the *Single-Density* and *Double-Density* images (Figure 17.2). Unlike the Figure 17.1 pseudocolor renditions, the retinal image contrasts are very similar to each other. Intraocular scatter reduces the 5.4 OD dynamic range of the *Double-Density* target to about 2.0 log units on the retina. It is only slightly darker than the *Single-Density* retinal radiance array.

The calculated retinal radiances show that intraocular glare limits the range of the image to roughly 2.0 OD units for these 50% White targets.

Figure 17.3 shows a different colorbar rendering of retinal image radiances from that in Figure 17.2. Here, we spread the same 64-colorbar elements over only 1.5 OD, instead of 5.4 OD. Rendering the range of 1.5 OD Target Contrast in 64 steps gives a range of 0.0234 OD per individual colorbar element.

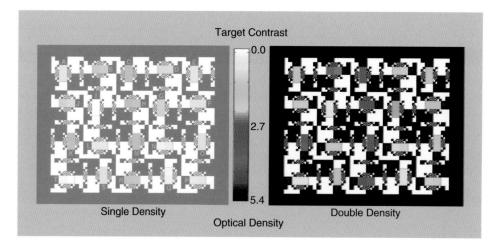

Figure 17.1   Target Contrast is a pseudocolor rendering of target luminances. The Double-Density target has a range of 5.4 (OD); the Single Density has a 2.7 (OD) range. The colorbar in the center identifies the color of each optical density over the range of 5.4 log units. The Single- and Double-Density targets are very different stimuli.

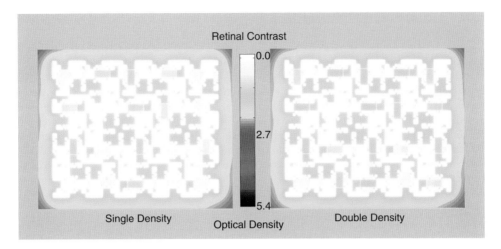

Figure 17.2   Retinal Contrast is a retinal image radiance using the previous pseudocolor LUT. Both the Single- and Double-Density retinal ranges are roughly 2.0.

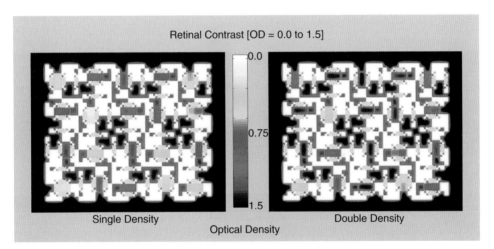

Figure 17.3   An expanded rendition of the range of retinal image radiances. To improve pseudocolor discrimination, the range of the colormap in this plot is only 1.5 (OD). It brings out the more subtle differences between Single- and Double-Density retinal images.

This rendition shows the after-scatter values of the 20 pairs of different gray patches in the targets. In the *Single-Density* target we see a range of different colors for the different transmissions. In the *Double-Density* target we see that intraocular scatter made many of the darker gray squares more similar.

17.4   Retinal Image Contrast

Veiling glare from a single pixel decreases with distance away from that pixel (Figure 16.2 shows the falloff of scattered luminance vs. distance). At large distances, this glare value is very small, but it is a small contribution from every pixel in the image. The sum of many small contributions is a significant number. The greater the % White in the scene, the more the glare, the lower the range of light of the retinal image.

The background in these images has a range of different sizes of uniform white or black squares. Each white pixel scatters a fraction of its light into surrounding pixels. In turn, each pixel receives a distance-dependent fraction of light from all other pixels. A pixel in the center of a large white square has the highest retinal radiance because there are many surrounding white pixels that contribute a larger fraction of their scattered light. Similarly, the lowest radiance pixels are found in the center of the largest black square since it is the furthest from white pixel scatter sources. The highest ratio of retinal radiances is the ratio of the center pixels in the largest white, or black squares. The same logic shows that the smallest white/black retinal radiance ratio is of the smallest, single pixel, retinal radiances. The results of the calculated retinal image show retinal radiance ratios from as high as 17.46 to 1 to as low as 1.45 to 1 (see Figure 17.4 and Table 17.1).

Table 17.1 Data and colormap rendering of white and black squares shown in Figure 17.4.

	DD white OD		DD black OD		WB contrast OD	W/B edge ratio
Target	0.00		5.4		5.40	251,188.64
DD 64×	0.08		1.02		1.24	17.46
DD 32×	0.09		0.89		1.08	12.02
DD 16×	0.17		0.61		0.86	7.31
DD 8×	0.24		0.43		0.59	3.93
DD 4×	0.25		0.51		0.32	2.11
DD 2×	0.35		0.47		0.22	1.64
DD 1×	0.46		0.53		0.16	1.45

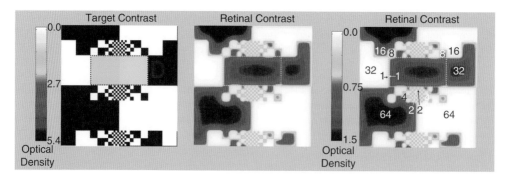

Figure 17.4 (left) The Target Contrast [colormap range = 5.4], and Retinal Contrast [colormap range = 1.5], (center) and (right) annotated retinal radiance.

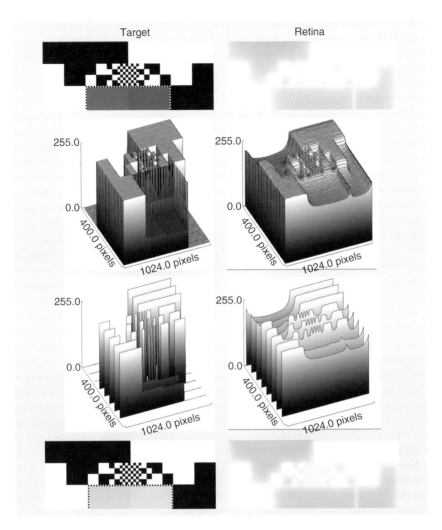

Figure 17.5 Comparison of target (left) with calculated retinal radiance (right), for a portion of the image in area D. These comparisons use the same pseudocolor scale (5.4 OD) for the left and right columns. The top row compares relative target and retinal radiances. The second pair of comparisons shows 100 horizontal plots of radiance. The next row is a surface plots with eight slices. The bottom pair is the pseudocolor map used in Figures 17.1 and 17.2. All plots show that the uniform high-dynamic-range [OD = 5.4] target is transformed by glare into a non-uniform low-dynamic-range retinal image. Observe the size-dependent effects on white/black edges.

In Figure 17.4, (right) the annotation numbers, ranging from 64 to 1, show the relative sizes of sides of the white and black squares. This figure shows only Double-Density Area D and its surround. It illustrates the effect of intraocular scatter on different size white and black squares.

Another feature of retinal images is the conversion of uniform target areas into gradients of retinal radiances. The target was designed to have uniform patches of white and black squares. At each white/black edge, intraocular scatter transforms the sharp edge into a much smoother transition with gradients. The slope of that gradient varies depending on the neighboring pixels. The simple uniform target elements have been transformed into a very complex array of gradients. Figure 17.4 (left) shows the gray area D in the target using colorbar = 5.4. Figure 17.4 (center and right) shows retinal radiances using colorbar = 1.5.

The retinal image is made of many complex gradients. The two different gray squares are very close in retinal luminance, in fact, they are almost indistinguishable in the pseudocolor rendering. Observer magnitude estimates of Area D were 11.9 ± 2.0 (left) and 6.9 ± 3.1 (right), on a scale of white = 100 and black = 1. Observers are just able to discriminate these small differences in retinal radiance.

All white surround squares have the same target luminances, and all blacks are the same. However, both white and black squares have variable retinal radiances depending on the size of the square, as well as its position in the surround. Therefore, constant target luminance does not ensure constant light on the retina. Nevertheless, the white squares appear to be uniform and have the same white appearance. Black squares appear uniform, and a constant black. Regardless of the retinal stimulus, the human spatial processing mechanisms make their appearances the same.

Table 17.1 shows the relative optical densities of typical retinal radiances for different size white/black pairs. It also includes the colorbar rendering of these retinal calculations. It lists the difference in OD, and the ratio of retinal radiances at the centers of the squares. Retinal radiance values vary considerably with the surrounding portion of the image. The size of white and black squares has considerable influence on the retinal radiances and white/black (W/B) ratios. The large white/black squares have retinal radiance ratios of 17.46 to 1, white the smallest square have a ratio of only 1.45 to 1. The ratios vary with square size.

Figure 17.5 uses four different visualizations to compare target and retinal images: scaled relative radiances, surface plots of radiance, eight slices of radiance, and pseudocolor. Intraocular scatter transforms high-dynamic-range, uniform, constant square targets (left) into low-dynamic-range, gradients with variable retinal radiance (right).

Human spatial-image-processing transforms this complex retinal image so that all "target whites" appear the same white and all "target blacks" appear the same black. HDR image processing techniques that attempt to mimic human vision need to differentiate the optical effects of intraocular glare from the neural spatial-image processing. Both optical and neural mechanisms show significant scene-dependent alteration of the image. These different scene-dependent processes tend to cancel each other out, making their presence less obvious (McCann & Rizzi, 2008; Rizzi & McCann, 2009). Veiling glare is seldom apparent. Neural contrast is so powerful that the effect we see, called Simultaneous Contrast, is the remainder after cancelation.

We see here two subtly different cancellations:

The shape of the sharp black-white edges were constant in the scene, but variable gradients on the retina.

The scene radiance contrast values for black-white were constant in the scene, but were different asymptotes on the retina.

Neural contrast transforms different spatial gradients with different contrasts into the same black and white appearances. Neither the glare nor the transform is apparent to the observer.

17.5 Summary

In Chapter 15 we saw that doubling the density of the 50% White target had a minimal effect on appearance. In Chapter 16 we saw that *Single- and Double-Density* targets have almost identical stimuli on the retina. In this chapter we inspected the images on the retina using pseudocolor techniques.

Further, the calculated retinal image shows that there is no simple relationship between the retinal radiance of a pixel and its appearance between white and black. Observers report that the appearances of white and black squares are constant and uniform, despite the fact that the retinal stimuli are variable and non-uniform. Human vision uses complex spatial-image-processing to calculate appearance from retinal radiance arrays.

17.6 References

McCann J & Rizzi A (2008) Appearance of High-Dynamic Range Images in a Uniform Lightness Space, *Proc. CGIV 08/IS&T*, Terrassa, **4**, 177–82.

Rizzi A & McCann JJ (2009) Glare-limited Appearances in HDR Images, *J Soc Img Display*, **17**, 3.

Rizzi A, Pezzetti M & McCann J (2007a) Measuring the Visible Range of High Dynamic Range Images, ECVP07, Arezzo: European Conference on Visual Perception, 36 ECVP.

Rizzi A, Pezzetti M & McCann J (2007b) Glare-limited Appearances in HDR Images, *Proc. IS&T/SID* New Mexico: Color Imaging Conf, **15**, 293–8.

18

HDR and Uniform Color Spaces

18.1 Topics

This chapter examines the effect of increased dynamic range on appearance in uniform color spaces. The experiments in Section C show that increasing the dynamic range of a transparent display by a factor of 500 has minimal effect on appearance. How does HDR imaging affect the data used to define a Uniform Color Space?

18.2 Introduction

There are many ways to study color vision. There is the study of light coming to the eye as described by physics (Radiometry and Photometry). There is the color response of the receptors, measured by Colorimetric Color Matching Functions (CMF). There are the neurophysiological measurements of neurons. There are the light absorptions of the visual pigments in the rods and cones in the retina. There are psychophysical measurements of color distance in Uniform Color Spaces (UCS).

Do they all combine to make the same simple model of how we see? Or, do these different techniques give different answers that lead to new paradoxes about vision?

In the 1860s, Maxwell made the first measurements of human color sensitivity. Since then, color has been studied using psychophysics to measure color matches and color appearances. Since 1964, physiological measurements of cone receptors' absorption spectra and electrophysiology have studied color vision at the cellular level. Since 1968 *High Dynamic Range* (HDR) image capture and image processing have studied color in the computer for processing and reproduction. Psychophysics, physiology and HDR imaging are three parts of today's color. However, they are often discussed independently without considering their inter-relationships. This chapter studies the points of agreement, points of possible disagreement, and the intellectual structure of how to combine all three disciplines.

18.3 Uniform Color Spaces – Psychophysics

As summarized in Chapter 2, human psychophysics uses two different data sets to describe color. The first set is Color Matching Functions (CMF), which predict whether two halves of a circle match, each half with a different spectral composition (Maxwell, 1860). Such color identities can be predicted by converting spectral radiance measurements (380 to 750 nm) into the tristimulus values X, Y, Z using CIE standard CMF functions (CIE Proceedings, 1931). Since psychophysics has no technique for measuring the peak wavelength sensitivities of individual L, M, and S cones, X, Y, Z are attempts at measuring linear transforms of cone sensitivities modified by pre-retinal absorptions. The X, Y, Z color matching functions can be used to predict whether patches will match, but cannot predict the color appearance of those patches, because X, Y, Z cannot take into account human spatial processing of other stimuli in the field of view (Wyszecki, 1973).

The second distinct set of color data describes uniform color spaces. The X, Y, Z values, from color matching experiments, form a 3D space. X, Y, Z space is not isotropic in appearance. Equal euclidean distances in X, Y, Z space do not predict equal changes in appearance. Munsell, Sloan and Godlove (1933) and others (Newhall, Nickerson & Judd, 1943) asked observers to find samples that appeared to be equally spaced (Chapter 2.6). Ostwald, OSA Uniform Color Space, NCS, and ColorCurve are examples of observations used to define other uniform color spaces. Munsell space is unique because it has no external restrictions imposed on the observers' judgements (McCann, 1999). For example, Ostwald constrains all maximum chroma to fall on a circle of constant value, and the OSA space assumes specific axes in the chroma plane. There are no constraints imposed on the Munsell space, other than achromatic value is the vertical axis and the space is a polar cylinder.

CIELAB, CIELUV, and CIECAM are examples of computational models of uniform color spaces. In this chapter we use CIELAB. We realize that there are other more recent color space models and that there are issues of accuracy of uniformity associated with L*a*b* (McCann, 1999). Nevertheless, L*a*b*, with its long history, and great computational expediency, has great popularity and common usage. The CIE 1931 standard for colorimetry calculates the tristimulus X, Y, and Z triplet from radiance, by integrating the light spectrum coming to the eye with the color matching functions $\bar{x}, \bar{y}, \bar{z}$. Using the CIE 1976 standard color space, we calculate L*a*b* from X, Y, Z (CIE, 1978). Lightness (L*) calculates appearances between white and black; a* calculates appearances between red and green-blue; b* calculates appearances between yellow and blue. The goal of these formulae is to be able to convert X, Y, Z to an isotropic color space, where all constant euclidean distances have approximately constant differences in appearances.

The formulae for converting linear (radiance) spectral (XYZ) to L*a*b* are as follows:

$$L^* = 116\left(\frac{Y}{Y_n}\right)^{1/3} - 16 \tag{18.1}$$

$$if\left(\frac{Y}{Y_n}\right) < 0.008856,\ then\ L^* = 903.3\left(\frac{Y}{Y_n}\right) \tag{18.1a}$$

$$a^* = 500\left[\left(\frac{X}{X_n}\right)^{1/3} - \left(\frac{Y}{Y_n}\right)^{1/3}\right] \tag{18.2}$$

$$b^* = 200\left[\left(\frac{Y}{Y_n}\right)^{1/3} - \left(\frac{Z}{Z_n}\right)^{1/3}\right] \tag{18.3}$$

where Xn, Yn, Zn are the integrals for the reference white (radiant power reflected from a perfect diffuser in the viewing illuminant).

For equations 18.2 and 18.3, when any of the ratios X/Xn, Y/Yn, Z/Zn is less than, or equal to 0.008856, then:

[(X/Xn)^1/3] is be replaced by [7.787*(X/Xn) 16/116],

[(Y/Yn)^1/3] is be replaced by [7.787*(Y/Yn) + 16/116],

[(Z/Zn)^1/3] is be replaced by [7.787*(Z/Zn) + 16/116]

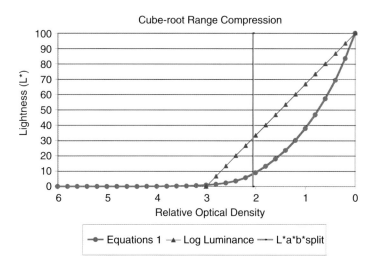

Figure 18.1 Plots of CIEL* equation 18.1 and 18.1a vs. relative optical density. The orange line delimits the range of each equation. L* reaches a value of 1 at OD = 3.0. For comparison, the triangles plot a linear log luminance function over 3 log units.

The way that L*a*b* handles High Dynamic Range (HDR) scenes has two components. The first uses a cube-root function in both lightness and chroma for high light levels (equation 18.1). The second uses a linear function to force the calculation to zero asymptotes (equation 18.1a).

In all cases, in evaluating this set of equations the first thing one does is to normalize the long-, middle- and short-wave integrals to the maxima in each channel; (X/Xn), (Y/Yn), (Z/Zn). Human vision normalizes appearances to maxima in L, M and S channels (Chapter 27, Land 1964; McCann, 1992; McCann, McKee & Taylor, 1976). This operation converts quanta catches to relative integrated radiances.

The next step in all calculations raises these normalized integrals to the power of 1/3, or cube root. This exponential step shapes the normalized X, Y, and Z to approach color space uniformity. In HDR terminology, this step scales the large range of possible radiances into a limited range of appearances.

Figure 18.1 shows calculated CIE Lightness L* (equations 18.1, 18.1a) vs. log luminance for a range covering 6 log units. Since we are concerned with the study of color over the range of HDR scenes, we will continue to plot luminance information as *relative optical density* (OD = log[1/(Y/Yn)]). The vertical orange line identifies the luminance that divides the regions used. On the right side of the orange line equation (18.1) applies; on the left side equation (18.1a) applies. L* describes white as 100. On this graph L* = 100 plots at 0 relative optical density, or 100% (Y/Yn). When we reduce relative luminance by *one half*, then equation (18.1) reduces L* by 24%. In order to get L* = 50, we have to reduce luminance to 18%.

The orange line delimits the ranges of L* equations and falls at 9% in apparent lightness and 0.9% in luminance. There is no cube root function in equation 18.1a. It controls the shape of the asymptote to 0 lightness. L* = 1 falls at OD = 3.0, or 0.1% luminance. In other words, L* suggests that 99% of possible apparent lightnesses fall in 3 log units of scene dynamic range. Since a* and b* use the same compressive cube root function on (Y/Yn), (X/Xn), (Z/Zn), then L* a* b* evaluates the very large range of X, Y and Z in the scene, over a 3 log unit cube. Uniform color spacing is achieved by the cube-root function, plus a linear segment below 0.9% relative luminance.

Figure 18.1 (blue triangles) plots a 3.0 log unit log10 function. By comparison, L* changes more quickly over the OD range 0.0 to 1.0, and more slowly over the OD range 1.0 to 3.0.

The Munsell Space data provides a description of color using all stages of processing. Since the data comes from asking observers to describe appearances, we can think of this as a "top-down" analysis of color space using the entire visual system. In comparison, as described by David Wright (Chapter 2), color matches probe the absorption of light by retinal receptors at the other end of the process. The CIE L* equation fits the Munsell Space lightness Value data very well (McCann, 1999).

18.4   Color Vision – Physiology

Although psychophysical measurements of human spectral sensitivities had a century-long head start, actual measurements of the absorption spectra of single rods and cones have been possible for 40 years. Marks, Dobelle, and MacNichol (1964) and Brown and Wald (1964), measured single cone cell spectral absorptions. About the same time, physiologists measured the electrical response of retinal summation cells to different levels of light. They reported that response to light was proportional to log luminance (Werblin & Dowling, 1969).

The study of physiology provides a second powerful description of human color vision. Since all data comes from biophysical measurements of single-cell preparations, we can think of this as a "bottom-up" analysis of light- and charge-responsive cells.

Do psychophysics and physiology agree?

The top-down and bottom-up data do not agree. The actual peak sensitivities of cone pigments are very different from the assumed peaks in color matching functions. The single cell response to light is logarithmic, while psychophysics describes a cube-root function of lightness vs. luminance. Are these serious discrepancies, or are these different results easily reconciled?

18.4.1   Spectral Sensitivity

The comparisons of spectral sensitivity of human cones, color matching functions, and camera spectral sensitivities have very interesting properties (McCann, 2005). If one expects that cones respond the same as color matching functions, then one would be surprised at just how different these sensitivity functions are. The peak sensitivity of the cones is 440, 540, and 565 nm (Brown & Wald, 1964; Marks, Dobelle & MacNichol, 1964). Dartnall, Bowmaker and Mollon (1983) have shown similar results with peaks at somewhat shorter wavelengths. The peak sensitivities of XYZ are 450, 555, and 610. The wavelength of CIE $\bar{x}$ peak sensitivity is 45 nm longer than the L-cone peak. The inclusion of pre-retinal absorptions helps somewhat to reconcile these discrepancies. They shift the peak sensitivities to longer wavelengths.

Since the CIE 1931 Committee did not know the actual peak sensitivities of cones, they had to assume one of a very large set of possible linear transforms of matching data. Smith and Pokorny (1975) argued that X, Y, Z are not inconsistent with actual cone sensitivities for color matching, as long as pre-retinal absorptions were included in the calculation.

18.4.2   Intensity Response

Werblin and Dowling (1969) showed that single cell recording of summation responses of horizontal cells are proportional to log light intensity. There is no evidence of a cube root function in retinal response to light.

Stiehl et al. (1983) studied images that contained equally spaced steps in lightness appearance. They measured the light coming from the display, eyepoint luminance, of these equally spaced lightness steps. Figure 18.2 plots Stiehl's 9-step Lightness data (triangles), rescaled to 100 to 1. Their data falls on top of the CIELAB and CIELUV L*function (circles).

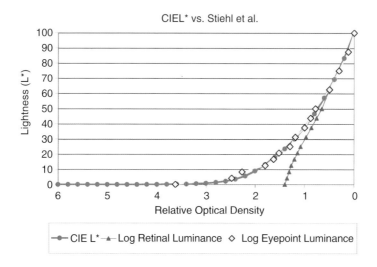

Figure 18.2   Plots of relative optical density vs. three sets of data: CIEL*; Stiehl scene luminance at the eyepoint; and Stiehl's retinal luminance after intraocular scatter.

Stiehl et al. (1983) also used the point-spread function of the human eye to calculate the luminance on the retina after intraocular scatter (Chapter 14). They found a different relationship between retinal luminance and appearance. Log luminance of the image on the retina is proportional to lightness appearance.

The cube-root function in CIELAB and LUV is caused by scattered light. We have just seen that pre-retinal spectral absorptions are critical to reconciliation of physiological and psychophysical sensitivity data. Pre-retinal veiling glare has to be included in evaluating lightness. Figure 18.2 plots L* vs. Stiehl et al.'s retinal and eyepoint luminances in the same manner as Figure 18.1. Stiehl's calculated retinal luminance covers a much smaller range of contrast (1.4 log units) and shows the same linear response to log luminance reported by physiologists.

Just as we saw that intraocular absorptions play a key role in understanding the spectral response of the eye, we see that intraocular scatter plays an equally important role in understanding the eye's dynamic range. Measurements of both spectral sensitivity and lightness show the significant impact of intraocular modifications of the target stimulus by absorption and scatter.

18.5   Accurate Transformations from CMF to UCS

The CIE UCS spaces are based on calculations that transform X, Y, Z to an approximate UCS value. CIELAB and CIELUV have similar, but different equations, using the cube-root for lightness. These conversions are suitable for a slide rule, or calculator. CIECAM uses many color-adaptation matrixes that are the equivalent of over 100 equations and suited for computer calculation.

The CIE L*a*b* space described above (18.3) is an approximate transformation of the CMF based X, Y, Z standards into a uniform color space (Fairchild, 1998). That raises two important questions:

How accurate is L*a*b* in converting X, Y, Z to Munsell color space?

Can we use color data to derive an error free conversion of CMF to UCS?

Since we know the spectra of all the chips in the Munsell Book, we can convert X, Y, Z individually to Munsell space, and evaluate the accuracy of the standard (McCann, 1999). We tested the uniformity of CIE L*a*b* and found surprisingly large discrepancies between CIE L*a* b* and isotropic observation-based color spaces, such as Munsell:

CIE L* predicts Munsell Lightness with great accuracy

CIE a* b* chroma exaggerate yellows and underestimate blues.

The average discrepancy between a* b* and accurate chroma prediction is 27%.

Chips with identical L* a* b* hue angles are not a single color.

There are many color evaluations in which we want to average color values. That is a problem if we calculate the average distance in a non-linear color space. For accurate color distance measurements we need to evaluate data in a uniform color space. Although it is very convenient to measure L* a* b* with many instruments, L* a* b* introduces large errors in color uniformity.

18.5.1   Data-based LUT Transformations from CMF to UCS

One way to minimize the L* a* b* errors is by using a *three-dimensional look-up table (3-D LUT)* to convert to position in Munsell space (Chapter 31.3.2). We have isotropic data in The Munsell Book. Computers using 3-D LUTS can convert instantly any measured L* a* b* to interpolated Munsell Book values. We call this space ML, Ma, and Mb, after Munsell. LUTs have been developed for both Lab2MLab and MLab2Lab. Each chip in the Munsell book is represented by the measured L*a*b*. With this zero-error, isotropic MLab space we can return our attention to the original color application problem (Marcu, 1998; McCann, 1999).

18.5.2   Data-based Fit for Transformation from CMF to UCS

A second approach is to fit the Munsell Book's data to a model of vision. You can solve for the transform that places all the chips in the Munsell Book in their defined locations.
    We know:

the selected location of each paper in the Munsell Color Space

the spectral reflectance of the paper that observers selected to occupy that location in the Munsell Book

the spectrum of the illumination

the LMS cone responses to the each paper

Indow and Romney have studied the properties of the transformation of cone responses to Munsell Space location (Indow, 1980; Romney & Indow, 2002a; b).
    D'Andrade and Romney (2003) began with the reflection spectra of the papers in the Munsell book, multiplied them by the spectrum of D65 illuminant, then multiplied by the short-, medium-, and long-wave cone sensitivity functions, then summed for each cone response. They then took the cube-root of integrated cone responses and normalized within cone type. We have seen above that the cube-root operation converts cone response to Munsell value for all papers. (McCann, 1999).
    Finally, in their last step, D'Andrade and Romney applied a color-opponent model based on neuro-physiological observations of the lateral geniculate. They converted cone response to the predicted Munsell location on the yellow to purple-blue (Y2PB), red to blue-green (R2BG), and white to black

axes (Value V). By applying different weights for their opponent processes they found the best fit to actual Munsell locations. They reported that:

$$V = 0.5 * \left[(Lcone)^{1/3} + (Mcone)^{1/3} \right] \tag{18.4}$$

$$R2BG = 7 * \left[(Lcone)^{1/3} - (Mcone)^{1/3} \right] \tag{18.5}$$

$$Y2PB = 2 * \left[-(Lcone)^{1/3} + 2*(Mcone)^{1/3} - (Scone)^{1/3} \right] \tag{18.6}$$

where V is lightness (equation 18.4); R2BG is one orthogonal axis in the chroma plane (equation 18.5); and Y2PB is the other (equation 18.6). The correlations for the fit between predicted and actual Munsell locations using these weights were V (0.997); R2BG (0.974), Y2PB (0.980). Their study evaluated the average statistics of the entire space and did not evaluate local regions of the Munsell Book.

The cube root played an essential role, in that it modified all the colors with a single, simple transform corresponding to the light on the retina after scatter. The central bracket V in this analysis is identical to the L* in L*a*b*. It uses only L and M cone responses as does Y (Chapter 2.5.2). Unlike L* it compresses the achromatic axis by a factor of 2.

The R2BG axis is very similar to a*. The R2BG uses broader spectral sensitivity functions (cones) leading to more crosstalk between receptors, and a very large amplification factor of 7. We need to multiply the small differences between L and M cone responses by seven times. Further, since the V has been weighted by 0.5, the combined result is that the difference in cone response needs to be amplified by 14! That is much larger than the five times required by L*a*b*. The Y2PB axis is very similar to b*. It also uses broader spectral sensitivity functions, and the amplification factor of 2, and a net amplification of 4.

The role of the color-opponent process is to counteract crosstalk (Chapter 27). Although the cones have very little difference in their spectral responses, humans' color vision stretches those small differences to be able to enhance differences in chroma.

18.6   Summary

Intraocular glare, or scattered light, controls the dynamic range of luminances falling on the retina. Lightness in scenes with uniform illumination correlates with log luminance on the retina. The most important mechanism in the transformation of the results from colorimetry to those from Uniform Color Space measurements is the cube-root of luminance. Stiehl et al. showed that the cube root function is the result of intraocular scatter. In normal scenes the cube-root function shows the effect of glare on areas below middle gray. The results in Section C support that finding.

One paradox, namely that colorimetry does not predict UCS appearance, has been resolved. It requires a two part solution.

First, pre-retinal scatter transforms scene radiances into the UCS lightnesses. Colorimetry makes the assumption that all appearances could be calculated from the matching properties measured by using spots in a black background. That assumption inhibited measurements of scenes that involved veiling glare. Without glare, color matches inform us of the spectral response of light receptors. Color matches in black surrounds fail to include scene-dependent spatial effects of glare, and hence they do not measure color vision for scenes.

Second, cone crosstalk severely limits the range of chroma output from cones. Neural image processing using spatially opponent color comparisons are needed to enhance extremely small differences in cone signals (Conway, 2010). D'Andrade and Romney estimate the chroma enhancement along R2BG color axis to be a 14 times stretch.

We need a combination of veiling glare and spatial opponent interactions to understand the complex transformation of light from the scene to Uniform Color Spaces.

18.7   References

Brown P & Wald G (1964) Visual Pigments in Single Rods and Cones of the Human Retina, *Science*, **144**, 45.

CIE Proceedings (1931) Cambridge University Press, Cambridge, 19.

CIE (1978) Recommendation on Uniform Color Spaces, Color Difference Equations, Bureau Central de la Paris: Psychometric Color Terms, Supplement 2 of CIE Publ. 15 (E-1.3.1).

Conway B (2010) *Neural Mechanisms of Color Vision: Double-Opponent Cells in the Visual Cortex*, Kluwer, Dordrecht.

D'Andrade R & Romney A (2003) A Quantitative Model for Transforming Reflectance Spectra Into the Munsell Color Space using Cone Sensitivity Functions and Opponent Process Weights, *PNAS*, **100**, 6281–86.

Dartnall H, Bowmaker J & Mollon J (1983) Human Visual Pigments: Microspectrophotometric Results from the Eyes of Seven Persons, *Proceedings of the Royal Society of London*, B **220**, 115–30.

Fairchild M (1998) *Color Appearance Models*, Addison Wesley, Reading MA.

Indow T (1980) Global Color Metrics and Color Appearance Systems, *Color Res & Appl*, **5**, 5–12.

Land E (1964) The Retinex, *Am Scientist*, **52**, 247–64.

Marcu G (1998) Gamut Mapping in Munsell Constant Hue Sections, *Proc. IS&T/SID Color Imaging Conference*, Scottsdale, **6**, 159.

Marks WB, Dobelle WH & MacNichol Jr. EF (1964) Visual Pigments of Single Primate Cones, *Science*, **143**, 1181.

Maxwell J (1860) *On the Theory of Compound Colours, in The Scientific Papers of James Clerk Maxwell* (1945), ed. Niven, Dover, New York, 410–44.

McCann JJ (1992) Rules for Colour Constancy, *Opthal. Opt.*, **12**, 175–77.

McCann JJ (1999) Color Spaces for Color Mapping, *J. Electronic Imaging*, **8**, 354–64.

McCann JJ (2005) The History of Spectral Sensitivity Functions for Humans and Imagers: 1861 to 2004, Color Imaging X: Processing, Hardcopy, and Applications, Bellingham, WA: *Proc. SPIE*, **5667**, 1–9.

McCann JJ, McKee S & Taylor T (1976) Quantitative Studies in Retinex Theory: A Comparison Between Theoretical Predictions and Observer Responses to "Color Mondrian" Experiments, *Vision Res*, **16**, 445–58.

Munsell A, Sloan L & Godlove I (1933) Neutral Value Scales, I. Munsell Neutral Scale, *J Opt Soc Am*, **23**, 394.

Newhall S, Nickerson D & Judd D (1943) Final Report of the OSA. Subcommittee on Spacing of the Munsell Colors, *J Opt Soc Am*, **33**, 385.

Romney A, & Indow T (2002a) Estimating Physical Reflectance Spectra from Human Color-matching Experiments, *P Natl Acad Sci*, **99**, 14607–10.

Romney A & Indow T (2002b) Model for the Simultaneous Analysis of Reflectance Spectra and Basis Factors of Munsell Color Samples under D65 Illumination in Three-dimensional Euclidean Space, *P Natl Acad Sci*, **99**, 11543–46.

Smith V & Pokorny J (1975) Spectral Sensitivity of the Foveal Cone Photopigments between 400 and 500 nm, *Vision Res*, **15**, 161.

Stiehl W, McCann J & Savoy R (1983) Influence of Intraocular Scattered Light on Lightness-Scaling Experiments, *J Opt Soc Am*, **73**, 1143–48.

Werblin F & Dowling J (1969) Organization of the Retina of the Mudpuppy, Necturus maculosus. II. Intracellular Recording, *J Neurophysiol*, **32**, 339–55.

Wyszecki G (1973) *Colorimetry, Color Theory and Imaging Systems*, ed Eynard R, Society of Photographic Scientists and Engineers, Washington, 38.

19

Glare: A Major Part of Vision Theory

19.1 Topics

This chapter reviews the experiments that isolate the role of optics from the role of neural image processing. It reviews the problems associated with pixel-based tone-scale image processing.

19.2 Introduction

In Sections B and C of this text we studied intraocular glare, or scattered light, that controls the dynamic range of light falling on the retina. Lightness in scenes with uniform illumination correlates with log luminance on the retina.

The paradox that colorimetry does not predict UCS appearance has been resolved. The usual hypothesis is that the post-retinal neurophysiology transforms the retinal response. In fact, we saw here that it is pre-retinal scatter that shapes the lightness. Further, we saw that the spatial comparisons of opponent-color processing are needed to overcome the overlap of cone spectral sensitivities.

19.3 Glare: Distorts Lightness below Middle Gray, More or Less

The analysis of lightness vs. light stimulus can be summarized by returning to the Stiehl et al. (1983; Chapter 14) data. The conventional psychophysical plot would use the amount of light on the horizontal (input) axis, and the appearance on the vertical (output) axis. But, here we are approaching the problem by starting with equally spaced lightnesses. Stiehl measured the scene luminances that made equal lightness steps. That is the cube-root data in Figure 19.1. As well, he calculated the contrast of the retinal image. That is, the straight-line log retinal luminance data in the same figure.

The two curves overlap above lightness 5.0, and are markedly different below this middle gray value.

The results in Figure 19.1 show that measures of discrimination are distinct from measurements of retinal dynamic range. Humans can discriminate between display blacks that are 1/1000th the white luminance, although the stimulus on the retina is limited by scatter to only 1/30th the white. Discrimination has to do with spatial comparisons. Cornsweet and Teller (1965) measured discrimination thresholds with different lightnesses. Using a pair of joined semi-circular test areas, with an adjust-

---

The Art and Science of HDR Imaging, First Edition. John J. McCann, Alessandro Rizzi.
© 2012 John Wiley & Sons, Ltd. Published 2012 by John Wiley & Sons, Ltd.

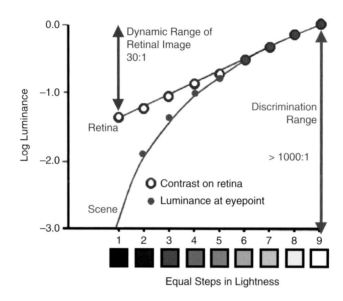

Figure 19.1 Comparison of the 30:1 range of contrast on the retina with the 1000:1 range of discrimination of scene luminances.

able surround, they measured observers' ability to detect edges. By changing the surround luminance around the test semicircles, they changed the test areas from light to dark. They showed that discrimination depends on the local, glare corrected, stimulus on the retina. The observers' discrimination was the same regardless of whether it appeared white or black.

Scatter, by its very nature, affects lower luminances more than higher ones. In order to study the spatial properties of vision one must analyze retinal contrasts. Earlier in Section C (Figure 16.5) we saw that different backgrounds had major effects on appearance. Scene content alters the visual response to light.

Veiling glare reduces the contrast of the scene. Spatial processing counteracts glare by increasing the rate of change of appearance with luminance. It not only cancels scatter, but overcompensates, leaving the examples of Simultaneous Contrast as the residual. A gray paper in a white surround appears 10% darker than the same gray on black.

Veiling glare masks the very powerful spatial processing in vision. Evidence for this spatial processing would be more dramatic without the influence of veiling glare. Regardless, spatial processing acts to minimize the effect of glare. Since both glare and spatial processing are complex, scene-dependent processes with different spatial properties, they do not exactly cancel. Rather, the spatial physiology reduces the effect of glare and contributes to the constancy of vision. To understand these phenomena in detail, we must study each process's complex spatial response to the entire scene. In this Section we have studied the spatial properties of veiling glare. In Section D, we will study the properties of neural spatial processing.

19.4 Pixel-based HDR Image Processing

When one measures the human response to spots of light we acquire valid data. The problem is that this data applies only to those specific test targets, their luminance distribution, their sizes, the relative position of white areas, and the viewing distance. One cannot generalize the results of these experiments and apply them to other images, with different luminance distribution, nonuniform illumination and different dynamic ranges. Thus, any function that attempts to correlate appearance with the luminance of a pixel is valid only for the images used in the measurements.

Over the past decade, HDR imaging research has devoted considerable study to the use of tone-scale maps that render HDR luminance and color data. (Tumblin & Rushmeier, 1993; Larson et al., 1997; Reinhard et al., 2006). Such studies assume that multiple exposures can capture the wide range of scene radiances (Debevec & Malik, 1997). Once the scene is captured, then one might want to select an appropriate tone scale to render the HDR scene digits for the human visual system (Reinhard et al., 2002). As we have seen from measurements of HDR test targets in this Section C, pixel-based HDR image processing has serious problems. Accurate capture is not possible, and the spot-based models of vision do not apply to any real scenes, with the possible exception of stars on a moonless night.

19.5   Summary

The interest in computer graphics HDR imaging in the past 10 years has stimulated research. Pixel-based image processing is very efficient. However, we need to be very careful about applying pixel based models of human vision to HDR imaging.

As we saw in Section C, intraocular glare helps to solve an important visual paradox. Uniform Color Spaces do not correlate with cone responses. Lightness correlates with the image on the retina after scatter. Equal steps in log retinal contrast generate equal steps in appearance. The contents of the scene determine the unique amount of glare added to each pixel. Glare is scene dependent. Neural contrast is also scene dependent, but tends to cancel glare. Both are substantial effects. Their large magnitude is concealed because they effectively cancel each other. Veiling glare is seldom apparent. Neural contrast is so powerful that the effect we see, called Simultaneous Contrast, is the remainder after cancelation. In the following section we measure the effects of post-retinal image processing. Spatial comparisons, instead of pixel response, are the basis of apparent lightness (Section D).

19.6   References

Cornsweet T & Teller D (1965) Relation of Increment Thresholds to Brightness and Luminance, *J. Opt. Soc. Am.,* **55**, 1303–8.

Debevec P & Malik J (1997) Recovering High Dynamic Range Radiance Maps from Photographs, *ACM SIGGRAPH'97*, 369–78.

Larson G, Rushmeier H & Piatko C (1997) A Visibility Matching Tone Reproduction Operator for High Dynamic Range Scenes, *IEEE Trans Vis & Comp Graphics*, **3**(4), 291–6.

Reinhard E, Stark M, Shirley P & Fewerda J (2002) Photographic Tone Reproduction for Digital Images, *ACM Transactions on Graphics*, **21**(3): 267–76.

Reinhard E, Ward G, Pattanaik S & Debevec P (2006) *High Dynamic Range Imaging Acquisition, Display and Image-Based Lighting*, Elsevier, Morgan Kaufmann, Amsterdam, 2006.

Stiehl W, McCann J & Savoy R (1983) Influence of Intraocular Scattered Light on Lightness-scaling Experiments, *J Opt Soc Am*, **73**, 1143–48.

Tumblin J & Rushmeier H (1993) Tone Reproduction for Realistic Images, *IEEE Computer Graphics and Applications*, **13**(6), 42–8.

Section D

Scene Content Controls Appearance

20

Scene Dependent Appearance of Quanta Catch

20.1 Topics

Appearance in High Dynamic Range images is controlled by intraocular glare and neural contrast. This section describes experiments that study how appearance is affected by the contents of the scene.

20.2 Introduction

We began the discussion of HDR imaging with some simple technology questions. Can better sensors and better displays make better images? We wanted to apply the principles set down by Hurter and Driffield, and embodied by Mees and Jones (Chapter 5). Their silver halide theory added the need for observers' subjective, as well as objective physical measurements. Their best compromise for color print photography was the use of a tone-scale curve designed by observers' preference. Their goal was to include in their photographic theory a model for vision.

HDR photography has demonstrated the limitations of silver-halide tone-scale imaging. Our analysis so far has followed the Mees and Jones approach in that we have measured, as they did, the range of light that can be captured by cameras. We have also measured the range of appearances of HDR test targets. The greatest difference between our conclusions and theirs is that models of vision require incorporation of the spatial characteristics of the scene, and the spatial image processing of human vision.

This chapter introduces a section that reviews the psychophysical measurements of human neural processing. We review a wide range of experiments that describe the properties of human spatial image processing.

20.3 Models of Vision – A Choice of Paradigms

The following chapters in Section D describe experiments that provide data for modeling vision. These experiments, using complex scenes, extend our understanding beyond what we can learn from the study

The Art and Science of HDR Imaging, First Edition. John J. McCann, Alessandro Rizzi.
© 2012 John Wiley & Sons, Ltd. Published 2012 by John Wiley & Sons, Ltd.

of spots of light. These experiments help us to refine our thinking about whether *silver halide*, or *neural contrast*, or *perceptual framework* paradigms will be the most fruitful in crafting models that mimic human vision.

Pixel-based Paradigms: Silver halide films work by counting photons at a pixel. Each small region of film reacts to all light falling on it, independent of the other regions in the image. Tone-scale management is all that one can do with such a silver halide system. One choice for a model of vision is that it too could be modeled at a pixel. It would be convenient for HDR system design, because it is so efficient; but it lacks the ability to mimic human's ability for constancy and HDR response to scenes.

Spatial Content – Neural Contrast Paradigm: If vision is fundamentally a spatial mechanism, then we must use spatial test targets to measure its response. If we can document the properties of neural contrast then we can write computer algorithms that perform the same operations on scene data.

Perceptual Framework Paradigm: Psychological studies of certain complex scenes have led some to conclude that simple *bottom-up* image processing mechanisms cannot account for appearances. These studies call for concepts of *belongingness, perceptual frameworks, and perceived illumination.* This provides a new third type of paradigm to consider. Does the experimental evidence require that we include *top-down* cognitive image processing to model the appearances of scenes. If humans recognize objects to influence their appearances, then do HDR models of vision have to include cognition?

20.4   Illumination, Constancy and Surround

Chapter 21 describes a systematic study of changes in uniform illumination on three images. The combined dynamic range of all these targets is 5.6 log units. The backgrounds are white, gray and black. By measuring the same targets in variable amounts of overall illumination we quantify the small departures from perfect constancy of appearance. By studying white, gray and black surrounds we include the effects of Simultaneous Contrast in constancy.

From this study we see that there is a small change in appearance with the overall level of light. The maxima in a scene have appearances that track one of the oldest psychophysical observations. The appearance of single spots of light and maxima in scenes show the same change in appearance with luminance. Parts of the image that are darker than the maxima show a very different behavior; they exhibit contrast. These darker areas show large changes in appearance with small changes in luminance.

Measurements show that white, gray and black surrounds cause different rates of change with luminance. Contrast depends on the contents of the image.

20.5   Maximum's Enclosure and Distance

Chapter 22 reviews studies of the spatial relationship of the contents of an image. Although a great many studies have measured appearance using uniform spots on uniform surrounds, we see that the spatial properties of the image components are important. This chapter studies the effect of a maximum luminance area (White) on a gray (Test Field). Here the white is adjacent to one, or two, or three or four sides of the test gray. It also studies the effect of separating the white from the gray on a black background. The results describe the visual system's response to enclosure and distance of the scene maxima.

20.6   Size of Maxima

Chapter 23 reviews studies of the spatial relationship of sizes of test patches and the sizes of maxima. This chapter introduces the idea of equivalent background for testing spatial models of human vision. It describes 50 different test targets of which 27 have identical pixel image statistics. It identifies sets of targets with equivalent backgrounds and analyzes the result using different approaches to modeling spatial vision. The results showed that despite a wide range of pattern types (*Snow, Corners, Sides Lines* and asymmetry) the observers' matches showed very high correlation with very simple spatial averages.

20.7   Assimilation

Chapter 24 reviews studies of *Assimilation*. In addition to the familiar examples of Simultaneous Contrast in which test gray areas are darker in white surround, there are other complex surrounds that make the same *test gray* appear lighter in an adjacent white surround. These *Assimilation* scenes add evidence to the complexity of the neural contrast mechanism. Chapter 24 reviews measurements of the appearance of *Assimilation* test targets.

20.8   Maxima and Contrast with Maxima

Chapter 25 summarizes Section D with its studies of how spatial content of images affect their appearance. It uses the results of spatial experiments to identify the properties of lightness appearance generated by complex scenes. It uses this data to merge the psychophysical *Aperture and Object Modes of Appearance* into a simple two step image processing mechanism. Although the matching data is complex there is no evidence that *top-down* perceptual mechanisms are necessary to account for changes in illumination, constancy, contrast and assimilation observations.

21

Illumination, Constancy and Surround

21.1 Topics

Nineteenth century psychologists segmented appearances into modes, such as Aperture and Object modes, based on observations that scene stimuli appear different in normal scenes than in black (no light) surrounds. The 19th century assumption was that the stimulus determines the mode, and perceptual feedback determines the appearance. A more modern view of vision is that appearance is generated by spatial interactions, or neural contrast. This chapter describes contrast and constancy experiments over a 5.6 log unit range of luminances. Matches are not consistent with discounting the illuminant. The observers' matches fit a simple two-step physical description: the appearance of maxima is dependent on luminance, and the appearance of less-luminous areas are dependent on spatial contrast. The need to rely on unspecified feedback processes, such as aperture mode and object mode, is no longer necessary. Simple rules of maxima and spatial interactions account for all matches in flat 2D transparent targets, complex 3D reflection prints and HDR displays.

21.2 Introduction

Our understanding of appearance is based on both experiments and the intellectual constructs, or logical frameworks, we build around the experiments. At the start of the 19th century Thomas Young suggested that the retina had three different types of spectral receptors. The problem of color is much more complicated because of observations that colored objects appear constant in different spectral illuminations. The simple, elegant theory relating color appearance to retinal receptor response was either wrong, or incomplete. Experiments showed that identical receptor quanta catches do not always generate the same color. The debate still continues about the nature of the missing elements in color constancy.

In 1953 the OSA Committee on Colorimetry (OSA, 1953) summarized an intellectual framework that included both the colorimetry and the ideas from Goethe, Hering, Helmholtz, Katz, Gelb, Evans and others. This framework describes five modes of appearance and 15 attributes and dimensions of color. The experimental fact of color constancy led to an assumption that humans inferred from the scene the spectral composition of the illumination. By this theory, appearance is significantly altered by scene inferences. Inferences determine the mode, which in turn affects the appearance of each pixel

The Art and Science of HDR Imaging, First Edition. John J. McCann, Alessandro Rizzi.
© 2012 John Wiley & Sons, Ltd. Published 2012 by John Wiley & Sons, Ltd.

in the scene. Katz (1935) described 11 modes of appearance. Evans suggested that Katz's modes be combined to three: Aperture or Film mode (without objective reference), *Object* mode, and *Illumination* mode. The OSA subdivided *Object* into *Surface* and *Volume*, and *Illumination* into *Glow* and *Fills Space*. All this complexity is a consequence of trying to fit experimental data into the framework that inferred modes that, in turn, control color appearances (OSA, 1953).

There is a different framework that can account for color constancy without modes of appearance, namely spatial comparisons. Starting in 1949 with Keffer Hartline's experiments on Limulus eyes (Hartline et al., 1964; Ratliff, 1974), followed by the neurophysiology of Kuffler (1953), Barlow (1953), Hubel and Wiesel (2005) and Zeki (1993), our mid-20th century understanding of visual mechanisms changed. Vision was no longer a biological array of pixels. It became a progression of spatial comparators. At the same time the work of Campbell, Gibson and Land provided psychophysical evidence for spatial mechanisms that did not require inference. It was no longer necessary to find the illumination of a scene to explain color constancy (Land, 1964; Land & McCann, 1971; McCann, 2004a). Quantitative measurements of color constancy were predicted accurately by spatial comparisons without determining the illuminant (McCann, et al., 1976; Section E). Modes of appearance were no longer needed to account for constancy.

If color constancy fails to provide experimental support for modes of appearances, is there still a need for these intellectual frameworks in achromatic experiments? Alternatively, is there a simple spatial mechanism that can be used to unify the thousands of experiments performed in trying to understand achromatic vision? (Gilchrist, 1994; 2006) This chapter describes lightness matches of areas in complex images in different levels of illumination. These results will help us evaluate the need for modes of appearance.

21.3   Hipparchus of Nicea

One of the first reported psychophysical experiments was the classification of stellar magnitude by Hipparchus of Nicea in the 2nd century B.C. Although his original manuscripts have been lost, Ptolemy documented them. After many centuries, stellar magnitude is still in common use today. It has been modified to be a photometric measurement starting with Pogson in 1856. The stellar magnitude changes by 100:1 when the measured luminance changes by 100000:1. In other words, stellar magnitudes have a slope of 0.4 in a plot of log luminance vs. log stellar magnitude (Zissell, 1998; McCann, 2005b). With achromatic stimuli there is excellent correlation between stellar magnitude and the aperture mode of appearance. Both study high luminance spots of light in very low-luminance surrounds.

21.3.1   Magnitude Estimation of Brightness

Many authors have measured the change in brightness as a function of luminance. This work is summarized in Wyszecki and Stiles (1982). The Bodmann et al. (1979) magnitude estimation experiments are somewhat similar to the conditions in stellar magnitude with a 2° test spot in an 180° surround. Figure 21.1 plots Bodmann's data. The solid blue line shows the plot of log estimates vs. log luminance. The straight line is similar to Pogson's formula, with a somewhat lower slope (0.3).

In a second experiment, Bodmann changed the surround luminance from 0 to 300 cd/m2. For spots greater than 300 cd/m2 the estimates are essentially the same as with no light in the background (dashed blue line). When the test spot is darker than the surround, then things become more interesting (red line). The plot of estimates has a much higher slope. Here small changes in luminance generate large changes in appearance. The high-slope behavior, called *lightness*, has been shown to fit the cube root of luminance (CIE, 1978, Chapter 18). The cube-root response function comes from intraocular scatter (Chapter 14).

The change from low- to high-slope behavior occurs when spot luminance equals surround luminance. We can generalize the data by saying the brightest area in the field of view tracks along the low-slope blue curve in Figure 21.1. Test spots with lower luminance than the surround exhibit

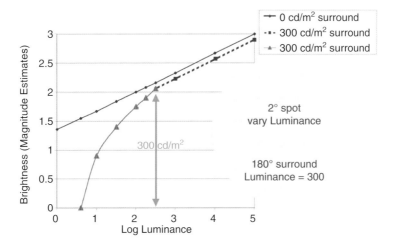

Figure 21.1 Estimated brightness vs. log luminance for a 2° test spot viewed on different surrounds. The solid blue line plots observed data for a 0 luminance surround. The dashed blue line shows spot > surround data for an 180° surround of 300cd/m2. The dotted red line shows spot < surround data for the same surround.

high-slope behavior, regardless of the actual luminance. The mechanism responsible for the appearance of the maxima in the field of view responds to the value of luminance. A second, independent, mechanism generates the high-slope behavior from spatial comparisons. It responds to relative, rather than absolute luminance values.

One can use an LCD display as a metaphor for vision. Conventional LCD displays have integrated two independent components: one is a uniform light source, and the other is a liquid-crystal transmission panel. The light source is analogous to the appearance of the maxima in the field of view. In the LCD, the light emitted is controlled by the number of lamps and their power supplies. The light source is the mechanism that controls the maxima in the display image. The liquid crystal panel with polarizers, color filters, transparent electrodes, and pixel signals reduces locally the amount of light to form the rest of the image. Analogously, the spatial interactions found in the human visual system control the appearance of everything, except the maxima. As shown above, these spatial interactions exhibit a high rate of change of appearance with luminance. The maxima exhibit a low rate of change, as observed in stellar magnitude.

This chapter's first study investigates the role of luminance and contrast in complex images. It uses eight different transmission patches in a white, a gray and a black surround. These patches are viewed at five different luminances and matched to a constant standard. The goal of the study is to understand the limits of constancy with overall changes in luminance. The experimental data will be compared to various hypotheses such as discounting the illuminant, simultaneous contrast, and the ancient low-slope change of appearance with luminance found in stellar magnitude. A further goal is to expand the range of luminances used in related studies (Steven & Stevens, 1963; Jameson & Hurvich, 1961, Bartelson & Breneman, 1967; Gilchrist, 1994) by using targets with a combined dynamic range of 5.6 log units, and variable surrounds.

The second study investigates the role of luminance and contrast in complex 3-D reflection scenes. It looks at matches of identical reflectances in bright light and in shade. The third study investigates the appearance of four identical gray scales in different luminances in complex images. It looks at the role of local maxima.

21.4   Flat-2-D Transparent Displays

The first experiment asked seven observers to match uniform luminance patches to a standard display (McCann et al., 1970). Both test and standard displays were transparent photographic films viewed on two high-luminance Aristo lightboxes. The lightboxes had light emitting surfaces of 20 by 24 inches, but only the most uniform central 10 by 12 inches were used to illuminate the transparent contrast test targets. The viewing distance was 21 inches. The two light boxes were mounted at right angles so that the observers turned their heads back and forth to make matches. The left eye was used for observing the contrast targets, and the right eye for observing the standard. Each observer had excellent color discrimination (McCann, 2007).

21.4.1   Experiments

This study uses eight different transmission test patches in a white, a gray and a black surround (Figure 21.2). These patches were viewed at five different luminances. Corresponding patches had the same transmissions.

Observers were asked to match these patches to a constant standard display. The standard was a series of 25 patches, each with a different luminance transmission surrounded by 1000 ft-L white (Figure 14.2). Observers were asked to report on the mixture of grays on a palette that would match what they saw.

In addition, a series of four different uniform transparencies (O.D. 0.37, or 42.7 %; O.D.0.75, or 17.65 %; O.D. 1.95, or 1.13 %; O.D. 3.25, or 0.056 %) were, in turn, placed behind the test target to control the illumination level of the lightbox. The dynamic range of the test targets covered 2.3 log units, and the dynamic range of the five illuminations covered 3.3 log units. Thus, the resulting dynamic range of the experiment covered 5.6 log units.

One advantage of this apparatus over electronic imaging displays is the accuracy of luminance values at all light levels. The luminances reported here were measured with a Gamma Scientific telephotometer. The ranges of the transparent contrast targets are larger because they are not limited by paper surface reflectances. The constancy of image content is far superior to the constancy possible using simulations on monitors and LCD displays. By using photographic transparencies, we are certain that the contrast displays are exactly constant at each illumination level. This property is essential to isolate contrast from overall luminance.

The three contrast targets are shown in Figure 21.2. The target called TSD-1 has a white surround (O.D. = 0.0) around the eight 2.7° by 5.5° patches. Their transmission optical densities were: 0.61, 0.32,

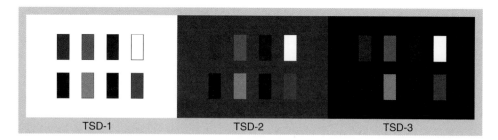

Figure 21.2   Three contrast targets. Transmissions of corresponding patches are the same. Observers matched these targets to the same lightness standard at constant illumination. The dynamic range of all test targets was 5.6 log units: the illumination varied over 3.3 log units and the test patches varied over 2.3 log units.

0.86, 0.00, 0.45, 1.16, 0.19 and 1.28 (clockwise), and TSD-2 has a gray (O.D. = 0.70) surround around the same patches. TSD-3 has a black (O.D. = 2.34) surround around the same patches.

21.4.2 Results

Figure 21.3 plots the average match for all observers for TSD-1 white surround vs. log luminance. The lines plot the eight contrast test patches in the five levels of illumination. The five data sets are parallel and exhibit high-slope behavior. That is, the matching value decreases rapidly with change in luminance. Figure 21.4 plots the average match for all observers for TSD-2 gray surround vs. log luminance. These data are parallel and exhibit high-slope behavior, but with a somewhat lower slope than with a white surround. Figure 21.5 plots the average match for all observers for TSD-3 black surround vs. log luminance. Again, these five curves are parallel and exhibit high-slope behavior, but with a lower slope than with white and gray. All three data sets show a departure from perfect constancy, in that the observers chose a less bright, or darker match for the white area as illumination decreased.

The comparison of Figures 21.3, 21.4 and 21.5 show that the surround luminance has a substantial effect on observer matches. We fit the data from each line in Figure 21.3 with linear least-square regression to calculate the linear slopes of the five fits for the data (decreases in overall illumination). The values were: 4.70, 4.75, 4.46, 4.34, 4.82. The average slope for a white surround is 4.62 ± 0.21. The linear fits for gray surround in Figure 21.4 were: 3.82, 3.77, 3.77, 3.83, 4.03 (Average = 3.85 ± 0.11). The linear fits for black surround in Figure 21.5 were: 2.59, 2.66, 2.32, 2.45, 2.83 (Average = 2.57 ± 0.20). These average slopes are plotted in Figure 21.9 (left).

Figure 21.6 replots the same data used in Figure 21.3. Only the plot lines are different. These graphs plot the changes in appearance of the white, gray, and black test areas. The previous graphs plot illumination. The lines plot the matches for single test patches. These eight lines are parallel and exhibit very low-slope behavior. Perfect constancy predicts all lines would be horizontal (slope = 0). Figures 21.7 and 21.8 replot the average observer match for TSD-2 and TSD-3. In each case the eight lines are parallel, and exhibit the same very low slope behavior.

Figures 21.6, 21.7 and 21.8 study the low-slope change in match with change in luminance. These Hipparchus lines, similar to stellar magnitude, plot the rate of change of match with overall changes in illumination. The comparison of these figures shows that the white, gray and black surrounds have no differential effect on the slope of observer matches with changing illumination. We used a least squares regression linear fit to the data from each line in Figure 21.6 to calculate the linear slopes of the eight patches. The average slope for a white surround is 0.50 ± 0.11. The fit for gray surround was 0.50 ± 0.10; the fit for black surround was 0.61 ± 0.07. These average slopes are plotted in Figure 21.8 (right). The average linear fit for all lines in Figures 21.6, 21.7, and 21.8 was 0.54 ± 0.10.

The average Hipparchus Line slopes are the same for all areas in all targets. Surrounds do not affect its slope. The average slope of 0.54 is in good agreement with the magnitude estimation data by Bodmann (slope = 0.3) using a 2° test spot on a black surround (Bodmann, Haubner & Marsden, 1979). It also is quite similar to the Pogson's standard for stellar magnitude that has a slope of 0.4.

21.5   A Simple Two-Step Physical Description

These results are very simple, and require only a simple physical description. As with color, constancy is not perfect (McCann, 2004b). As with color, departures from constancy have simple physical correlates (McCann, 2005a). We can model this data with a two-step process.

The first step is to find the area with the maximum luminance in the fields of view. Appearance of the maximum is nearly constant, but with sufficient change in the value of the maximum luminance, changes in appearance are obvious. The maximum's luminance predicts that area's match on the low-slope Hipparchus line.

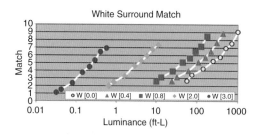

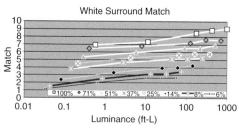

Figure 21.3 Matches for eight test areas in TSD1 in a white surround with variable illumination.

Figure 21.6 Matches for each of the eight test areas in white.

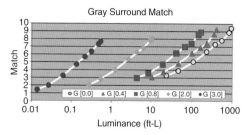

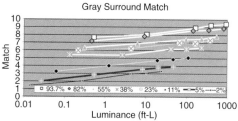

Figure 21.4 Matches for eight test areas in TSD2 in a gray surround with variable illumination.

Figure 21.7 Matches for each of the eight test areas in gray.

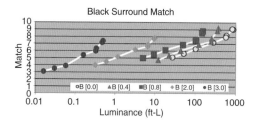

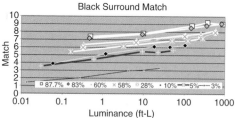

Figure 21.5 Matches for eight test areas in TSD3 in a black surround with variable illumination.

Figure 21.8 Matches for each of the eight test areas in black.

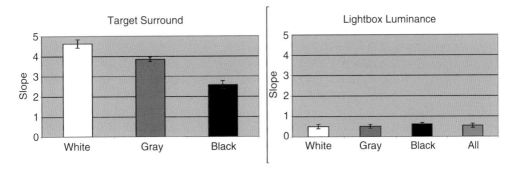

Figure 21.9   (left) Average slopes and standard deviations for surrounds in Figures 12.3, 12.4, 12.5; (right) average slopes for five illuminants in Figures 12.6, 12.7, 12.8.

The second step can calculate the match for nearby less-luminous areas. The composition of the surround determines the slope: White = 4.6; gray = 3.8; Black = 2.6. The ratio of the darker area's luminance to the maxima's luminance determines the distance of the match along the high-slope line.

The data does not show perfect object constancy. If the Hipparchus line data fit a slope of 0, then all luminances would be have exactly the same appearance in a black surround. The data does not correlate with the quanta catch of receptors (luminance).

The data show quite good fit to the simple two-step physical description of luminance and contrast (ratios). Matches to maxima are dependent on absolute luminance, and less-luminous areas are dependent on neural contrast interactions. These results measure the value of the Hipparchus-line slope for complex images. We can use the simple two-step physical description of maxima and spatial contrast to evaluate complex scenes using direct and shadow illumination.

Figure 21.10 illustrates the two-step process. It shows four sets of scene-dependent high-contrast lines for four different values of uniform overall illumination. Detailed plots of the data and this summary are found in McCann (2007).

21.6   Complex 3-D Scenes

These 3-D experiments study the influence of uniform direct illumination and shade in real 3-D scenes. The experiments use printed-paper targets that are folded. The images on each side of the fold are the same. One side was in direct illumination and the other in shade. A separate card has standard reference reflectances printed on the direct illumination side (Figure 21.11 left). Observers matched these patches in both direct and shade illumination. This data can be used to assess whether real 3-D illuminations of scenes influence the appearance of matches. If humans discount the illumination, then these matches will differ from those reported in flat transparent experiments. If the observers behave the same as with 2-D targets, then we have no need for discounting the illumination in complex scenes.

In trying to tease apart the mechanism of appearance it is essential to control the question presented to observers. McCann and Houston (1983) used a photograph of a float to illustrate that observers give different answers to different questions about the same stimulus. (Described below in Chapter 29: Figure 29.1.) When asked to recognize the paint on the sides of the raft, observers say it is white, even though one side is in sunlight and the other is in shade (perception). When observers are asked to hypothetically mix paints to render the appearance in a painting of the scene, they select a whiter, more yellow match for the sunny side, and a darker, more blue match for the shady side (sensation). Arend and Goldstein (1987) documented this observation in experiments using CRT display Black and White

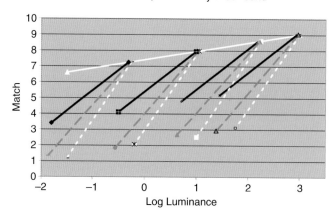

Figure 21.10 The two-step physical description of matches for four illuminants. The upper (white solid) line plots the low-slope Hipparchus line (slope 0.54). Dotted white lines (slope 4.6, white surround); dashed gray lines (slope 3.8, gray surround); and solid black lines (slope 2.6, black surround) fit observer matches. The display's maximum luminance determines the starting point of these high-slope contrast lines. The target's background controls the slope of the high-slope line. Luminance predicts the match of the scene maxima, and spatial contrast (luminance/max) predicts darker areas.

Mondrians. Bloj and Hurlbert (2002) also make this distinction in the analysis of Mach Card experiments. In the experiments described below the observers were asked to report on the mixture of colors on a painter's palette that would match what they saw. They were instructed not to try to guess the reflectance of the patch they were matching.

21.6.1 Experiments

A 4.4 by 2.8 inch folded card was placed on a table on a black cloth. The light falling on the paper came from two fluorescent lamps (distance = 18 inches). Care was taken to make the illumination of each side of the card uniform. The side directly illuminated had a 99.3 ft-L luminance from the white part of the card. The other side had 4.86 ft-L from the white paper. The shade was 4.9 % of the bright side. The standard card was slightly wider than the test targets. The 13 standard patches covering the range of 9.0 to 3.0 were calibrated to have the same % reflectance as the % transmission in the standard in the flat-2D transparent study. The reflectance standard had lower luminance and lower dynamic range than the transmission standard.

Measurements of the luminances from the cards showed that the illumination was uniform over the entire card. The black cloth under the paper plays an important role in uniform illumination. Without a black cloth, it is very difficult to control unwanted light reflected from the table surface, leading to higher luminances along the bottom of the card. A large white surface (2 by 2 feet) was parallel to the lamps on the far side of the folded paper target. It acted as a reflector to control the uniformity and illuminance falling on the paper on the darker side of the fold. A black card (3 by 3 feet) was placed perpendicular to the lamps to reduce reflected light onto the card from that direction.

The observers' eyepoint was to the right of the photograph of the cards and lamps (Figure 21.11 left). The box that contained the lamps masked the lamps from the observers' field of view. The observers viewed the folded card and the matching palette from a distance of 21 inches using both eyes. Binocular vision prevented perceived reversal of depth for the cards.

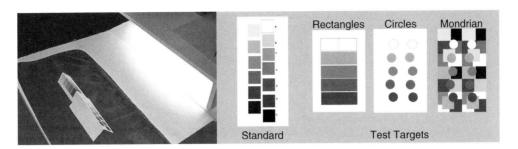

Figure 21.11   (left) 3-D display of folded card on black cloth and fluorescent tubes; (right) shows the Standard and the three test targets. The test targets were folded along the vertical centerline. Matches for the five test areas were measured using the Standard.

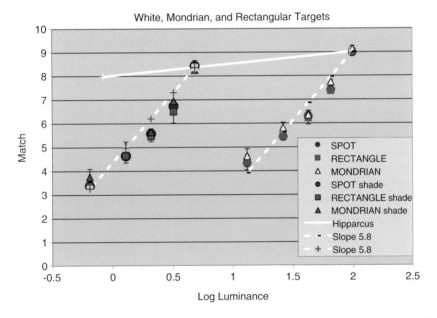

Figure 21.12   Matches for grays in three different backgrounds. Hipparchus line predicts matches for whites or local maxima. The dashed white line is slope 5.8.

The experiments used three different folded cards shown in Figure 21.11 (right). Observers were asked to match test areas in the Rectangles, Circular Spots and Mondrian targets. In addition, they were asked to make additional matches within areas of uniform reflectance of the Rectangles target.

21.6.2   Results

Figure 21.12 plots the average of the matching data for all targets. It plots the average of eight trials (two observers) ± the standard deviation for three different backgrounds around the matching patches. The linear regression fit for circular spots with a white surround has a slope of 5.2 for direct and 6.2 for shade. The fit for the Mondrian surround has a slope of 5.0 for direct and 6.1 for shade. That is not very different from the slope of 5.1 for direct and 6.8 for shade in the Rectangles experiment. The average slope is 5.8. The lines are the same as predictions made in simple physical description in Figure 21.10. The solid white line plots the Hipparchus line that predicts the matches for whites or local maxima. The dashed white line plots slope 5.8.

The data show that there is very little difference between matches with the center of the Rectangles, the Circular Spots, (both in white surrounds), and a Mondrian. Further, the data fits very well the physical relationship described for the flat transparent experiments. The whites in the direct illumination have matches averaging 9.0. The average match in shade was 8.4 and falls on the Hipparchus (slope 0.54 line).

Observer matches are the same as in uniform illumination. The observer seems to follow the simple physical rule that the local maxima respond to luminance. Areas less luminous than the local maxima in white surrounds are controlled by the same high-slope contrast mechanism.

It is interesting to note that the Mondrian background, despite its lower average luminance, has the same slope as the other two white backgrounds (Circular Spot and Rectangular patches). This fact is important because uniform gray surrounds do not have the same visual effect as complex images. Standard matching targets should use white surrounds.

The targets in Figure 21.11 are binocular variants of the Mach Card demonstration that has considerable variability with spatial content (Hurlbert, 1998). The appearance matching experiments in this chapter differ in three properties from Mach's perceptual experiment. First, the displays have three different complex spatial patterns printed on them. Second, the displays were seen in binocular vision that inhibited the depth reversal necessary for change in the perceived reflectances in the Mach card experiments. After all matches had been completed, observers were asked if they could make the card reverse using binocular vision. They said they could not. Third, we asked observers of the artist's palette the question of finding a value in the standard that looked like the patch. Again, after all matches had been made, observers were asked to use monocular vision to see if they could make the card reverse in depth. They could. They were then asked if the reversal changed the matches of the gray patches on the card. They said that the reversal did not. Although there are physical similarities to Mach card, there are also important differences. These results are not in conflict with perceptual studies of Mach's original experiment, they are different answers to different questions using different stimuli.

21.6.3   Do Uniform Stimuli Appear Uniform?

Additional matches were made along the rectangular patches as illustrated in Figure 21.13 (left). Observers were asked to match all along the gray stripe that started near the table at the bottom of the

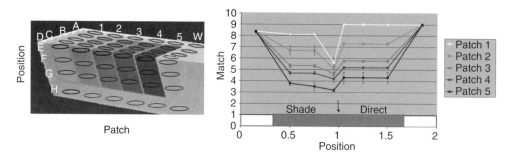

Figure 21.13   (left) Diagram given to observers to identify the spatial position of matches; (right) matches along the rectangular patch for five target patches. The horizontal axis is the distance along the paper from white surround in shade, to the fold at the center, and on to the white surround in the direct illumination.

card, traveled to the top and continued down the shade side. At the top the observer was asked to fixate on the edge created by the shadow at the very top. Here the observers were asked to match the edge they observed at the transition.

Figure 21.13 (right) plots the average matches for image segments along the rectangular patches. The plot starts at the bottom of the card in the shade (Position 0). The first matches are for the white surround below the rectangles H (position 0.15); the last matches are in the white surround below the rectangles A (position 1.85).

The results in Figure 21.13 (right) show that uniform luminances do not always appear uniform. The right side of the graph shows that the gray rectangles appear nearly uniform in direct light. At the edge created across the rectangle by the change from direct to shade illumination there is a significant decrease in match followed by higher values further along the strip. The match at the center of the shade portion of the strip falls on the Hipparchus line. The spatial comparisons at the illumination edge report a much darker patch. All the rest of the spatial comparisons report a lighter patch consistent with the spatial comparisons to the local maxima. This non-uniform spatial appearance is an important piece of data for any computational model using multi-resolution computations. This data can be used to identify the contributions of different size spatial components.

21.7 Local Maxima

In Chapter 12 we described observers' magnitude estimation data for 40 pie-shaped test sectors in white and in black surrounds. The plots of appearance estimate vs. luminance are replotted separately in Figure 21.14. The displays were made of four identical transparencies with 10 gray steps with different neutral density filters (0.0, 1.0, 2.0, 3.0). With a no-light surround the observers estimated that the highest luminance in each circle fell on a low-slope line (left). Adjacent, and nearby sectors, change more rapidly with luminance. Observers can discriminate all 40 test sectors.

With a maximum-light surround on the right, the appearance estimates are more complex, involving both contrast and veiling glare. Observers can discriminate only 28 test sectors. The relevant facts here are that the white surround makes the higher density circle segments darker, even though glare made them have more light on the retina.

The results of Figure 21.14 show that the same pair of descriptive rules found in previous experiments applies to HDR images. Low-slope changes for maxima; higher-slope ratio-dependent changes for less-luminous sectors. The new result here is that the Hipparchus line behavior is seen for local maxima in the same field of view, for the black surround.

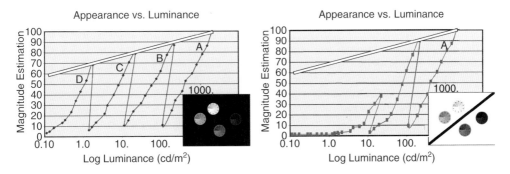

Figure 21.14 (left: 4scalesBlack) Magnitude estimation of all the local maxima fall on the Hipparchus line in the black surround; (right: 4scalesWhite) adjacent white surround makes local maxima in C & D gray and black.

21.8   Review

One principal goal of these experiments was to measure the extent of constancy with changes in luminance. We studied flat-2-D transparent targets and complex 3-D shapes in direct light and shade. To make these measurements we must use an experimental design that quantifies constancy accurately. We chose to instruct the observer to make the "artist's palette" judgment to match the appearance of the test patch. We told the observer not to guess the reflectance of the patches. Guessing the patch's reflectance does not quantify the appearance accurately. Two patches that look different could be assigned the same apparent reflectance (Beck, 1972). Such surface estimates are appropriate for measuring how well humans do at reflectometry (Katz, 1935; Robilotto & Zaidi, 2004), but do not help quantify the accuracy of appearance constancy matches.

These experiments were designed to measure the magnitude of the phenomenon called *discounting the illuminant*. We measured the changes in match with overall, uniform changes in illumination (flat-2-D transparent targets). In 3-D, we measured identical reflectance patches in direct and shadow illumination. We found no difference in observer matches in uniform illumination and matches in non-uniform illumination in real 3-D scenes. We found similar behavior for local maxima in magnitude estimates of scale targets in an HDR image. We found no evidence for perfect constancy. We found no evidence for high-level recognition of illuminants and mechanisms that discount those illuminants. Instead, we found that that the maxima in the field of view, and the local maxima, follow the low-slope decrease in match with luminance. This is the simplest form of response to light. Response is proportional to log luminance. Areas less luminous than the maxima show a rapid change in match due to contrast [slope 5.4 (flat-2-D transparent targets) and slope 5.8 (complex 3-D targets)]. These two simple low-level mechanisms are all that are necessary to explain luminance constancy in flat-2D and complex 3-D targets as long as observers were asked the painter's palette question.

> As discussed above, stellar magnitude (Hipparchus line) is the same as aperture mode of appearance. In the past this data required the separate category of sensation, namely *brightness*. The experiments with 2-D transparencies and 3-D folded prints show that the maxima in these complex images are no different from aperture mode. The shaded areas in the folded card and the HDR *4scaleBlack* experiments (Chapter 12) show that local maxima also act in so called aperture mode.

> Nearby, lower luminance segments in all these images show spatial-dependent neural contrast. This is the same data that historically required the category of *lightness*. There is no difference between neural contrast and object mode of appearance.

Both intellectual frameworks, modes of appearance and neural contrast, attempt to explain the same data. The traditional modes framework assumes recognition of either objects or illumination. The spatial processing framework builds the appearance from the scene content, with a dependence on scene maxima.

This analysis leads to the same pair of alternatives discussed for color in the introduction. There is a 150-year-old framework for vision that assumes the stimulus determines the mode, and then, unspecified mechanisms control feedback that determines appearance. An alternative framework combines the Hipparchus line from stellar magnitude with spatial contrast. Leonardo introduced the study of contrast which is the 500 year-old legacy of spatial interactions. This framework suggests that observations requiring "top-down" modes of appearance can be more easily predicted by the "bottom-up" combination of the Hipparchus line and neural contrast.

21.9   Summary

Objects in complex images appear almost constant. This is true with changes of the overall level of illumination. The small departures from perfect constancy provide important evidence describing the underlying mechanisms:

First, we measured the departures from constancy with changes in overall luminance.

Second, we measured the effects of contrast using white, gray, and black surrounds.

Third, we compared flat, transparent displays with results from 3-D shapes.

Fourth, we reviewed measurements of appearances in HDR targets.

In all cases, we found that the appearance of local maxima is dependent on luminance, and other, less-luminous areas, are dependent on spatial contrast. In all the complex images studied we found no need for modes of appearance to explain the observer results.

The appearance of the maxima and local maxima depends on their luminance, as they fall on the Hipparchus line.

The appearance of the rest of the image depends on the area's relative luminance to the maxima.

The paradox of modes of vision can be greatly simplified. Vision does not need separate *modes* with different *aperture* and *object* mechanisms, requiring perceptual analysis. *Aperture mode* targets are only a singular uniform light stimulus, so it is the maximum. It always tracks the *Hipparchus Line*. *Object Mode* describes complex images with many stimuli. Only the local maxima track the *Hipparchus Line*. The other luminances show neural-contrast interactions and track the high-slope line determined by image content.

21.10   References

Arend L & Goldstein R (1987) Simultaneous Constancy, Lightness and Brightness, *J Opt Soc Am*, **4**, 2281–85.
Barlow H (1953) Summation and Inhibition in the Frog's Retina, *J Physiol*, **119**, 69–88.
Bartelson C & Breneman E (1967) Brightness Perception in Complex Fields, *J Opt Soc Am*, **57**, 953–57.
Beck J (1972) *Surface Color Perception*, Cornell University Press, Ithaca.
Bloj M & Hurlbert A (2002) An Empirical Study of the Traditional Mach Card Effect, *Perception*, **31**, 233–46.
Bodmann H, Haubner P & Marsden A (1979) A Unified Relationship Between Brightness and Luminance, *CIE Proc. 19th Session*, Kyoto, **50**, 99–102.
CIE (1978) Recommendation on Uniform Color Spaces, Color Difference Equations, Bureau Central de la Paris: Psychometric Color Terms, *Supplement 2 of CIE Publ.*, **15** (E-1.3.1).
Gilchrist A (1994) *Lightness, Brightness and Transparency*, Hillsdale: Lawrence Erlbaum Associates.
Gilchrist A (2006) *Seeing Black and White*, Oxford University Press, Oxford.
Hartline H, Wagner H & Ratcliff F (1956) Inhibition in the Eye of Limulus, *J Gen Physiol*, **39**(5), 651–73.
Hubel D & Wiesel T (2005) *Brain and Visual Perception*, Oxford University Press, Oxford.
Hurlbert A (1998) Editorial: Illusions and Reality-checking on the Small Screen, *Perception*, **27**, 633–6.
Jameson D & Hurvich L (1961) Complexities of Perceived Brightness, *Science*, **133**, 174–9.
Katz D (1935) *The World of Color*, Trans. MacLeod RB & Fox CW, Kegan Paul, London, Trench, Truber.
Kuffler S (1953) Discharge Patterns and Functional Organization of the Mammalian Retina, *J. Neurophysiol.*, **16**, 37–68.
Land EH (1964) The Retinex, *Am Scientist*, **52**, 247–64.
Land EH & McCann JJ (1971) Lightness and Retinex Theory, *J Opt Soc Am*, **61**, 1–11.
McCann JJ, Land EH & Tatnall S (1970) A Technique for Comparing Human Visual Responses with a Mathematical Model for Lightness, *Am J Optom*, **47**, 845–55.
McCann JJ (2004a) Capturing a Black Cat in Shade: Past and Present of Retinex Color Appearance Models, *J Electronic Img*, **13**, 36–47.
McCann JJ (2004b) Mechanism of Color Constancy, *Proc. IS&T/SID Color Imaging Conference*, Scottsdale, Arizona, **12**, 29–36.
McCann JJ (2005a) Do Humans Discount The Illuminant? *Proc. SPIE* **5666**, 9–16.
McCann JJ (2005b) Rendering High-Dynamic Range Images: Algorithms that Mimic Human Vision, *Proc. AMOS Technical Conference*, Maui, 19–28.
McCann JJ (2007) Aperture and Object Mode Appearances, in *Proc. SPIE*, 6292–26.
McCann JJ & Houston K (1983) *Color Sensation, Color Perception and Mathematical Models of Color Vision*, *Colour Vision*, ed. Mollon, and Sharpe, Academic Press, London, 535–44.

McCann JJ, McKee S & Taylor T (1976) Quantitative Studies in Retinex Theory: A Comparison Between Theoretical Predictions and Observer Responses to "Color Mondrian" Experiments, *Vision Res*, **16**, 445–58.

OSA Committee on Colorimetry (1953) *The Science of Color*, Thomas Y. Crowell, New York.

Ratliff F, ed (1974) *Studies on Excitation and Inhibition in the Retina: A collection of Papers from the Laboratories of H. Keffer Hartline*, Rockefeller University Press, New York.

Robilotto R & Zaidi Q (2004) Limits of Lightness Identification for Real Objects Under Natural Viewing Conditions, *J Vision*, **4**, 779–97.

Steven JC and Stevens SS (1963) Brightness Function: Effects of Adaptation, *J Opt Soc Am*, **53**, 375–85.

Wyszecki G & Stiles W (1982) *Color Science: Concepts and Methods Quantitative Data and Formulae*, 2nd edn, John Wiley & Sons, Inc, New York, 486–513.

Zeki S, (1993) *Vision of the Brain*, Blackwell Science Inc, Williston, Vermont.

Zissell R (1998) Evolution of the "Real" Visual Magnitude System, *JAAVSO*, **26**, 151.

22

Maximum's Enclosure and Separation

22.1 Topics

We used quantitative lightness matching to measure changes in achromatic lightness as a function of the separation between a bright White Maximum (1000 mL) and a nearby 3 mL, or a 300 mL Test Gray. The experiments varied size, angular subtend and separation of the Maximum White area. All variables showed some influence on the matching lightness to the Test Gray. Of particular interest are substantial changes in lightness from Maximum White areas that have 7.5 degrees separation. This chapter describes the quantitative results, and their implications for models of lightness.

22.2 Introduction

Unlike photographic films and electronic sensors, vision exhibits a number of mechanisms that demonstrate significant departures from simple correlation between the quanta caught and the sensation observed. One of the most fascinating properties of human vision is that it has both local and global properties (McCann, 1987). Many experiments clearly demonstrate the powerful influence of local changes (Wallach, 1948; Albers, 1963). The area surrounding a gray can influence its appearance, even if it does not change the global properties of the image.

The goal of this chapter is to measure both global and local spatial influences of a variable Maximum White on the appearance of a constant gray area (Test Gray). In particular, the goal is first, to maximize the changes due to global mechanisms, and second, to measure the local influence of very bright areas in the field of view with variable separations and sizes.

The largest change in appearance in uniform illumination is an experiment described by Gelb (1929). A black paper appears white when an intense light is projected only on the black paper. When Gelb added a white paper to the field of view, the observer reports that the white paper is white and the black paper is black. Gilchrist (2006) has demonstrated this experiment by introducing a series of new maxima in the field of view.

Heinemann's (1972) excellent review article summarizes many measurements on luminance, induction field size, test field size, distance between the test and the inducing field, and luminance differences.

The Art and Science of HDR Imaging, First Edition. John J. McCann, Alessandro Rizzi.
© 2012 John Wiley & Sons, Ltd. Published 2012 by John Wiley & Sons, Ltd.

In addition, papers by Walraven (1977), Cicerone et al. (1986) have measured the influence of local surrounds. Reid and Shapley (1988) have measured the effects of surrounds in the presence of a gradient and in the context of a Mondrian. Gilchrist's books (1994; 2006) describe many lightness experiments, and present them in the context of top-down perceptual frameworks.

22.3   Experimental Design

In this chapter we study the influence of a maximum white area on gray test patches. We study the effect of separating the white from the gray and the effect of enclosing it in white.

The experimental design was described in McCann, Land, and Tatnall (1970). The apparatus is the same as the one described in Chapter 21. The observers' task was to match the appearance of all areas in the Test Display. The observers looked with their left eye at the Test Display; they looked with their right eye at the Standard Display to assign a number to the appearance of the test patches. The observers looked back and forth to find a match in the Standard Lightness Display (McCann & Savoy, 1991).

The Test Displays were a series of transparency targets that had spatially different arrangements of, at most, three different luminances. The objective was to study the properties of position; thus we severely limited the number of different luminances in the displays. The three patches were called the *Maximum White*, the *Test Gray*, and the *Background*. The *Maximum White* varied in size and position, but always had the same luminance of 1000 mL. The *Test Gray* was always in the center of the light box. It was square and subtended 2.5 degrees. In the first experiment the *Test Gray* had a luminance of 300 mL, and in all other experiments had a luminance of 3 mL. The *Background* was a uniform low luminance of 2 mL, that covered 25 by 30 degrees. In other words, it filled in any area not covered by either the White, or the Test Field.

22.4   Lightness Matches – Light Gray on Black

These results were first presented by Savoy and McCann (1975). These experiments matched the relatively simple Test Targets to the same complex Standard Lightness Display.

The results (Figure 22.1) show three phases in the continuum between the extremes of the Gelb experiment. First, a rapid increase in Lightness as maximum moves from 0 to 1.25 degrees. Second, a region where the matched lightness increases very slowly with substantial increases of separation from the maximum. Third, a jump in lightness when the *Maximum White* is removed.

Figure 22.1 is a plot of observer RLS data. It is made up of nine different displays. Eight displays have a concentric Maximum White square annulus surrounding the *Test Gray* [T] at different eccentrici-

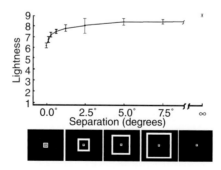

Figure 22.1   Diagrams of the targets along with the separation angle between the constant central gray and the maximum. For each separation distance we plot the matching lightness in the standard.

ties. The ninth display has no Maximum White. The first and last slides in this series are Gelb's experiment. Without the presence of the Maximum White, the 300 mL *Test Gray* matches a "white" (slightly less than 1000 mL) in the Standard Lightness Display.

When the Maximum White is adjacent to the 300 mL *Test Gray* with no intervening *Background*, the observer matches it to a lightness slightly over 6.0. In the next four displays (separations up to 1.25 degrees) the 300 mL *Test Gray* increased to a lightness of 7.6. It gets only slightly lighter in the next three displays (separations from 2.5 to 7.5 degrees). Without the Maximum White, the lightness of the *gray* jumps to 9.0.

In summary, *Maximum White* darkens the *Test Gray* over a wide range of separations and causes a rapid change of lightness at close separations. It remains to be seen if these two rates of change with separation are caused by the same mechanism. When the Maximum White is removed, there is an increase in lightness.

22.5   Lightness Matches – Dark Gray on Black

The second series of experiments repeats the first with a darker 3 mL *Test Gray*. In addition, it contains three additional sets of experiments that vary the circumferential extent of the Maximum White surround.

Figure 22.2 reports the observer KLH data. It contains the results from four sets of targets each with a different degree of enclosure (four sides, three sides, two sides and one side). The leftmost graph is made up of the data from six different displays. Five displays have a concentric Maximum White square annulus surrounding the *Test Gray* at different eccentricities. The sixth display has no White. Again, the first and last slides in this series are Gelb's experiment; this time with much higher contrast between the *Maximum White* and the *Test Gray*.

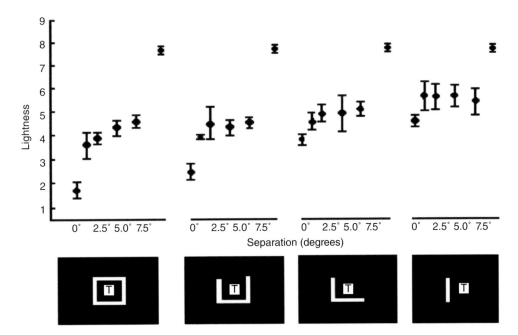

Figure 22.2   Diagram of the four enclosures, each with six matches. This repeats the extension of Gelb's experiment with variable enclosure and separation.

22.5.1   Dark Gray on Black – White on Four Sides

In Figure 22.2 the leftmost graph has six targets. Five have Maximum White on four sides. Without the presence of the Maximum White, the 3 mL *Test Gray* matches 7.7 "light gray" (about 800 mL) in the Standard Lightness Display. When the Maximum White is adjacent to the 3 mL *Test Gray* with no intervening *Background*, it matches 1.7. In the next display (separations of 1.25 degrees) the 3 mL *Test Gray* had a lightness of 3.5. In the three displays (separations from 1.25 to 7.5 degrees) the 3 mL *Test Gray* increased in lightness only slightly to 4.5 .

The results show again three phases in the continuum between the extremes of the Gelb experiment. First, a rapid increase in lightness as the separation from maximum grows from 0 to 1.25 degrees. Second, a region in which the matched lightness increases only slightly with substantial increases of separation. Third, a jump in lightness when the Maximum White is removed. These results are qualitatively like those in Figure 22.1; but, quantitatively they show much larger changes in matching lightness.

22.5.2   Dark Gray on Black – White on Three Sides

In Figure 22.2 the left of center graph has three sides of maximum. When the maximum White is adjacent to the 3 mL *Test Gray* with no intervening *Background*, observers match it to a lightness of 2.1. With separations of 1.25 degrees, the 3 mL *Test Gray* had a lightness of 3.9. Increasing separations from 1.25 to 7.5 degrees, the 3 mL *Test Gray* increased in lightness only slightly to 4.5. With only three sides of Maximum White, the *Test Gray* is lighter than with four sides. The results show again three phases, with similar characteristics.

22.5.3   Dark Gray on Black – White on Two Sides

In Figure 22.2 the right of center graph has two sides of maximum. Without the presence of the Maximum White, the 3 mL *Test Gray* was matched by the observer to 7.7. When the Maximum White is adjacent to the 3 mL *Test Gray* with no intervening *Background*, the observer matched it to a lightness of 3.7. With a separation of 1.25 degrees, the 3 mL *Test Field* grew slightly lighter to a lightness of 4.2. In the three displays (separations from 1.25 to 7.5 degrees) the 3 mL *Test Gray* increased in lightness only slightly to 4.9. With only two sides of Maximum White, it is lighter still. Again, the results show the same three phases.

22.5.4   Dark Gray on Black – White on One Side

Figure 22.2 right shows one side of maximum. When the Maximum White is adjacent, the gray matches 4.6. With separations of 1.25 degrees, it had a lightness of 5.5. In the three displays (separations from 1.25 to 7.5 degrees) it decreased in lightness very slightly to 5.3.

In summary, Maximum White anywhere in the 25 by 30 degree field of view darkens the *Test Gray*. Over a wide range of separations from 7.5 to 1.25 degrees, lightness changes gradually with separation. Maximum White between 0 and 1.25 degrees exhibits rapid change of lightness with changing separation.

22.5.5   Dark Gray on Black

Figure 22.3 combines all the matching data in a 3-D plot. Without the maximum the 3 mL target appears a very light gray of 7.7 (off the graph). When the maximum is in the field of view it varies from 5.6 to 1.7. Both enclosure and separation influence the lightness matches.

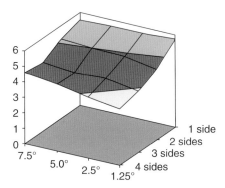

Figure 22.3 Joint influence of enclosure (number of sides) and separation (degrees). The lightness match for the central Test Gray is darkest surrounded by four sides of Maximum White with no separation. As black areas increase in the vicinity of the Test Gray, it appears lighter.

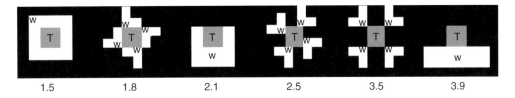

Figure 22.4 Target diagrams; the numbers refer to lightness matches.

22.6 Dark Gray on Black: Varying White's Position

Figure 22.4 shows the results of six different experiments in which the 3.0 mL *Test Gray* is touched by the same area of 1000 mL Maximum White (McCann & Savoy, 1991). Here we studied the influence of circumferential extent. All displays have an equal area of Maximum White. That means that all displays have the same global, or *GrayWorld* statistics. They vary only in local image properties.

The placement of the Maximum White varied from touching the Test Gray along the entire perimeter, left-most display, to only touching on one side, right-most display. Here we reposition a constant area of Maximum White. The central Test Gray (T) appears darkest when surrounded by Maximum White (W) on all four sides. It is lightest when Maximum White is on only one side.

The comparison of the two left-most experiments as well as the comparison of the two right-most experiments shows that having a common border with a Maximum White is not the only constraint on the contrast mechanism. The two lightnesses, 1.5 and 3.9 show dramatic change with only small changes in the local positions of the maximum areas.

Let us think of the Maximum White as influencing the neural contrast mechanism so as to drive the lightness of the *Test Gray* towards black.

> **Left-most display:** the Maximum White encircles the *Test Gray* around the entire circumference. In this experiment we see that when the Maximum White touches all four sides (left-most); the *Test Gray* has the lowest lightness (1.5).

> **Second from the left:** the Maximum White still touches all four sides, but is less homogeneously distributed around a circumference of the *Test Gray*, then the *Test Gray* has slightly higher lightness (1.8).

Third from the left: the Maximum White touches three sides; the *Test Gray* has slightly higher lightness (2.1) than the first two.

Fourth from the left: the Maximum White still touches all four sides, but only for half the length; the *Test Gray* has lightness 2.5.

Fifth from the left: the Maximum White touches only at the four corners; the *Test Gray* has a higher lightness (3.5).

Right-most display: the Maximum White touches on only one side; the *Test Gray* has significantly higher lightness than the first five (3.9).

This data shows that the absence of Maximum White in the near periphery affects the outcome. A model for lightness needs to respond to the influence of the presence of a Maximum White in the near vicinity of the *Test Gray*.

22.7 Review

Although you can characterize film in a physical pointalistic calculation in which all pixels react to light in exactly the same manner, human vision is fundamentally different. The problem of appearance in human vision requires a more complex class of model because all pixels in the image can influence the appearance of all the others. Human vision is a field phenomenon. Appearances are determined relative to all other pixels in the field.

There seem to be four different observations about the mechanisms that control the lightness of a dark *Test Gray* when influenced by a very bright Maximum White.

First, is the Gelb phenomenon. Whenever the Maximum White is removed from the field of view, all appearances are reset to the new lightest area in the field of view.

Second, the circumferential angle is important. At all separations the greater the percentage of the circumference encircled by Maximum White, the darker the sensation of the Test Gray.

Third, the lightness of the Test Gray is nearly constant beyond 1.25 degrees separation for a constant circumferential angle (Figure 22.2).

Fourth, in the region between 0 and 1.25 degrees separation, the appearance is controlled by the physical image on the retina. Intraocular scattered light controls the contrast of the retinal stimulus. The 3 mL Test Gray is indistinguishable from the 2 mL surround. As the Maximum White areas moves further away from the dark Test Gray, then the Test Gray rapidly becomes lighter, between 0 and 1.25 degrees.

As calculated by (Stiehl et al., 1983; Chapter 14) an opaque patch surrounded by a 1000 mL area, has a retinal luminance of 30 mL at the center. The luminance image of the opaque patch on the retina is entirely due to intraocular scatter. The scatter contribution is U-shaped across the opaque area, 30 mL at the center. The 3 mL *Test Gray* simply adds a constant luminance to the greater than 30 mL scattered luminance. At 0 degrees separation scattered light into the dark Test Field is substantially larger than the real light from that area of the target. As the separation increases, the scatter decreases with distance from the Maximum White and the luminance from the Test Field becomes a larger percentage of the light on the retina.

Table 22.1 shows the effects of scattered light on the edge ratio of the *Test Field* and *Background*. The first column lists different scattered light values. The value 30 mL comes from Stiehl's calculations

Table 22.1   Estimates of glare's influence edge ratios.

Scatter [S]	Test Field [T]	Background [B]	Test Field on Retina [S + T]	Background on Retina [S + B]	Ratio [S + T]/[S + B]
30	3	2	33	32	1.031
3	3	2	6	5	1.200
0.3	3	2	3.3	2.3	1.435
0.03	3	2	3.03	2.03	1.493
0.003	3	2	3.003	2.003	1.499

of an opaque area surrounded with white. As the white moves away, the scatter decreases (column 1). When scatter is added to constant target luminances the estimated edge ratio varies from 1.03 (no change) to an asymptote of 1.5 (scene luminance ratio) with decrease in scattered light.

As the ratio of luminances on the retina increase, the Test Field becomes visible against the Background and the Test Field increases in lightness. As the scatter approaches a very low number the ratio of radiances on the retina at the boundary between the 2 and 3 mL areas approaches a constant; and the appearance of the Test Field approaches a constant appearance.

The experiments by Reid and Shapley studied the region for 0.0 to 0.7 degrees separation. They conclude that decline of influence is so great from 0 to 0.7 degrees that it is unlikely that one edge of the Mondrian can influence the next. Our data, however, shows a dramatic influence from stimuli 7.5 degrees away. Informal experiments indicate similar results for 15 degrees. Reid and Shapley's experimental data is most likely in closer agreement to our data, than their conclusion. The region from 0 to 0.7 degrees is highly influenced by scattered light. We see dramatic change in appearance from 0 to 1.25 degrees. Our conclusion, however, is that this data supports the long-distance interaction hypothesis rather than discredits it.

In Ratio, Product, Reset models of lightness (see Chapter 32), reset accounts for the change in lightness when the Maximum White is removed and for the asymmetry associated with the Gelb phenomenon. Variable probability of encountering a Maximum White, and hence having a Reset, can account for the circumferential extent lightness dependence. Long-distance integrations (Ratio Product) account for nearly constant lightness from 1.25 to 7.5 degrees separation. The most difficult data to model is the region controlled by scattered light. Here each target must be evaluated for the real stimulus and the scattered light image so as to determine the real luminance on the retina. Here at least the qualitative argument seems to hold. As described above, with 0 degrees separation the retinal stimulus is virtually all due to scattered light. The *Test Gray* appears darkest when it has its highest luminance. As the separation increases, the *Test Gray* has a lower luminance since the scatter is less. Nevertheless, it appears lighter. At very small separations the observer cannot differentiate the 3 ml *Test Gray* from the 2 mL *Background*. As the separation increases further, the *Test Gray* appears still lighter from a still lower luminance. At about 0.5 degrees separations the observer can differentiate the 3 ml *Test Gray* from the 2 mL *Background*. As described above, the magnitude of the ratio between the Test Field and the *Background* increases as the scatter contribution decreases. The *Test Gray* must look lighter because the edge ratio between Test Field and the *Background* is greater. After the influence of scatter has passed, the Ratio-Product model generates a nearly constant lightness up to 7.5 degrees.

22.8   Summary

The appearance of a *Test Gray* varies with the presence and position of maxima in the field of view. These results cannot be explained by image statistics, or by very simple center-surround models. As well, they cannot be explained by cognitive *top-down* percepts. They show again that the maximum in the scene behaves in a different manner than segments less than the maximum. The influence of the maximum on darker areas has both global and local properties.

22.9 References

Albers J (1963) *Interaction of Color*, Yale University Press, New Haven.

Cicerone C, Volbrecht V, Donnelly S & Werner J (1986) Perception of Blackness, *J Opt Soc Am A*, **3**, 432–36.

Gelb A (1929) Die "Farbenkonstanz" den Sehdinge. *Handbuch der normalen und patholgischen Physiologie*, 12, Springer, Berlin.

Gilchrist A (1994) *Lightness, Brightness and Transparency*, Lawrence Erlbaum Associates, Hillsdale.

Gilchrist, A (2006). *Seeing Black and White*, Oxford University Press, Oxford.

Heinemann E (1972) Simultaneous Brightness Induction, *Handbook of Sensory Physiology VII/4, D. ed. Jameson and Hurvich*, Springer-Verlag, Berlin, 146–69.

McCann JJ (1987) Local/Global Mechanisms for Color Constancy, *Die Farbe*, **34**, 275–83.

McCann JJ, Land EH & Tatnall S (1970) A Technique for Comparing Human Visual Responses with a Mathematical Model for Lightness, *Am J Optom*, **47**, 845–55.

McCann JJ & Savoy RL (1991) Measurements of lightness: dependence on the position of a white in the field of view, *Proc. SPIE* **1453**, 402–11.

Reid R & Shapley R (1988) Local Contrast and Spatial Dependence of Assimilation,*Vision Res.*, **28**, 115–32.

Savoy R & McCann JJ (1975) The Changes of Test Fields by Varying the Shape, Distance and Size of Brighter Fields, ARVO Spring Meeting.

Stiehl WA, McCann JJ & Savoy RL (1983) Influence of Intraocular Scattered Light on Lightness Scaling Experiments, *J Opt Soc Am*, **73**, 1143–48.

Wallach H (1948) Brightness Constancy and the Nature of Achromatic Colors, *J Exptl Psychol*, **38**, 310.

Walraven J (1977) Color Signals from Incremental and Decremental Light Stimuli, *Vision Res*, **17**, 71–6.

23

Maxima Size and Distribution

23.1 Topics

Simultaneous Contrast is the well-known observation that a central Test Gray in a white surround appears darker than in a black surround. This chapter measures the gray appearance as influenced by 2112 white pixels in 25 different spatial configurations. The set of different spatial patterns of white pixels that generate the same matching lightness for gray are defined as equivalent backgrounds.

The chapter then analyzes the spatial properties of equivalent backgrounds. A Test Gray square appears darkest when the solid white surrounds the gray and is contiguous with it. In the case of the distributed white pixels, the gray appears lighter. This chapter presents an analysis of the spatial properties of intermediate surrounds that give the gray center equal visual appearances.

23.2 Introduction

Spatial processing in human vision makes identical retinal stimuli (quanta catch) appear as different sensations (Benary, 1950; Land & McCann, 1971; White, 1979; DeValois & DeValois, 1986; Gilchrist (1994). Different spatial configurations of surrounds can make substantial changes in appearance. Studies of Simultaneous Contrast showed that changing the size of both center and surround together does not change the observed lightnesses. However changing the size of the center relative to surround does. With very large surrounds and small centers the Gray in White is 20% darker, while it is only 10% darker with large centers (McCann, 1978).

The goal of this chapter is to measure the appearances of a wide range of different spatial arrangements of an identical set of white, gray, and black pixels. By comparing the results of observer matches, we can identify different patterns of surround that have the same effect on the human spatial processing, namely when the observers' matches are consistent. These patterns that generate equal matches for a constant test patch are defined as equivalent backgrounds.

Sets of displays having equivalent backgrounds can be used for analyzing different spatial models of vision (McCann, 2004). A model that mimics the human visual process will be able to correctly predict the spatial patterns that are equivalent.

The Art and Science of HDR Imaging, First Edition. John J. McCann, Alessandro Rizzi.
© 2012 John Wiley & Sons, Ltd. Published 2012 by John Wiley & Sons, Ltd.

23.3   Experimental Procedure

This chapter reports experiments using 25 different spatial patterns all composed of a central test patch (made up of 1024 light-gray pixels) on a background (made up of 2112 white pixels, and 62 400 black pixels). These targets were computer generated and were displayed on a CRT monitor viewed at a distance of 38 inches. The square central Test Gray square subtended a visual angle 0.75°. The entire 256 by 256 display subtended 5.9°. Each pixel subtended 1.4 minutes of arc. Each pixel was slightly above resolution threshold and clearly visible (McCann & Rizzi, 2003).

The rest of the monitor screen was covered with opaque material in a darkened room. Observer matches were made using a paper Munsell chart with samples at 0.25 Munsell Value increments. This matching target was placed in front of the observer in an opaque box, so that no light illuminating the standard papers fell on the computer monitor. The observer looked down to see the matching Munsell Value target and looked up to see the test display.

23.4   Controls

Figures 23.1 and 23.2 illustrate two control experiments. In this experiment the observer's only task was to find a match in the Munsell Value scale for the central Test Gray patch. The first control patch varied the pixel value of the Test Gray patch in a constant white surround (pixel value 255). Obviously, when the experimenter decreased the digital value of the central patch, the observer matched it to a darker Munsell Value.

The second control experiment used a constant pixel value of 140 for the central Test Gray and varied the value of the uniform surround.

With a white surround (digital value 255), the observer match was Munsell Value 5.2 ± 0.4. With darker surrounds the observer matches were lighter. Below digit 100, matches reached an asymptote of Munsell Value 8.0. The surround can influence the observer's choice of match over one-third of the range from white to black.

23.5   Dispersion of White ("Snow")

Figure 23.3 shows the beginning of the series of constant average displays. The target on the top left has a 32 by 32 pixel central Test Gray patch (digit value 140). It is surrounded by a 12 pixel band of white. The rest of the target is black. The observer match was 6.50 ± 0.39.

Figure 23.1   The change in lightness with luminance of the central test square in a uniform white surround.

Figure 23.2   demonstrates the change in lightness with luminance of the surround around a constant test square. The experimenters set the constant central patch (digital value 140) and varied the surround.

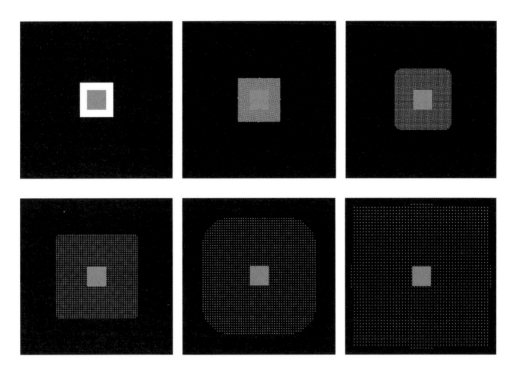

Figure 23.3   The Snow targets with constant image statistics.

This experiment tests whether the global average of the scene controls the lightness of the test patch (Figure 23.3). These *Snow* displays were made with the identical pixel elements. They all have a central Test Gray patch (32 by 32 pixels) with pixel value of 140. They all have a black background of 62,400 pixels (value 0); and a white surround of 2112 pixels (value 255). Each successive target disperses the white pixels in the black surround:

(top-left) 1/1 white pixel fraction; 12 bands of white pixels adjacent to Test Gray

(top-middle) 1/2 white pixel fraction; checkerboard of white and black pixels

(top-right) 1/4 white pixel fraction.

(bottom row), 1/9, 1/16 and 1/25 white pixel fractions

Figure 23.4 plots the Munsell Value matches for the central Test Grays for these targets.

The solid white band adjacent to the Test Gray patch is matched by 6.50 ± 0.39, while the 1/25 white pixel fraction was matched by 7.85 ± 0.49. All displays have the identical global average, or *GrayWorld* value, yet the matches appear lighter with greater dispersion of white pixels.

23.6   Sides and Corners

Figure 23.5 continues the series of constant average displays. Here the 2112 white pixels form four squares each 32 by 32 pixels. The 32 by 32 pixel central Test Gray patches have a digit value of 140.

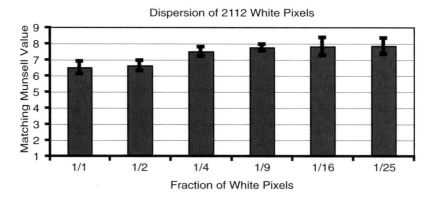

Figure 23.4   Munsell Value for the test patches in Figure 23.3.

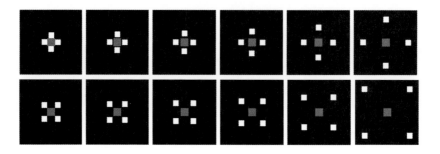

Figure 23.5   The Sides targets (top), and the Corners target (bottom).

In the top row the four squares are adjacent to the sides of the Test Gray square (*Sides*), while in the bottom the four squares are on the diagonal of the gray square (*Corners*).

In both Sides and Corners, the left targets have the white adjacent to the Test Gray. Moving to the right, the white squares are separated by 4, 8, 16, 32, and 64 pixels. Separating the white squares from the Test Gray patch makes the matches appear lighter (Chapter 22).

Figure 23.6 plots the matching lightnesses for four sets of data. Starting on the left, we see Snow, Sides, and Corner targets. On the right we see the Lines target set described below.

On the far left of Figure 23.6, we see the matching Munsell Values for this Test Gray central square in control solid white [matches 8.0]; and in solid black surround [5.2]. We used these values to select the range of the Munsell Value axis, from 8.5 to 5.0.

All other displays in this Figure have identical image statistics, (1024 gray; 2112 white; 62 400 black pixels) and hence the same global average values. The orange bars plot the Munsell Value for the *Snow* targets in Figure 23.6. The red bars plot the matches for the *Sides* and *Corners* targets. The *Sides* have greater influence (darker match) than the *Corners,* as seen in Figure 22.4.

23.7   Lines

Figure 23.7 continues the series using lines parallel to the sides of the Test Gray square. The targets include white lines that are separated by 1, 3, or 7 black pixel lines (2, 4, 8 pixels per cycle). The effect

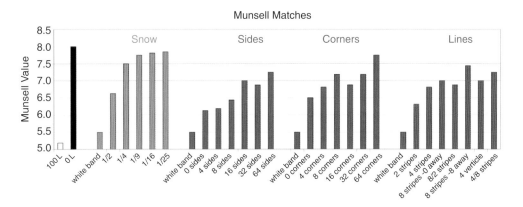

Figure 23.6   Munsell matching data for all test targets: range varies middle gray (5.0) to white.

of separating the white lines was that matches were lighter. Figure 23.6 plots these matches with magenta bars on the far right (McCann & Rizzi, 2003).

23.8   Equivalent Backgrounds

In the all white surround, the Test Gray was matched by a lightness of 5.2, while in the all black surround it was matched by 8 (Figure 23.2). The range of observer matches for the 25 patterns with identical white, gray, and black pixel counts is from 5.25 to 7.85. This is 2.6 lightness units on a scale in which white is 9.6 and black is 1.5. Hence the surround with identical average statistics can manipulate the matching lightness one-third of the range between white and black.

One of the goals of this chapter is to understand the underlying mechanisms of these large changes in appearance. Obviously, the spatial pattern of the white pixels is influencing the appearance. The mechanism, however, is not at all obvious. Figure 23.8 displays the set of eight different surrounds that generated Munsell Value matches of 7.0. We chose the value of 7 because it fell very close to the median of observer matches for the 25 patterns.

The top left shows that in a white surround, the central test patch that matched by 7.0 had 71% of the luminance of the white (100%).

The second Test Gray test patch had a digital value of 140, (50% maximun luminance) surrounded by a uniform gray background, 256 pixels on a side. From Figure 23.2 a background luminance of 63% surrounding 50% makes the Test Gray match Munsell 7.0.

The rest of the Test Gray patches had a digital value of 140, and 2112 white pixels on a 62 400 black pixels background. They are:

    asymmetrical single square of white,

    1/2 dispersion fraction of snow (checkerboard),

    white squares on corners with 16 pixel separations,

    white squares on sides with 16 pixel separations,

    stripes with 8 pixel cycle

    stripes with 8 pixel cycles vertical and 2 pixel cycles horizontal

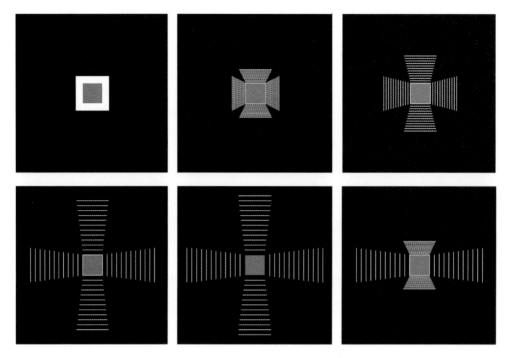

Figure 23.7 Relocation of the 2112 white pixels into parallel stripes with variable spacing.

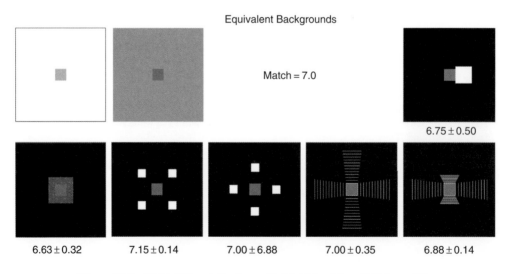

Figure 23.8 Eight different displays with Matching Munsell Value = 7.0.

23.9   Equivalent Backgrounds and Models of Vision

Equivalent background patterns provide a challenge to spatial models of vision. Namely, models that are designed to calculate the appearance of lightness need to generate identical predictions for the Test Gray patch from these diverse spatial input targets.

Computational models can be based on:

Perceptual frameworks with depth planes and illuminants

Global GrayWorld

Spatial frequency filter models using single MTF

Multi-channel spatial frequency models

Multi-resolution spatial processing

It is difficult to propose meaningful cognitive frameworks for either perceived illumination, or depth planes for these displays. The ideal model is one that only requires the array of pixel data, without additional interpretation of image segments. It is difficult to imagine a perceptual framework that predicts variable lightness from the different dispersions of *Snow, Sides, Corners* and *Lines*.

It is also easy to see that global-average models e.g., *GrayWorld*, cannot predict the results of 25 experiments with identical image statistics. As well, normalizing by the single maximum value pixel in the entire image will not vary with these targets.

Further, a simple model employing a single spatial-frequency filter convolved with the spatial-frequency energy distributions of the targets will not account for observer data. The spatial-frequency transforms of eight equivalent background targets is not constant. It is not obvious how a single filter could transform the equivalent background in Figure 23.8 into equal outputs.

That leaves multi-resolution spatial models and multi-channel spatial frequency models as likely candidates. The design of these experiments was to collect data for future use in designing models. The intent was to extend our understanding of how white pixels influence appearance.

There are many uses for *Whites* in imaging. Whites have been assigned different roles by photographic densitometry, photographic quality control, Retinex image processing, global normalization, perceptual frameworks, clues for identifying the illuminant, and normalization in CIE appearance models.

If we consider the *pixel based* vs. *spatial comparison* paradigms, we could use the single pixel with the highest value to define white, or we could define it to depend on scene content. These measurements of the size and distribution of whites are the signature of our neural contrast mechanism. They show that a Snow distribution has the same effect on the lightness of the Test Gray area as a uniform gray background. Although they appear white, small white *Snow* pixels behave the same as uniform gray. Although these targets have many different characteristics (*Snow, Lines, Corners, Sides*), they have the same effect on the Test Gray area. The presence of small white pixels affects the image, but that effect is not the same as the presence of large white areas. This observation shows that algorithms that normalize to the single pixel with maximum value do not mimic vision. They may be appropriate for instrumentation, but not for models of vision.

23.10   Summary

These experiments were designed to identify the underlying spatial information that is important to the human visual system. This chapter introduces the idea of equivalent background for testing spatial models of vision. It describes 25 different test targets that have identical pixel image statistics, but different matches. It identifies the targets with equivalent backgrounds, and analyzes the result using different approaches to modeling spatial vision. Section F will be devoted to HDR models with special attention given to the influence of vision.

23.11   References

Benary W (1950) *The Influence of Form on Brightness, A Source Book of Gestalt Psychology*, W, Ellis trans. Kegan, Paul, Trench, Truber & Co. Ltd, London, 104–8.

DeValois R & DeValois K (1986) *Spatial Vision*, Oxford University Press, New York.

Gilchrist A (1994) Lightness, *Brightness and Transparency*, Lawrence Erlbaum Associates, Hillsdale, 136.

Land EH & McCann JJ (1971) Lightness and Retinex Theory, *J Opt Soc Am*, **61**, 1–11.

McCann JJ (1978) Visibility of gradients and low-spatial frequency sinusoids: Evidence for a distance constancy mechanism, *J Photogr Sci Eng*, **22**, 64–8.

McCann JJ ed (2004) Retinex at 40, Journal Special Collection, *J Electronic Img*, **13**, 1–145.

McCann JJ & Rizzi A (2003) The Spatial Properties of Contrast, *IS&T/SID Color Imaging Conference*, Scottsdale, Arizona, **11**, 51–8.

White M (1979) A New Effect of Pattern Lightness, *Perception*, **8**, 413–16.

24

From Contrast to Assimilation

24.1 Topics

This chapter reports experiments using segmented black and white surrounds. Observer matches showed marked dependence on the surround's spatial pattern: adjacent pixels have much more influence than diagonal pixels. The appearance of grays with segmented surrounds having constant average luminance depending on the spatial pattern. Some of these test targets exhibit the assimilation phenomenon, while other very similar ones do not.

24.2 Introduction

In real scenes and in complex Mondrians the appearance of two identical colored papers in different locations is remarkably constant. Changing the position, and hence the surround, does not usually alter appearance. In simple displays, grays vary in lightness with surround. The spatial arrangement of the surround can make appearance more similar to (called *Assimilation*), or more different from (called *Simultaneous Contrast*) the immediate surrounding area. In describing lightness effects, *Simultaneous Contrast* refers to the fact that a large white surround makes a gray center appear darker than a black surround. Following the discovery of spatial opponent ganglion cells by Barlow (1953) and Kuffler (1953), it is generally believed that the white surround stimulates inhibition of the center for certain cells. That inhibition could make gray centers in white look darker.

Assimilation is the name of the observation with the opposite effect. When the surround is smaller, it produces a different effect called Assimilation (Gilchrist, 1994). Grays with small adjacent whites look lighter than the same gray with small adjacent blacks. Examples are Benary's Cross, White's Effect, Checkerboard and Dungeon Illusions. These effects have been used to suggest a top-down analysis of the scene, implying mechanisms based on the recognition of illumination, objects or junctions. Shevell and colleagues have measured the effect of checkerboard backgrounds on color chromatic displays (Schirillo & Shevell, 1996; Shevell and Wei, 1998; Barnes, Wei and Shevell, 1999; Shevell, 2000). Shapley and Clay Reid (1985) report strong individual differences in the strength of assimilation.

Experiments demonstrate that *Simultaneous Contrast* is much more complex than predicted on the basis of inhibition by average luminance in the surround (chapter 23). In this chapter we describe displays

The Art and Science of HDR Imaging, First Edition. John J. McCann, Alessandro Rizzi.
© 2012 John Wiley & Sons, Ltd. Published 2012 by John Wiley & Sons, Ltd.

with a square gray central element and eight square surround elements that demonstrate sensitivity to
the placement of white and black surround elements. Equal-average surrounds in the target do not give
equal gray appearances. We then present variants of the classic checkerboard assimilation pattern, that
represent a further challenge for any simple explanation of the assimilation phenomenon.

24.3   Segmented Surrounds

The experiment uses segmented black and white surrounds for a gray central patch (McCann, 2003).
Figure 24.1 illustrates one of the possible test target combinations, named 1346 from the number of
the squares set to black (Figure. 24.1, right).

   The square gray center element subtends 1.25° with eight square surrounding elements (four adjacent
– four diagonal). There are 256 combinations of white and black elements in eight locations. A single
black segment positioned at the top center (target 1) is assumed to be the same as all the other single
adjacent black squares (targets 3 [east side], 5 [bottom], and 7 [west side]). Target 1 was tested and the
stereoisomers, targets 3, 5 and 7 were not. Removing all the stereoisomers leaves 56 unique spatial
surround tests.

   Figure 24.2 is a diagram of the target with angular sizes, and an illustration of the tasks for the
experimenter and the observer. The observers saw the 3 × 3-segment (3.75°) test target on an
18.75° × 11.25° gray background. The top of the illustration shows a target with all eight surround
squares set to white (Target W, or 0). The right side of the display had a fixed white surround and a
variable center. The observers' task was to adjust the center on the right to match the center on the left.

   The experimenter controlled the pattern of the segmented surround on the left. In the bottom illustration (Target B, or 12345678), the experimenter set all eight surrounding squares to black. The observer
varied the intensity of the right matching gray center in the white maximum luminance surround. With
eight white surround elements, grays matched 17.5 % maximum luminance; with eight black surround,
68.2 % maximum luminance. The results are analyzed using log luminance axes with 100 % scaled to
1.0. The matching value on this scale for the white surround is 0.23, and for the black surround is 0.85.

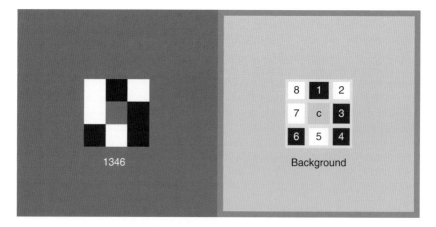

Figure 24.1   (left) The "1346" segmented test target; (right) diagram of the nomenclature . This display
has 3 % luminance in the 1, 3, 4, 6 segments, hence its name. The center [c] and the background were
17 % maximum luminance. The surround was segmented into eight elements, numbered clockwise
starting at the top center. Each surround segment was either 100 % or 3 %.

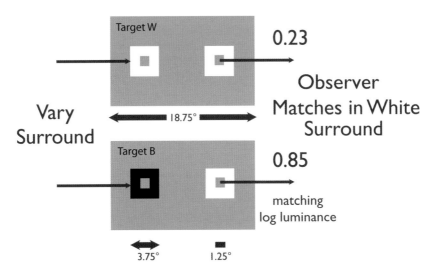

Figure 24.2   Segmented surround experiment.

24.3.1   56 Combinations

Figure 24.3 shows matching data for all 56-surround arrays. The vertical axis plots the relative log matching luminance (LML). The icon for the all white surround is placed at 0.23, and the icon for the black surround match is placed at 0.85 on the axis to illustrate the range of possible matches. The horizontal axis identifies the segmented surround. The data have been sorted so that the number of black elements increases from left to right. In each group the data is ordered by average log matching luminance. Figure 24.3 shows the results for two observers: each with 56 targets, eight trials per target. The matches showed little correlation with the number of black segments, or the average luminance.

If the number of black elements, or a surround average, controlled appearance, then we might expect a series of flat steps with vertical risers at the change in number of black segments. Instead we found a marked dependence on the surround's spatial pattern:

Target 2 (LML = 0.24) with one diagonal black segment is the same as target 0 (LML = 0.23) with an all white surround.

Target 1 with one adjacent black segment is lighter (LML = 0.32).

Among the six targets with two black sectors, the average log matching luminances vary from 0.22 to 0.53.

Among the 12 targets with three black sectors, the average log matching luminances vary from 0.26 to 0.56.

Among the 14 targets with four black sectors, the average log matching luminances vary from 0.26 to 0.67.

Among the 12 targets with five black sectors, the average log matching luminances vary from 0.29 to 0.74.

Among the six targets with six black sectors, the average log matching luminances vary from 0.31 to 0.75.

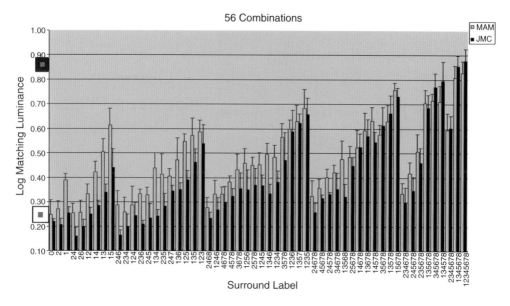

Figure 24.3 Matching data for all 56-surround arrays. The vertical axis plots the relative log luminance of the observer's matches. The icon for the all-white surround is placed at 0.23, and the icon for the black surround match is placed at 0.85 on the axis to illustrate the range of possible matches. The horizontal axis identifies the segmented surround. The data have been sorted so that the number of black elements increases from left to right. In each group the data is ordered by average log matching luminance.

Target 2345678 (LML = 0.83) with one diagonal white segment is the same as the all black target 12345678 (LML = 0.85). However, target 2345678 with one adjacent white segment is darker (LML = 0.59).

Instead of flat steps correlating with number of black sectors, we find that there is a very wide range of matches for each set of constant number of black sectors. The matching data is inconsistent with the hypothesis that appearance is controlled by the average luminance of the surround.

24.3.2   Combinations with Four White and Four Black Squares

Figure 24.4 is a plot of Segment Pattern vs. Log Matching Luminance for all 14 patterns with four white and four black elements in the surround (McCann, 2003). They are sorted from left to right in order of increasing average log matching luminance. The two lowest LML values have zero adjacent blacks. The next two patterns have one adjacent black, and the following seven LML values have two adjacent blacks. The remaining five LML values increase with the number of adjacent blacks, but with more variability than previous patterns. The adjacent segment has more influence than the diagonal on matching luminance. The data from the 14 test targets with four white and four black elements are more consistent with the number of gray-black edges/gray-white edges than with the average luminance of the surround.

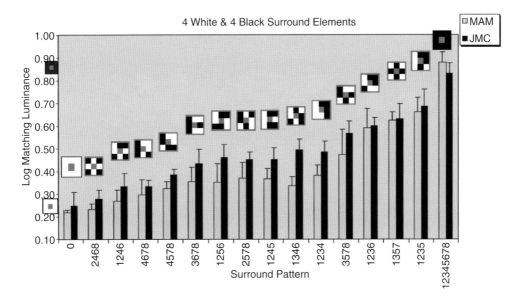

Figure 24.4   Matches for all the 4-white/4-black surround targets.

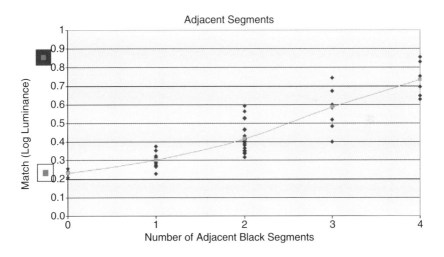

Figure 24.5   Matches for the number of adjacent black edges for all 56 displays.

All 14 targets in Figure 24.4 have four white surround elements. The vertical axis plots the relative log luminance of the observer's matches. Matches vary from 0.26 to 0.67 LML depending on the placement of the white and black surround elements (McCann, 2003).

Figure 24.5 is the plot of number of gray-black edges vs. log matching luminance for all 56 targets. The graph shows good correlation in that the data are clustered around the solid-line plot of average log luminance values for 0, 1, 2, 3, 4 black segments. Nevertheless, the range of data around the average

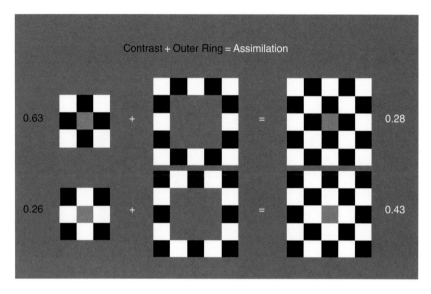

Figure 24.6   Central gray squares change from Contrast to Assimilation with the addition of more checkerboard squares.

suggests that number of black segments is not a sufficient explanation of the data. A more complex spatial analysis is required.

Adjacent elements have much more influence than diagonal elements, although adjacent elements alone cannot account for the matching data. The appearance of grays with segmented surrounds having constant average luminance shows dependence on spatial pattern. *Simultaneous Contrast* effects with segmented surrounds are much more complex than predictions of models in which the radiance of a central area is compared with a spatial average of a concentric surround.

On the left of Figure 24.6, observers reported *Simultaneous Contrast*. The gray with four adjacent black squares had a matching log luminance of 0.63, while the one with four adjacent white squares had a matching log luminance of 0.26. The center shows the checkerboard ring of 16 square rings added to the *Simultaneous Contrast* display to make the "Assimilation" display. Observer matches for the gray squares in the center of displays on the right were 0.28 for four adjacent black squares and 0.43 for four adjacent white squares.

In *Simultaneous Contrast*, adjacent large white areas make grays look darker; in *Assimilation*, small adjacent whites, and the rest of the scene, make grays lighter (McCann, 2001a, 2001b, 2001c).

The antagonism is apparent when you compare the Assimilation found in the checkerboard and *Simultaneous Contrast* found in the very similar 1357 and 2468 segment pair (Figure 24.6 left). The only difference between the *Simultaneous Contrast* displays on the left and the *Assimilation* display on the right is the outer ring of black and white segments. It is as if this outer ring shuts off *Simultaneous Contrast* and lets *Assimilation* be apparent.

One important feature is that the outer ring is a periodic addition to the nine segment inner area. If the outer ring is replaced with equal areas of white and black in an aperiodic pattern, then *Simultaneous Contrast* remains and *Assimilation* is not apparent. McCann (2001c) reports that if the outer ring has been shifted by half a width of a surround square that change is enough to disrupt the mechanism that caused *Assimilation*. Periodicity plays an important role in these checkerboard test targets, but there are other examples of assimilation, that are not periodic.

24.4   Checkerboard Variants

Here we present some variants on basic assimilation patterns to understand the *Contrast-Assimilation* antagonism. Starting with Rus and Karen De Valois's (1988) checkerboard assimilation pattern (Figure 24.7 left), we designed some variants (Rizzi et al., 2001).

The right side of Figure 24.7 shows the variant of the checkerboard effect. The gray squares are smaller and surrounded by a white and black frame. This target has some of the characteristics of assimilation and some of contrast. In this case, the relative appearance of the two gray squares changes is reversed: the upper square appears darker than the lower. The relative lightness of the gray squares in the variant reverse, but, since the immediate surrounding area changed as well, the *assimilation* name for the effect remains. The white and black squares (Figure 24.7) were 64 pixels on a side (subtending 1°) with the frame of the first variant that was four pixels wide (0.06°, or 3.8′).

Different width frames were tested: 4, 8, 16, 24 pixels on each side leaving the test gray patches of 56, 48, 32 and 16 pixels, respectively. The results are presented in Figure 24.8. As the frames enlarge, and gray squares shrink, the darker, top gray gets lighter until it is slightly lighter than the bottom. In the classic assimilation target (Figure 24.7, left), the observers reported that the top gray was lighter 100 % of the time.

Regardless of the name Simultaneous Contrast or Assimilation, the appearance of the gray patch responds to scene content. Figure 24.7, and many other experiments in this series (Rizzi et al., 2001), show that the mechanism that generates the assimilation phenomenon responds to the spatial content of the surround.

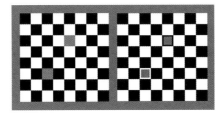

Figure 24.7   The checkerboard pattern (left) and the variant (right). The relative lightnesses of the gray square in the variant are reversed.

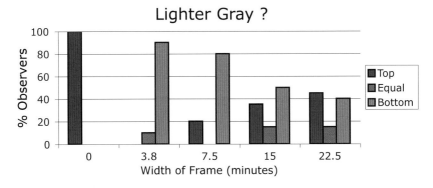

Figure 24.8   Which gray appeared lighter; top, bottom, or equal?

The analysis of assimilation targets has another complication. These experiments are usually conducted with papers and low-dynamic range displays. However, Assimilation targets work best with smaller scene segments that are close to the resolution threshold (McCann, 2001c). As seen in Section C, glare has a dramatic effect on the retinal image, changing uniform, square scene elements into complex *retinal image* gradients. Assimilation targets add optical blur because their image segments are near the resolution limit. Both glare and blur transform the scene into a very different image on the retina.

The data shown here makes it very clear that the appearance of test gray cannot be predicted by simple rules that just simply identify adjacent white and black elements in the surround. The problem is much more complex than dependence on adjacent areas. The whole scene and its multi-resolution, spatial-domain characteristics play a role, as seen in the sensitivity to periodicity. Many image characteristics such as size, relative size, shape, position, and periodicity of scene segments influence the appearance of test gray patches. One thing is clear: Assimilation uses very complex scene arrays. The complete analysis of these scenes requires spatial calculation of the retinal image (after blur and scatter), and the scene characteristics in many different layers of its multi-resolution pyramid. How the scene's multi-resolution spatial content stimulates the multi-resolution cortical cells is an important first step in understanding the data. This is an area that requires further research.

24.5   Summary

Contrast is a complex spatial mechanism that is evident with simple center-surround displays. It can be shut off, and reintroduced by the spatial pattern of outer-surround elements.

Assimilation in periodic displays is a mechanism found in the absence of contrast. Its explanation is still an object of research.

24.6   References

Barlow H (1953) Summation and Inhibition in the Frog's Retina. *J Physio.*, **119**, 69–88.

Barnes C, Wei J & Shevell S (1999) Chromatic Induction with Remote Chromatic Contrast Varied in Magnitude, Spatial Frequency and Chromaticity, *Vision Research*, **39**, 3561–74 .

De Valois R & De Valois K (1988) *Spatial Vision*, Oxford University Press, New York.

Gilchrist A (1994) *Lightness, Brightness and Transparency*, Lawrence Erlbaum, Hillsdale, 136.

Kuffler S (1953) Discharge Patterns and Functional Organization of Mammalian Retina. *J Neurophysiol*, **16**, 37–68.

McCann JJ (2001a) Calculating the Lightness of Areas in A Single Plane. *J Electron Img*, **10**, 110–22.

McCann JJ (2001b) Image Processing Analysis of Traditional Gestalt Vision Experiments, *Proc. AIC Conference*, Rochester, NY.

McCann JJ (2001c) Assimilation and Contrast, *Proc. IS&T/SID Ninth Color Imaging Conference*, Scottsdale, 91–6.

McCann JJ (2003) Calculating appearances in complex and simple images, *Normal and Defective Colour Vision*, Mollon J, Pokorny J & Knoblauch eds, 231–9.

Schirillo JA & Shevell SK (1996) Brightness Contrast from Inhomogeneous Surrounds, *Vision Res*, **36**(12), 1783–96.

Shapley R. & Clay Reid R. (1985) Contrast and assimilation in the perception of brightness, *Proc. Natl Acad Sci USA*, **82**, 5983–6.

Shevell SK & Wei J (1998) Chromatic Induction: Border Contrast or Adaptation to Surrounding Light? *Vision Research*, **38**, 1561–66.

Shevell SK (2000) Chromatic Variation: A Fundamental Property of Images, *Proc. IS&T/SID Eighth Color Imaging Conference*, Scottsdale, Arizona, 8–12.

Rizzi A, Gatta C & Marini D (2001) Experiments on new contrast patterns, ECVP2001, European Conference on Visual Perception, Kusadasi (Turkey), 26–30.

25

Maxima and Contrast with Maxima

25.1 Topics

This chapter reviews the common conclusions about the lightness experiments described earlier in Section D. These measurements provide a data set that can be used to evaluate models of the appearance of lightness.

25.2 Merger of Aperture and Object Modes

Traditional 19th century psychology segmented appearance into different *Aperture* and *Object Modes*, because constant stimuli generated variable responses. Vision is not limited to a single appearance from a single set of receptor quanta catches. The 19th century approach was that observers inferred a *Mode* from the stimuli and the Mode mechanism generated appearances appropriate for the scene content. Later, 20th century thinking showed that top-down *Modes* can be replaced with bottom-up spatial processing. The same data that stimulated Aperture and Object Modes can be used to measure the spatial properties of the visual process.

25.2.1 Appearance of Maxima

We began with a set of measurements of the effects of changing the illumination on a group of gray test areas in three different surrounds. The scene maxima behaved the same regardless of the surround. The maxima matches became slowly darker with large changes in illumination. The data was consistent with observations of stellar magnitude and magnitude estimates of brightness. As well, the appearance of shadows in 3-D displays was consistent. We called this slow change of appearance of the maxima the Hipparchus line.

25.2.2 Contrast with Maxima

The appearances of other scene segments follow different rules. In all complex images, segments with less luminance than the maxima get darker quickly with small changes in light. The rate of getting

The Art and Science of HDR Imaging, First Edition. John J. McCann, Alessandro Rizzi.

darker varies with scene content. For the same decrement in luminance, lightness decreases most rapidly in a white surround, less rapidly in a gray surround, still less with grays on a black surround, and finally falls on the Hipparchus line with single spots of light, or local maxima.

The responses to complex images are evidence for post-receptor processing. The *neural contrast* mechanisms respond to scene content to calculate lightness appearances.

25.3   Influence of the Maxima

Lightness constancy is impressive, but not perfectly constant. As well, objects do not change in lightness as they move through real scenes, or in complex Mondrian displays. Many of the everyday changes of appearance are so small that they are essentially unnoticed, or regarded as unimportant.

25.3.1   Take Away the Maximum

To make a dramatic visual demonstration we need to introduce a new maximum, or remove an old one. In Section D we described a number of different ways to modify and remove maxima from a complex scene. We made modifications of the maximum by:

varying the separation between *maximum white* and *test gray*

varying the enclosure of *test gray* by *maximum white*

varying the size and distribution of the *maximum white*

varying the relative size of *test gray* and *maximum white*

varying the adjacency of *maximum white*

All of these parameters have significant influence on the appearance of the *Test Gray* areas. These studies reveal a great deal about the influences of the *neural contrast* mechanisms that generate our spatial vision.

25.3.2   Phenomena Names vs. Scene Content

Many of the test targets shown in Section D have been studied in previous experiments, but have been described by a wide range of different names, implying that they are different phenomena, with different neural mechanisms. In this section we have discussed Simultaneous Contrast and Assimilation, Aperture and Object Modes, lateral inhibition and global frameworks.

Our goal was to concentrate on the data. We also wanted to present the data as simply as possible. We avoided the extreme hypotheses, such as over-simplified lateral inhibition, and over-complex perceptual frameworks. Neither the simple bottom-up, nor the perceptual top-down mechanisms can model this collection of data. Simple bottom-up hypotheses, (using local lateral inhibition, or global averages) cannot account for separation and enclosure experiments. Perceptual frameworks cannot account for asymmetric assimilation, and separation. Currently there is no single model that can predict this collection of data and those in the literature. However, we can make a number of conclusions:

Scene maxima have their own independent function of appearance vs. luminance.

The smaller the extent of maxima in images, the slower the darkening of appearance with decrease in luminance.

Maxima with small extent, large separations, and mismatched relative sizes have less influence on test grays.

Image statistics, global averages, maximum values, *GrayWorld*, and local averages cannot account for the set of appearance measurements presented here.

The data continues to support the hypothesis that appearances of both *Aperture* and *Object Modes* can be translated to the *Hipparchus line* behavior of maxima (spots of light) and *neural contrast* response to image content below the maxima.

There is a critical need for future research to integrate all of the data above into a comprehensive model of lightness appearance. Given a complex scene, with all sizes, shapes, and retinal contrasts we need an analytical description that predicts appearance. Such a model, that can calculate reliably appearance from *retinal radiance*, will be of great help in identifying the best image processing algorithms, as discussed in Section F.

25.4 Summary

The results found in Section D parallel those in Section C, namely that, the greater the presence of white maxima in images, the greater the glare, the lower the retinal contrast, and the greater the apparent contrast.

Section D described experiments that provide data for modeling images. Using complex scenes, *these* experiments, extend our understanding beyond the limits of spots of light experiments. These experiments help us to refine our thinking about whether the *silver halide*, the *neural contrast*, or the *perceptual framework* paradigm will be the most fruitful in crafting models that mimic human vision.

Section E expands the discussion to include the study of complex color images such as Mondrians, real scenes, color constancy and 3-D Mondrians. These studies will continue to show the critical role of scene maxima. As well, the neural contrast mechanism is an essential part of color appearance.

Section E
Color HDR

26

HDR, Constancy and Spatial Content

26.1 Topics

Scene content affects the colors we see. The study of Color Constancy provides a rich set of vision experiments that parallel and intertwine with the study of HDR. Section E reviews these experiments and points out similar conclusions to those found in the previous HDR experiments. We describe color in the natural image, color constancy in Mondrians, searches for evidence of color adaptation in complex scenes, and 3-D Color Mondrians. These experiments show the spatial nature of human color image processing.

26.2 Introduction

When John McCann started to work in the Polaroid Research Labs, while a freshman at Harvard, there was a great deal of discussion about making color photographs of Land's *Red and White projections*. If you make a standard Ektachrome transparency slide of a projected Red and White image on a screen, it does not look like the original. Why? Slides work well at reproducing what we see in everyday scenes. Why don't they work for red and white projections? Additionally, if you take a slide using daylight film of a scene in tungsten light, it is very yellow. While the actual scene appears slightly yellow, the film reproduction has distorted colors. What controls whether the film makes an excellent reproduction some of the time, and very poor ones at other times? If the film reproduced the light coming from the scene perfectly, the original and the reproduction must appear the same because they would be identical stimuli over the entire field of view.

The answer was that tungsten illumination increased the scene's dynamic range beyond the camera film's range. The long-wave emulsion had more exposure from tungsten light than the middle-wave emulsion; and much more than the short-wave emulsion. If white and black papers were exposed properly for the long-wave emulsion; they would be underexposed for the middle-waves; and

The Art and Science of HDR Imaging, First Edition. John J. McCann, Alessandro Rizzi.
© 2012 John Wiley & Sons, Ltd. Published 2012 by John Wiley & Sons, Ltd.

severely underexposed for the short-waves. Photographic films have an S-shaped, nonlinear response to light (Chapter 5). The reproduction recreated the long-wave original adequately, but distorted, and severely distorted, the other wavebands because of their nonlinear response. The limited dynamic range of the film altered the spectral reproduction. Severe underexposure replaces variable scene radiances with the uniform asymptotic black. Moderate underexposure distorts different scene radiances by making them disproportionally darker. Such distortions in the middle- and short-wave records change the film's color responses. Its limited dynamic range has substituted different colors in the reproduction, so that it does not reproduce the light coming from the original.

Vision is different. Three independent spatial channels normalize the cone responses. A white paper in tungsten light, with its L > M > S radiances, plots as different radiances on the Hipparchus line. They have different radiances, but similar lightnesses. The spatial information in the scene in middle- and short-wave light is not distorted, as in film. Vision's HDR spatial imaging uses neural contrast mechanisms, and hence has better renditions. Spatial image processing does not introduce the color distortions found in films.

Human color constancy experiments always involve HDR imaging. As we discussed in Chapter 2, objects have a range of reflections of roughly 1.5 log units. In order to do a color constancy experiment we need to increase, or decrease, the radiance in part of the spectrum. These changes in illumination change the dynamic range of the scene. In experiments that use papers and lights, the range varies with changes in illumination. Objects' relative reflectances are preserved, and we observe constancy. Once we introduce reproduction technologies (prints, displays, etc.), we have to be certain that the range limitations of the media do not introduce distortions, as found in film.

26.3   Red and White Projections

Maxwell's experiments on color matching were done early in his career. After that he turned his attention to electromagnetic fields (Maxwell, 1861). Land often quipped that the study of color vision would have been very different if Maxwell studied color matching after he had studied electromagnetic fields. Maxwell analyzed colors with the assumption that color vision works, as film does, using a pixel-based mechanism, as shown by his Royal Institution demonstration of color photography. Maxwell never discussed the idea that color was a field phenomenon.

For Land it was the opposite. Although his first published paper was on retinal rivalry (Land, 1936), his scientific interest had been focused on polarized light, polarized images, and instant photography until he was 48. He accidentally observed colors in a red and white projection that he could not explain. By then, he had several hundred patents on how to make images. He was fascinated because vision did not work at all like photography; it was spatial in nature and could not be explained by pixel-based logic (McCann & McCann, 1994).

26.3.1   Maxwell's Three-Color Projectors

In the mid-1950s Edwin Land, while developing instant color film, made a startling observation. Din often told the story about repeating the famous experiment that James Clerk Maxwell presented at the Friday Evening Discourse to the Royal Institution on May, 17, 1861. Maxwell demonstrated the first color photograph using three superimposed projections. Each projector had either a red, green or blue filter. Each projector had a black-and-white photograph taken with the same filter. The photographs were reported to be records of the image information from the long-, middle-, and short-wave visible light. Maxwell's lecture used this superimposed photographic image to support the assertion of Thomas Young's trichromatic theory. However, this event is remembered by photographic historians as the invention of color photography (Evans, 1961; Coote, 1993).

26.3.2 Colors from Red and White Projections

Land wanted to see the colors possible in three-color additive projections as a part of his research in developing an instant, three-color subtractive print film. At the end of reviewing a variety of three-color projections, while putting things away, a colleague, Meroe Morse asked Land why the colors were still there with only red and white light. Land replied, "Oh, that is adaptation". Land frequently attended Optical Society of America meetings and was familiar with the work of Selig Hecht and George Wald on dark and light adaptation of the retina.

Land reported that, at 2 o'clock in the morning, he sat up in bed, and said "Adaptation, what adaptation"? Polaroid's research in color vision was a direct result of the fact that Land got up, and returned to the lab at 3 o'clock in the morning to see whether these colors could be explained by adaptation.

It is interesting to recall that Meroe Morse was the Director of Black and White Photographic Film Research at Polaroid. She was a relative of Samuel F. B. Morse who, while promoting his telegraph in France, met Daguerre and brought photography to the United States.

In the late 1950s Land was well on his way to inventing 350 US patents, the majority on the topic of image capture and reproduction. Land recognized instantly that the human visual system was fundamentally different from all existing image reproduction systems. Land (1993) reported his research on red and white photography in a number of papers (Land, 1955), and lectures from 1955 to 1962, and developed a prototype Red-and-White television system with Texas Instruments. McCann, Benton, and McKee (2004) wrote an annotated bibliography of the Polaroid Red and White papers. Following Red and White Photography, Land started experimenting with real papers and controlled illumination.

26.4 Color Mondrians

Color Mondrians provide a large amount of quantitative data on spatial interactions. Chapter 27 reviews many different experiments that led to the following conclusions:

Color is a spatial process, not a pixel-based process such as film.

There are three channels that build three independent lightness images from long-, middle-, and short-wave scene content (Retinex).

As with achromatic lightness, the maxima play a special role in each channel.

Areas darker than the maxima show rapid changes in appearance with small radiance changes.

Experiments failed to find a significant influence of adaptation on Mondrians' appearances.

Chapter 27 extends the search for the influence of visual pigment adaptation in complex color images. Despite many attempts to measure it, we can find no evidence that visual pigment adaptation plays a role in the color constancy of complex scenes.

26.5 Constancy's On/Off Switch

Identical stimuli everywhere in the field of view have to look the same. Chapter 28 describes experiments, proposed by Vadim Maximov, in which two sets of carefully selected papers in two different illuminations combine to make two identical visual stimuli. Although these pairs of papers are difficult to make, it is possible, if one can manufacture exactly the colored papers required. The results show that color constancy in complex scenes can be shut off. The experiments study whether additions to the scene can restore color constancy.

26.6   Color of 3-D Mondrians – LDR/HDR Illumination

So far, the study of Color Mondrians has been limited to flat displays in uniform illumination. Chapter 29 describes experiments that use two identical 3-D Mondrians made with only 11 paints. One Mondrian is in nearly-uniform (LDR illumination), the other in highly directional, mixed spectral light (HDR illumination). They are viewed in the same room at the same time. Observers are asked to measure the appearances of constant reflectance in variable illumination.

26.7   Color Constancy is HDR

Chapter 30 reviews the literature on the only Retinex that can be completely isolated by controlling the content of the illumination. The rods allow us to observe the appearances of a single set of receptors.

Further, the rods are perfectly good color receptors when combined with light appropriate for the long-wave cones near their threshold. We see object constancy in HDR imaging from observations in the dark at night to bright daylight. We see the same constancy in color experiments. In this section we review color constancy and compare human color mechanisms with those found in HDR experiments.

26.8   References

Coote J (1993) *The Illustrated History of Colour Photography*, Surrey: Fountain Press.

Evans R (1961) Maxwell's Color Photograph, *Scientific American*, CCV(11), 118–20.

Land E (1936) The Use of Polarized Light in the Simultaneous Comparison of Retinally and Cortically Fused Colors, *Science*, 83, 309.

Land E (1955) The Case of a Sleeping Beauty or a Case History in Industrial Research, in *E. H. Land's Essays* (1993) Vol. **III**, ed. MA McCann, IS&T, Springfield, 3–4.

Land EH (1993) *Edwin H. Land's Essays*, ed. MA McCann, IS&T, Springfield.

Maxwell JC (1861) On the Theory of Compound Colours, in *The Scientific Papers of James Clerk Maxwell*, (1945), ed. Niven, Dover, New York, 445–50.

McCann J, Benton J & McKee S (2004) Red/White Projections and Rod-Lcone Color: An Annotated Bibliography, *Journal of Electronic Imaging*, 13(01), 8–14.

McCann M & McCann J (1994) Land's Chemical, Physical, and Psychophysical Images, Optics & Photonics News, 34–7.

27

Color Mondrians

27.1 Topics

This chapter integrates a wide variety of psychophysical experiments using Mondrians to measure color appearance. By carefully measuring the accuracy of color constancy we learn a great deal about the mechanisms of color vision and HDR in color.

27.2 Introduction

There are two major threads in color constancy. One follows the traditional 19th century idea that human vision discounts the illumination. The other, a 20th century idea, suggests that vision uses the spatial information in the image content to synthesize appearances. Observations of color constancy showed that color appearance does not correlate with the radiance of light coming to the eye from objects in the scene. That radiance is the product of the illumination and the object's reflectance. In general, appearances correlate more with reflectance than with illumination, however constancy is rarely perfect. This chapter studies complex Mondrian scenes with particular attention to measuring and analyzing the small departures from perfect constancy. These departures are the signature of human color mechanisms, and help us to evaluate different theories of color.

Figure 27.1 illustrates the alternative approaches. Starting with the radiance at each wavelength of light coming from the scene:

> *Spatial comparisons* build up appearance by analyzing the response to scene content.

> *Discounting the illumination* attempts to calculate the illumination from the response to light, in order to remove it, so as to determine an object's reflectance.

There are a number of different versions of discounting illumination. They have different mechanisms, and resulting properties:

> *Adaptation* is a "bottom-up" explanation that assumes that increased long-wave illumination initiates a decrease in response to that light.

The Art and Science of HDR Imaging, First Edition. John J. McCann, Alessandro Rizzi.
© 2012 John Wiley & Sons, Ltd. Published 2012 by John Wiley & Sons, Ltd.

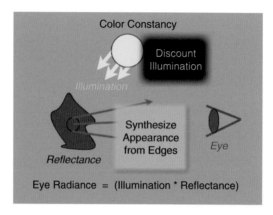

Figure 27.1   Two different approaches to color constancy: discounting the illumination, or synthesizing appearance from the spatial content of the scene.

> *Perceptual frameworks* is a "top-down" explanation that assumes humans can recognize the illumination and remove its influence.

The chapter uses a variety of Color Mondrians to measure human color constancy and evaluate these different hypotheses of how color constancy works.

27.2.1   Land's Study of Complex Images

In 1963 Edwin Land decided to take his research on vision in a different direction. He joked that many critics thought that his conclusions about human vision were suspect, because the experiments were photographic, and that these critics thought that he (Land) could do anything with photographic film. He set out to do experiments using just papers, lights and a very good telephotometer. Red and White experiments (Chapter 26) had shown the importance of complex images, so Land started with displays using about 100 papers (Figure 27.2). He asked Lucretia Weed to make a display resembling Mondrian's painting in the Tate Gallery, London. As it turned out, Lucretia finished the display before we could find a color photograph of that painting. Years later at a conference in Amsterdam, Hank Spekreijse pointed out that his countryman Mondrian had never used high-chroma greens. While these experimental displays are higher in chroma and contrast than that Mondrian painting, they are better visual test targets than their inspiration because they have a much larger range of colors.

Nigel Daw's (1962) experiments with afterimages convinced Land to avoid regular arrays of same-size squares, or other identical shapes. Daw's experiment had two parts: First, he had observers make a strong color afterimage of an object by fixating at a point in a color image, say a square pillow, which formed a diamond afterimage on the retina. Second, he asked observers to describe the afterimages as they moved their gaze to different fixation points on a black and white image of the same scene. When observers fixated on new points in the image, the mismatch of contours between external scene and the internal afterimage inhibited the visibility of the afterimage. However, when the observers fixated on the original point, thus registering the color afterimage with the contours of the black and white image on the screen, the color afterimage became more visible. For several minutes observers could make the color afterimage of the pillow appear by looking at the original fixation point, and make it disappear by looking at another point in the image. The conclusion was that afterimages are inhibited by mismatched contours on the current image on the retina. Hence, Land avoided the problem of afterimages,

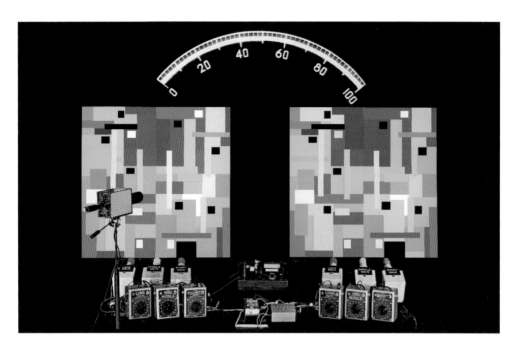

Figure 27.2 Land's Double Mondrian Experiment. Two identical sets of matte colored papers, with separate L, M, S illuminating projectors with voltage transformers for control of the amount of light. Telephotometer readings were projected above the Mondrians. The experimenter measured the *L, M, S* radiances separately from a green paper in the left Mondrian. Then, he adjusted the *L, M, S* radiances from a red paper on the right Mondrian to be the same. Observers reported different red and green colors produced by identical light stimuli.

by making each color in the Mondrian a different size, or shape. Regular arrays of constant size patches are subject to the adaptation problems caused by color afterimages of the last square that fits the contours of the new area of interest. The adaptation problems literally disappear when looking at complex scenes.

Land set out to study the appearance of colors by varying the spectra, intensity and the duration of illumination falling on Mondrians.

27.3 Color Mondrians

We have discussed the Black and White Mondrian which showed that a particular luminance at the display, and a particular quanta catch at the retina, can appear any lightness (white to black) (Chapter 7). Here we describe a series of Color Mondrians that show the need for independent spatial processing of L-, M-, S-cone responses. These experiments also show that the popular mechanism of color adaptation does not correlate with measured appearances.

27.3.1 Land's Original Color Mondrian

The Color Mondrian experiments showed that a particular triplet of radiances from a display can appear as any color. All color sensations can result from a single triplet of quanta catches by Long-wave (L), Middle-wave (M), and Short-wave (S) cones in the retina.

The first Color Mondrian experiments (Land, 1964; Land & McCann, 1971; Land, 1974a, 1977) used two identical Mondrians made of color papers, and three different, non-overlapping spectral illuminants (long-, middle-, and short-wave visible light). In this experiment observers reported the colors of papers in the Mondrians. Two different reflectance papers were shown in two different illuminant mixtures, such that both had equal quanta catches by the L, M, S cones. In order to get the same quanta catch from a green paper as a red paper, the experimenter measured separately the L radiance from the red paper in *ILL* illumination alone. He repeated the process to measure M radiance in *ILM*, and S radiance in *ILS*. For the second Mondrian, the experimenter aimed the meter at the green paper, and separately adjusted three new illuminant intensities *ILL2, ILM2* and *ILS2* to get L, M, and S radiances from that paper. When all six illuminants were turned on, observers reported that the red paper looked red, and the green paper looked green, despite the fact they were identical stimuli in the center of the areas.

Land repeated this experiment with all the papers in the Mondrians. He made two important observations:

A constant L, M, S radiance triplet could generate any color sensation.

With single illuminants, such as *ILL* alone, the appearance of the Mondrian changed very little with large changes in the amount of uniform illumination. The red paper was light in *ILL* and in *ILL2*. The green paper was dark in *ILL* and in *ILL2*.

If we compare the two Mondrians in only long-wave illumination with different amounts of light on each, the appearances of the papers were nearly constant. Figure 27.3 shows the pair of Mondrians in only long-wave light with more light on the left, than right. Observers report only small changes in appearance, with large changes in the amount of uniform illumination. This is the same observation as found with achromatic test targets in Chapter 21 measuring the Hipparchus line. In fact, we observe in spectral narrow-band illumination all the same properties we measured for achromatic HDR targets in Sections C & D. The signature behavior of lightness, namely the Hipparchus line and neural contrast, are found in colored light.

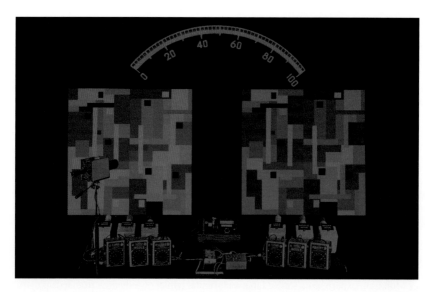

Figure 27.3   Pair of Mondrians in only long-wave light. The left Mondrian has more illumination than the right. Observers report nearly the same set of lightnesses, with the left set detectably lighter than the right.

If we look around the pair of Mondrians in the same color light, it is difficult to estimate how the cones adapt to the light from this scene. We have light and dark areas in each Mondrian, and we have global shifts in illumination left and right. We have very complex patterns of eye movements. Clearly visual pigments in the cones adapt to the history of light falling on them, but it is very difficult to imagine how that non-uniform state of adaptation of the array of cones can account for color constancy of Mondrians.

When the Mondrian is viewed in only middle-wave light the lightness appearances of the papers changed. The red paper appeared dark and the green paper appeared light, this time in a greenish wash over the scene. The scene content is different. It appears to be a different set of papers. It is what photographers call a different color separation record.

We see the same result with short-wave illumination. In summary, if the two Mondrians are side by side in the same band of wavelengths, but different intensities, the observers report nearly the same set of lightnesses, with one set detectably lighter than the other. With large uniform changes in illumination, we see nearly constant lightnesses of the colored papers.

We use the illuminations described in Figures 27.3 and 27.4. The piece of green paper on the left Mondrian has more long-wave and less middle-wave illumination. The red paper on the right Mondrian has less long-wave and more middle-wave illumination. When adjusted, those changes in illumination make the red and green papers have identical radiances. Those changes do not significantly alter the lightnesses of the areas in separate illumination. When viewing the Mondrian in combined illumination, in color, those changes in illumination do not change the color appearances of the red and green papers.

These observations led Land to propose the Retinex theory. If the apparent lightnesses of the papers in long-wave light are constant with large changes in overall illumination, and the different lightnesses of the papers in middle-wave light are also constant with similar large changes, then the colors are constant because the lightnesses are constant. The triplet of apparent lightnesses, not radiances, determines the color appearance. Constant lightnesses generate constant colors.

That hypothesis led to a study of color appearances in L, M, S bands of light. Did all red colors have the same triplet of lightness appearances? Did a red color always look (light, dark, dark) in L, M, S light? Did a green always look (dark, light, dark)? Did color appearance always correlate with the triplet of L, M, S lightnesses?

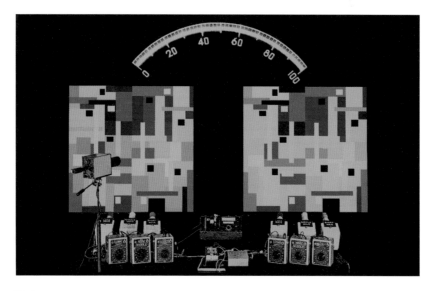

Figure 27.4   Pair of Mondrians in only middle-wave light. Now, the left Mondrian has less illumination than the right.

Table 27.1 Correlation table of color appearances and the apparent lightnesses in L, M, S illumination.

Color Appearance	Appearance in L- light	Appearance in M- light	Appearance in S- light
Red	light	dark	dark
Yellow	light	light	dark
Green	dark	light	dark
Cyan	dark	light	light
Blue	dark	dark	light
Magenta	light	dark	light
White	light	light	light
Black	dark	dark	dark

The experiment is easy. Find a red, a green and a blue filter. Be sure that the filters exclude the other two-thirds of the spectra. With a green filter you should just see greens with different lightnesses. You should not see a mixture of greens, and yellows, and blues. If you do, you need a filter with a narrower band of transmission.

Identify a group of red objects. Look at then sequentially through the L, M, S filters. Look at them in different ambient illuminations. Look at them at different times of day. Bright red colors are always (light, dark, dark) in L, M, S light (Table 27.1).

Retinex theory predicts that the triplet of L-, M-, S-lightnesses determines color. Colors are constant with changes in illumination because the triplet of lightnesses is constant. How does human vision generate constant lightnesses in the different wavebands? We have been discussing such a mechanism for the last 100 pages: spatial comparisons. If each spectral channel made its own spatial comparisons, then the lightness in long-wave light would be independent of changes in overall illumination. The same holds for each spectral channel.

The last remaining step is that the spatial comparisons described in sections C and D need to work independently for each spectral channel. Making that theoretical assumption makes understanding color constancy easy. Land proposed the word *Retinex*, a contraction of retina and cortex, as the name of these independent spectral channels. Chapter 32 provides a full discussion of the changes in usage of the word Retinex over the past 50 years.

According to Retinex theory, first, the rod and cone receptors responded to their quanta catch. Since color appearances correlate with lightness, and lightness is a spatial comparison, then color vision must have three independent spatial channels. Figure 27.5 is a simplified block diagram that illustrates the spatial nature of color comparisons, different from pixel based colorimetry. It should not be considered as a wiring diagram of the retina.

The physiology of color vision is a large and complicated topic that is beyond the scope of this text. Here we limit our discussion to psychophysical observations using the system as a black box. The physiology has many fascinating problems to solve:

Capture quanta over 10 log units of dynamic range

Generate Hipparchus line responses to maxima

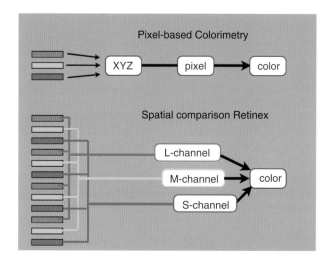

Figure 27.5   Spatial organization of Retinex compared with Colorimetry.

Implement neural contrast responsive to scene content

Amplify color separation information from overlapping spectral sensitivities

Account for color constancy

The Retinex diagram in Figure 27.5 should not be mistaken for a cone wiring diagram. Color physiology is much more complicated. Opponent color processing was suggested by Hering (1872). Ladd, Franklin, and Konig's suggested Helmholtz trichromacy was followed by Hering opponency in the neural pathway. Dorothea Jameson and Leo Hurvich added quantitative modeling of opponent processes. Rus DeValois's (1986) neurophysiology experiments brought opponent processing into the spotlight of a field dominated by trichromatic theory at the time. Color opponent processes are found in the optic nerve. Color opponent processes amplify color differences (Chapter 18.6.2). Color opponent processes are also spatial. Jameson points out that Hering never suggested that opponency was spatial, as described by DeValois. Nigel Daw (1984) and Bevil Conway (2010) have studied spatial double opponency. Semir Zeki (1980; 1993) made intracellular recordings from V4 from alert monkeys, and has shown that V4 cells exhibit color constancy using Color Mondrians. These results were consistent with three independent channels of color information reaching that part of the cortex. Despite the importance of integrating psychological experiments and mathematical models into the fabric of physiological data, it is far too big a subject to discuss here.

27.3.2   Quantitative Color Mondrians

McCann, McKee, and Taylor (1976) used a simplified 18-Area Mondrian in narrow-band illumination (630, 530, 450 nm). They repeated and extended the original Color Mondrian experiment (Land, 1977). They:

asked observers to match the color appearance of each paper to the Munsell Book

studied five illuminants: *L,M,S* from gray, red, yellow, green, and blue papers

measured *Scaled Integrated Reflectance* for each paper in each illuminant compared *Scaled Integrated Reflectance* to observed color matches

calculated color appearance using spatial comparisons

McCann, McKee and Taylor selected five papers (gray, red, yellow, green and blue). They adjusted five triplets of illumination falling on these papers so that the same *L, M, S* radiances reached the eye from each. Sequentially, observers matched each of the 18 papers to chips in the Munsell Book.

In addition, they used a Gamma Scientific telephotometer to measure the radiances of each paper in each illuminant. The meter had three spectral filters that converted the sensor's inherent spectral sensitivity to *L*-cone, *M*-cone, and *S*-cone sensitivities, used sequentially. The filters were selected to match Paul Brown and George Wald's (1964) measurement of cone sensitivities. That was the best spectral data at the time (Land, 1974b). These measurements gave them the cone responses to all papers in five illuminations. They compared the matches to the Munsell Book with a variety of hypotheses of how vision uses cone quanta catches. Triplets of L, M, S cone quanta catch did not predict the observer match data, as expected in a color constancy experiment. They tested whether color matches correlated with reflectances of papers in the Mondrian.

The definition of reflectance in physics is the ratio of incident to reflected light. It is not possible to measure that directly because the light meter gets in the way of the illuminating light. In practice reflectance is measured using a two step process:

Measure the light coming from a paper in appropriate illumination

Measure the light from a white, or known reflectance standard

Physical reflectance is the ratio of the two measurements, or percentage of the light from the standard (Figure 27.6, left).

Vision does not use an independent reflectance standard with known properties. If vision estimates something like "reflectance" it has to do it using the light falling on the retina. The Retinex theory proposed that spatial comparisons can be used to calculate the relative "reflectances" of papers in the Mondrian (Figure 27.6 right). If vision does this, then it must use cone quanta catches as the input to the calculation. The extremely broad, overlapping spectral sensitivity curves of cones introduce important differences between reflectance, as measured in physics, and "reflectance" estimated by spatial vision.

In*tegrated Reflectance* is the ratio of [a cone's response to a paper of interest] to [that cone's response to a white paper] in a particular illuminant. Thus, *Integrated Reflectance* is a physical measurement of

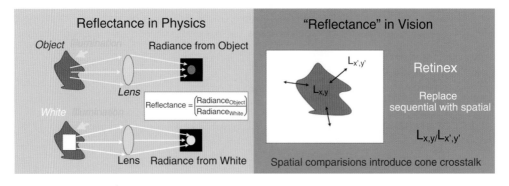

Figure 27.6   (left) Two step process for measuring reflectance; (right) estimating "reflectance" using spatial comparison.

light using the very broad-band sensitivity curves found in cones. These broad-band sensitivities intro-duce *crosstalk* into the measurements that allows the spectra of the illuminations to affect the measured ratios (McCann et al., 1976; McCann, 2004; 2005).

Scaled Integrated Reflectance takes the Integrated Reflectance measurements and scales them using a standard lightness function such as Glasser et al. (1958), or L* (CIE,1978). Linear reflectance over-emphasizes high-reflectance values, logarithmic reflectance over-emphasizes low-reflectance values, so we used equally-spaced apparent reflectances in the evaluation of our data (Chapter 18).

The McCann McKee and Taylor results showed:

Scaled Integrated Reflectance data predicted color matches

Spatial-comparison computational model predicted color matches

Small changes in matches in different illumination correlated with cone crosstalk

Small overall lightness changes from illumination intensity (Hipparchus slope 0.3)

Physical reflectance is the ratio of the light reflected from a paper to the light from a reflectance standard, or white. *Physical reflectance* is always measured with the narrowest waveband possible so that the sensor sensitivity and illumination spectra play a minimal role in the measurement of the paper's ability to reflect light.

Scaled Integrated Reflectance differs from *physical reflectance* in four important ways:

It replaces the narrowest possible waveband with broad band cone-sensitivity curves, causing crosstalk. Thus, *integrated reflectance* values change with changes in illumination.

It replaces ratio of reflected light to a standard: it uses spatial comparison to scene maximum.

It replaces sequential pairs of measurements (sample, then standard); it uses simultaneous spatial comparisons of scene areas, and does not need a standard.

It uses cube root scaling to compensate for intraocular scatter and makes L, M, S values proportional to lightness.

These four important changes make it possible to look for a physical measurement that correlates with human color constancy. Human vision uses very broad spectral sensitivities, has no control of the spectral content of the illumination, and has to build appearance out of the spatial content of the image on the retina.

27.3.3   Physical Correlate of Color Constancy

The McCann, McKee, and Taylor (1976) results compared the physical measurements of Scaled Integrated Reflectance with observed color matches for *L, M, S* components of all five experiments, each using 18 color patches. Observers matched each color paper in each illuminant to a chip in the Munsell Book in constant illumination. The data showed nearly constant appearances, with subtle changes in different illuminants. However, those shifts in appearance correlated with changes in *Scaled Integrated Reflectance*. Figure 27.7 plots the physical measured scene *Scaled Integrated Reflectance* vs. observer matches for 18 areas, five illuminants and three channels. Retinex color constancy predic-tions and observer matches are available on the Wiley web site http://www.wiley.com/go/mccannhdr

The very broad-band L-cone sensor responds primarily to the L illuminant. However, cone crosstalk is the result of L-cone's very large response to the M illuminant, and a smaller response to the *S*

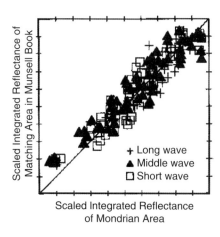

Figure 27.7   Scene vs. matching papers' Scaled Integrate Reflectances.

illuminant. When the experimenter changes the amounts of the *ILL, ILM, ILS* illumination, the ratio of radiances from the colored sample to the white changes. The L-cone radiances are not simply illumination (*ILL*) times reflectance (*RL*), but rather the sum of integrals of L-cone responses to the three illuminants. For analysis, we describe the L-cone response as the sum of three responses to three illuminants in equation 27.1.

$$Lcone = \int_{\lambda=400}^{700} \left(ILL_\lambda * R_\lambda * LS_\lambda \right) d\lambda + \int_{\lambda=400}^{700} \left(ILM_\lambda * R_\lambda * LS_\lambda \right) d\lambda + \int_{\lambda=400}^{700} \left(ILS_\lambda * R_\lambda * LS_\lambda \right) d\lambda \quad (27.1)$$

where *LS* is the spectral sensitivity of the long-wave cones.

The McCann, McKee, and Taylor experiments used a very-narrow-band interference filters to simplify the analysis. We can make a very close approximation to the L-cone integral with the following sum of three products in equation 27.2.

$$TLR = \left[(ILL * RL * LSL) + (ILM * RM * LSM) + (ILS * RS * LSS) \right] \quad (27.2)$$

where *TLR* is the Total L-cone Response to narrowband IL at 630, 530, and 450 nm light, and R and LS are the values for those wavelengths.

So, despite the fact that the paper's surface is unchanged, we can measure changes in the *Integrated Reflectance* values with changes in illumination. McCann, McKee, and Taylor (1976) measured these changes and reported them in detail. The purpose was to try to correlate color matches with *Scaled Integrated Reflectance*. The data shows that a substantial part of the small, but measurable, color change observed in color constancy experiments correlates with *Integrated Reflectance* – a phenomenon in the domain of physics. Most of these color changes have nothing to do with adaptation mechanisms attributed to the domain of biology. In other words, the cause of much of the "departures from perfect constancy" is simply the physics of the broad-band spectral sensors found in cones. If the *Scaled Integrated Reflectance* ratios change, then appearance changes.

These observations are central to models of human color constancy. This subtle difference between *physical reflectance* and *Scaled Integrated Reflectance* defines a watershed that distinguishes different models. Ebner (2007) describes a number of studies of constancy as a problem in physics, i.e. separating physical reflectance from actual illumination. He reviews the work of Horn, Stockham, Drew, Funt, Finlayson, Hordley, and others, whose techniques attempt to calculate *physical reflectance,* rather than

Scaled Integrated Reflectance. This important distinction is that these studies changed their goal to a different one than the study of human vision. *Physical reflectance* does not correlate with human color constancy, while *Scaled Integrated Reflectance* does only in uniform illumination (Chapter 29). This subtle difference is the key signature of the underlying color constancy mechanism of human vision, as we will see below.

There is a second important distinction between color appearance and object recognition. A number of psychological studies ask observers to report the perceived reflectance and perceived illumination. These studies ask a different question of observers, and get different answers than those described here. We are studying the question: What does the area look like, rather than what do you think the object is made of (Chapter 29). Although these distinctions between models of appearance, reflectance estimation, and object recognition may seem to be a small detail, they are very important because each have markedly different ground truths for their calculations. They may seem to be indistinguishable, for these flat uniformly lit Mondrians, but as we will see in Chapters 29 and 32, they are very different in real scenes, with real illumination.

27.3.4   Predicted Color Constancy

The McCann, McKee, and Taylor (1976) results included computer model predictions of color matches. Using the Ratio-Threshold-Product-Reset Model (Chapter 32) they build the L, M, S lightness predictions for the five experiments using 18 color patches. Although these experiments were very early examples of calculating color appearances in images, using only 20 by 24 pixels (large for the time), the results were quite good. Spatial comparisons of three independent L, M, S channels predicted observer color matches, including departures from perfect constancy.

27.4   The Signature of Color Constancy

Color Constancy experiments show that the appearance of objects is almost constant with large spectral changes in illumination. Von Kries suggested that changes in spectral illumination caused relative changes in receptor sensitivities, now called adaptation (Wyszecki & Stiles, 1982). McCann, McKee, and Taylor pointed out that the departures from perfect constancy correlated with the spectral crosstalk between broadband retinal receptors. Nayatani (1997) proposed that incomplete-adaptation explained these departures from perfect constancy. Nayatani's mechanism is illumination based; it is the result of imperfect adjustments to changed illumination. McCann, McKee, and Taylor's mechanism is reflectance based. Cone response ratios of two papers change with illumination because of crosstalk. Recent experiments (McCann, 2004; 2005) compared the predictions made by the illumination and reflectance hypotheses. Those experiments provided an unequivocal test of which mechanism controls color constancy.

The use of narrow band illumination is key to measuring departures from perfect constancy. Using similar broad band illuminants, such as tungsten and daylight minimizes the amount of crosstalk, and thus limits the size of the subtle changes caused by illumination.

The experiment included four papers; white, yellow, purple and gray; illuminated by three narrowband light emitting diodes (LEDs); 625, 530 and 455 nm. The Total Long-wave Response (TLR) is:

$$TLR = \left[\left(I_{625} * R_{625} * LS_{625} \right) + \left(I_{530} * R_{530} * LS_{530} \right) + \left(I_{454} * R_{455} * LS_{455} \right) \right] \tag{27.3}$$

where *I* is the incident illumination, *R* is the % reflectance and *LS* is the long-wave channel's sensitivity for each wavelength used. The long-wave sensors' response to 625 nm light is a primary response. The long-wave responses to 530 nm and to 455 nm are both crosstalk. The Total Medium-wave and Total Short-wave Responses are similar. The experiment used four identical LEDs for each wavelength. The experimenter switched on one, two, or four LEDs of each wavelength to vary the spectral content of the illumination. The I_{625}, I_{530}, I_{455} values were: 4,4,4; 4,4,2; 4,4,1; 4,2,4; 4,2,2; ... 1,1,4; 1,1,2; 1,1,1. In total 27 different combinations of three LEDs at three intensities were tested.

Spatial-comparison models hypothesize that human vision builds independent L, M and S lightness images by taking spatial ratios of different image areas. The long-wave output calculation uses the ratio of the yellow paper to the white paper. This ratio changes with relative changes in 625, 530 and 455 nm illumination because the proportions of crosstalk contributions change. This argument holds for colored papers, but not for achromatic ones. By definition, both white and gray papers have the nearly constant reflectance for all wavelengths. When the crosstalk component is the same as the main component, the ratio of gray to white is constant for all changes in illumination. Consider the ratio of the *Total Long-wave Radiance* (*TLR*) for the neutral gray paper (Reflectance *Rn*) to the *TLR* of the white paper (Reflectance *Rw*):

$$Ratio = \frac{[(I_{625} * Rn_{625} * LS_{625}) + (I_{530} * Rn_{530} * LS_{530}) + (I_{455} * Rn_{455} * LS_{455})]}{[(I_{625} * Rw_{625} * LS_{625}) + (I_{530} * Rw_{530} * LS_{530}) + (I_{455} * Rw_{455} * LS_{455})]}$$

$$when: \quad Rn_{625} = Rn_{530} = Rn_{455} = c1 \quad and \quad Rw_{625} = Rw_{530} = Rw_{455} = c_2$$

$$Ratio = \frac{[(I_{625} * c1 * LS_{625}) + (I_{530} * c1 * LS_{530}) + (I_{455} * c1 * LS_{455})]}{[(I_{625} * c_2 * LS_{625}) + (I_{530} * c_2 * LS_{530}) + (I_{455} * c_2 * LS_{455})]} \tag{27.4}$$

$$Ratio = \frac{c_1}{c_2} \frac{[(I_{625} * LS_{625}) + (I_{530} * LS_{530}) + (I_{455} * LS_{455})]}{[(I_{625} * LS_{625}) + (I_{530} * LS_{530}) + (I_{455} * LS_{455})]}$$

$$Ratio = \frac{c_1}{c_2}$$

As shown in equation 27.4, crosstalk for the gray paper is canceled by the crosstalk from the white. The ratio of neutral gray papers to white is completely unaffected by the spectral composition of the illuminants. This is not true for colored papers. The ratio of yellow to white depends on the crosstalk contributions controlled by the overlap of sensitivity functions and variable proportions of spectral illuminants. The independence of grays and the dependence of colored papers on spectral illumination is an important signature of crosstalk.

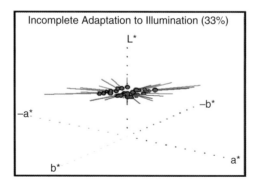

Figure 27.8   The effect of incomplete adaptation. L*a*b* plot of the 27 (625, 530 and 455 nm) LED illuminants. The ends of the lines represent illuminants. Assuming 33 % incomplete adaptation we get compression to the volume of the 27 solid spheres. All lines converge towards theoretical perfect color constancy at a point on the L* axis.

27.4.1   Incomplete Adaptation Predictions

As the explanation of constancy, the von Kries adaptation changes receptor sensitivities in response to spectral illumination changes. Incomplete-adaptation hypotheses assume that the departures from perfect constancy are the result of imperfect adjustments to changes in illumination. Figure 27.8 is a 3-D diagram of incomplete adaptation in L*a*b* space. The 27 illuminants form a cloud of points. L*a*b* stretches the radiances into an elliptical volume with a* as the longest axis. Using 33% incomplete adaptation shrinks the ellipsoid to the volume shown by the solid balls.

Since the incomplete adaptation hypothesis is controlled only by illumination changes, it predicts that gray and yellow papers will have identical constancy departures. In other words, the lack of complete adaptation should cause color shifts in the same direction in color space and they should have the same magnitude. Asking the observer to look at a different paper in the same field of view does not change the illumination predictions. Figure 27.9 (top left) shows L*a*b* plots of the incomplete adaptation predictions for three papers in the same field of view: Yellow (Sunburst), purple (Weeping Wisteria) and gray (Steel Gray). Each paper generates a cloud of 27 different points representing 33% incomplete adaptation for each illuminant.

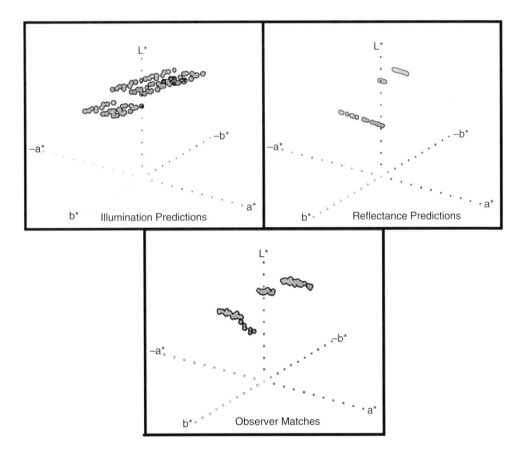

Figure 27.9 (top) Predicted matches based on: (left) Nayatani's illumination hypothesis; (right) Integrated Reflectance crosstalk hypothesis. Observers' matches (bottom) are consistent with the crosstalk predictions.

27.4.2   Spatial-Comparisons Predictions

Let us analyze color constancy assuming that vision is controlled by a number of independent channels using spatial comparisons. Here we calculate the ratios of Total Responses: (TLR yellow/TLRwhite); (TMR yellow/TMRwhite); (TSRyellow/TSRwhite) for all 27 illuminants. Similar calculations are made for the purple and gray papers. Figure 27.9 (top right) plots the spatial-comparisons predictions in L*a*b* for the three papers in 27 illuminants. Here the results of crosstalk fall along a long line for yellow, a shorter line for purple, and a cluster around a point for the gray. (The measured reflectance values for the gray paper were 64%, 60%, and 56% (L, M, S); hence the departure from a single point.)

27.4.3   Predictions vs. Observed Color Constancy

Since the incomplete adaptation hypothesis is controlled only by illumination changes, it predicts that all papers will have identical constancy departures, as long as the field of view is constant. Since the crosstalk hypothesis is based on the ratio of radiances sensed by broadband retinal receptors, it is paper dependent. Here the model predictions are unequivocally different.

The results of observer matches are shown in Figure 27.9 (bottom). The predictions and matching data were displayed and analyzed using 3-D rotating plots of all 27 illumination points for yellow, purple and gray papers. Illumination prediction plots show three nearly identical elliptical clouds of data. These clouds are the compressed colorimetric distribution of the 27 illuminations. The purple and gray illumination predictions partially overlap. These matches are inconsistent with *incomplete-adaptation* predictions.

The *spatial-comparisons* predictions show no resemblance to the colorimetric distribution of illuminants. They fall on lines for yellow and purple, and occupy a very small volume for gray. The observer matches are very similar to the spatial-comparisons predictions. They support the idea that departures from perfect constancy are the result of crosstalk between receptors. Crosstalk is a property of spatial comparisons. There is no crosstalk for pixel-based color comparisons.

The illumination and reflectance hypotheses that attempt to predict the departures from perfect constancy have markedly different predictions. Illumination mechanisms are indifferent to crosstalk and predict identical discrepancies for all colors of papers. The crosstalk hypothesis predicts no discrepancy for grays, and variable discrepancies with different hue and chroma papers. The matches are inconsistent with the incomplete-adaptation predictions that gray-paper matches must change as much as colored ones. Humans do not discount the illuminant in color constancy. Humans build constancy from a spatial comparison of scene content.

The color constancy experiments above are unusual in that they provide clearly different predictions for illumination models and spatial-comparison models. That was not the case with either Land's, or McCann, McKee, and Taylor's Mondrian experiments. We need a number of such experiments with unequivocally distinct predictions if we are to make progress in our understanding of the mechanisms of vision. We need additional unambiguous analysis of appearance in order to move forward.

Rotation 3-D plots in Quicktime® movie files are available on the Wiley web site http://www.wiley.com/go/mccannhdr for a better view of the three results in Figure 27.9. Their similarities and large differences are more clearly seen in the 3-D plots.

27.5   Search for Evidence of Adaptation – Averages

A more interesting version of the Color Mondrian are those done with colored surrounds. (McCann, 1987; 1989; 1992; 1997a; 1997b) The original McCann, McKee, and Taylor experiments did not try to discriminate between the various normalization techniques, such as maximum, average (*GrayWorld*) and local average. Figure 27.10 (left and center) illustrates the original color Mondrian experiment. The left panel shows that a gray paper with *ILL, ILM, ILS* illumination sends radiance of *L, M, S* to the observer, and the average of the field of view is *AVL, AVM, AVS*.

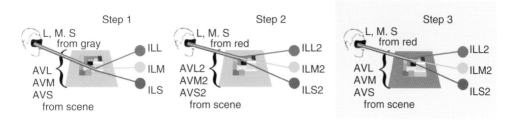

Figure 27.10   Three steps in searching for the effects of adaptation.

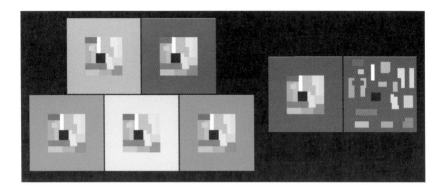

Figure 27.11   (left) Five Mondrians with different color surrounds that compensate for changes in illumination; (right) Global and Local Surround Mondrians.

In order to get constant *L, M, S* radiances from a different red paper, they changed the illuminants to *ILL2, ILM2, ILS2*. That action changed the average radiances to *AVL2, AVM2, AVS2* (Figure 27.10, center). That change in average radiances leaves the door open to *GrayWorld* explanations of color constancy. A better color constancy experiment is one that includes a surround around the Mondrian. We can use the color of the surround paper to cancel the changes in *AVL, AVM,* and *AVS* measured in Figure 27.10 (center).

27.5.1   Search for Global Adaptation

Three steps are necessary to perform color constancy experiments that can separate the underlying mechanism controlling color constancy. In order to get *L, M, S* to come from the red paper, the experimenter decreased *ILL* and increased *ILM* and *ILS*, because the red paper reflected more long-wave and less middle- and short-wave light than the gray. That meant that the *AVL2* is lower than *AVL*; *AVM2* is higher than *AVM*; and *AVS2* is higher than *AVS*.

Figure 27.11(right) adds a third step to return *AVL2, AVM2 AVS2* to the original *AVL, AVM AVS* values. The experimenters searched though 1000 colored papers and found one that reflected the proper amount of more long- and less middle- and short-wave light. The experimenters changed the color reflectance of the surround area around the Mondrian to move the *AVL2, AVM2, AVS2 GrayWorld* averages back to be equal to the averages in the original illumination. (McCann,1987; 1989; 1992; 1997a; 1997b)

Now, with this Step 3 Color Mondrian, there is no change in G*rayWorld* input. As well, there is no change in the overall scene to cause any adaptation. If either *GrayWorld*, or adaptation play a role in color constancy, this test target should shut off color constancy. There is no signal to cause any adaptation. The red paper must look gray because the *L, M, S* radiances match, and the *AVL, AVM, AVS* values all match the original gray paper case.

Figure 27.11 shows the McCann, McKee, and Taylor Mondrian surrounded by five different papers viewed in five different illuminations. These reflectance and illumination combinations gave constant *L, M, S* radiances and *AVL, AVM, AVS* average radiances.

27.5.2   Search for Local Adaptation

One can hypothesize that adaptation is a local mechanism, and not global one. It might be that the local area around each Mondrian area is more important than the average over the entire image. McCann tested the local influence by making a set of five *Local Surround Mondrians*. The Global and Local Surround Mondrians are shown in Figure 27.11.

27.5.3   No Evidence of Adaptation in Mondrians

A pixel-based alternative to spatial processing is the adaptation hypothesis. Unfortunately, the term adaptation has many conflicting definitions:

Dark adaptation – slow chemical migration of retinene (Pirenne, 1962);

Light adaptation – fast neural responses (Dowling, 1987);

von Kries – phenomenon driven model;

CIECAM models (Hunt, 2004).

One, or all, of these adaptations could play a role. We see adaptation when entering a movie theater, and when returning to daylight. The rods and cones adapt their responses to the level of light.

If human response to pixels in complex scenes is controlled by adaptation, then we should be able to find evidence of changes in appearance with changes in adaptation. In principle, the introduction of Step 3 (Figure 27.11) is an attempt to shut off color constancy. By changing the average radiance back to the original AVL, AVM, AVS we removed the hypothetical cause of adaptation to the new illuminant. If the controlling mechanism of constancy was the average radiance that caused adaptation, then having no change in average radiance means no adaptation. With no adaptation, then there cannot be any color constancy.

Observer matches show the same color constancy as it did in Step 2. Table 27.2 shows the matching Munsell chips for McCann, McKee, and Taylor (MMT) experiment, the *Surround* experiment and the *Local Surround* experiment.

Table 27.2 shows that color constancy is not changed significantly by the adaptation canceling surrounds. Neither large *Surrounds*, nor *Local Surrounds* disturbed the color constancy mechanism. The MMT matches were close to the chips' actual values. *GrayWorld* and adaptation hypotheses predict that all matches must be the identical; they are identical quanta catches, and identical averages. The matches do not fit this prediction. The color matches were very close to those for the MMT experiment and model results. We designed experiments to measure the influence of *GrayWorld*, or adaptation, and were unable to find evidence that it has an effect on color constancy experiments using Mondrians.

27.5.4   Change Adaptation by Changed Average Radiance in Mondrians

To further elaborate on this point we performed a third set of Color Mondrian experiments (McCann, 1997b). In the previous experiments, we balanced the change in illumination with a canceling change in reflectance of the surround. There, *GrayWorld,* and adaptation, are reduced to zero.

Table 27.2 Munsell matches for five papers. There are five sets of results: actual papers, MMT, surround, local surround, and computer model results.

Test Target	Actual Paper	Matching MMT chip	Matching Surround Chip	Matching Local Chip	Calculated MMT Chip
Gray	N 6.75/	5 YR 6/1	N 6.0/	N 5.75/	N 6.0/
Red	10 RP 6/10	5 R 6/6	2.5 R 7/4	5 RP 7/4	5 RP 5/4
Yellow	5 Y 8.5/10	5 Y 8/8	5 Y 8.5/8	7.5 Y 8.5/12	7.5 Y 7/8
Green	2.5 G 7/6	10 G 7/4	7.5 G 7/4	2.5 G 8/2	5 G 6/6
Blue	2.5PB 6/8	2.5PB 4/6	10B 6/2	10B 6/4	7.5B 6/6

Using new surround papers, we introduced a change in *GrayWorld* as large as that introduced by a change in illumination, but we kept the original illumination constant. As above we change the *ILL, ILM, ILS* illuminant to get the desired *L, M, S* radiances at a particular point. The consequences are shifts in the *AVL, AVM, AVS* averages as big as the illuminant changes. We measure the amount of this shift. Now, we change the color reflectance of the surround area around the Mondrian to move the averages as much as the illumination change would have. In other words we shift the average as much as the MMT experiment would have, but we view the display in the original illumination. Any *GrayWorld*, and any adapting mechanism, must introduce large changes in appearance. The matching data (McCann, 1997b) show that appearance does not change. Changes in *GrayWorld*, and changes in adaptation, do not cause changes in color appearance. As well, we made spaced Mondrians in which the surround changed the local average while viewing in the original illumination. These surrounds had very little effect on color matches, showing that local adaptation is not involved in Mondrian color constancy.

27.6   Transparency in Mondrians

When we hold a transparent filter over a portion of a Mondrian, we recognize the property of transparency. Ripamonti et al. (2004) studied the spatio-chromatic constraints for achromatic and chromatic transparent displays. They showed that transparency correlates with constant spatial edge ratios for cone types. They found that the greater the number of surfaces in the display that are partially covered by a transparent filter, the stronger the impression of transparency. These results and other studies show the importance of spatial ratios of cone responses within each spectral channel (Westland & Ripamonti, 2000; Ripamonti & Westland, 2003; Foster & Nascimento, 1994; Khang & Zaidi, 2004; Smithson & Zaidi, 2004).

27.7   Color Assimilation

Land's observation still stands: The triplet of apparent lightnesses correlates with color. The observation is important because a variety of different phenomena can influence lightness, such as Simultaneous Contrast, the Cornsweet effect, assimilation, and spatial blur of the retinal image. Regardless of the cause of the lightness changes, when two identical physical objects look different, color appearances correlate with their *L, M, S* lightnesses.

Land (1974b) reported on Smitty Stevens test of Retinex Theory with studies of the Cornsweet's spinning disks in color.

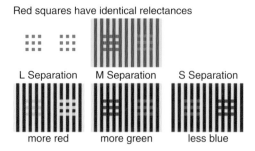

Red squares have identical relectances

L Separation   M Separation   S Separation

more red   more green   less blue

Figure 27.12 (top) Color squares and Color Assimilation changes their appearance;(bottom) L, M, S separation images. Color appearances changes correlate with L, M, S lightness changes.

McCann (1990) studied color produced in Jay Hannah's paintings in which appearance changed dramatically with distance.

McCann (2002) studied color assimilation.

In all cases the colors correlated with the triplet of lightnesses.

In an effective color assimilation display (Figure 27.12), there are two sets of nine red squares that are the same (top left). When we place the same red squares on yellow and blue stripes they look different (top center). The left patches appear a purple red, while the right ones appear a yellow orange. In other words, the left patches appear more blue and the right ones more yellow.

In the L separation (Figure 27.12, bottom) the corresponding patches are lighter on the right; in the M separation these patches on the right are lighter; in the S separation they are darker on the right. Land's retinex predicts that whenever L and M separations are lighter and S separation is darker, then that patch will appear more yellow. Whenever S separation is lighter and L and M separations are darker, then that patch will appear more blue. Colors correlate with *L, M, S* lightnesses. Color assimilation displays exhibit larger color effects than color contrast (Shevell, 2000). McCann showed that the very large color changes are associated with color channel crosstalk (2002).

27.8   Summary

Land made Mondrians with irregular shapes to inhibit afterimages. He demonstrated that identical *L, M, S* radiances generated any color sensation. Land's Retinex theory describes colors as the triplet of L, M, S lightnesses. That means we can transfer all we learned about lightness and its Hipparchus-line and neural-contrast behavior to color.

McCann, McKee, and Taylor asked observers to match 18 areas in five different spectral illuminants. They showed that colors matched *Scaled Integrated Reflectance* predictions. Spatial comparisons using cone sensitivity curves introduced crosstalk, not found in measurements of physical reflectance. Crosstalk accounts for the departures from perfect constancy with changes in spectral illumination.

McCann's experiments looked for the effects of *GrayWorld* and adaptation. He could not find adaptation effects in color constancy experiments using Color Mondrians. Adaptation of rods, cones, and neurons is a fundamental property of vision. All visual cells adapt. However, our color vision mechanisms process the light falling on the retina in such a way that it removes almost all of the influence of adaptation in rendering complex images.

The experiments in this chapter carefully separate and compare the ideas of *discounting the illumination* with those of building *color from spatial content*. Since these mechanisms are very different, we designed experiments that have different predicted results.

Discounting illumination has the properties:

It is responsive to light from the scene (average of light)

An adaptation hypothesis allows an efficient pixel-based strategy

Constant L,M,S and AVL, AVM, AVS must shut off constancy

Spatial Content has the properties:

It is responsive to scene content (maxima and spatial contrast)

Constancy is unresponsive to adaptation

Constancy is unresponsive to averages.

The first Color Mondrian experiments by Land and McCann, McKee, and Taylor demonstrated that the Retinex model works quite well. Crosstalk is the result of spatial comparisons between single-cone-type receptors. There is no crosstalk from the comparison of *L, M, S* radiances at a pixel. Departures from perfect constancy correlate with crosstalk.

The search for a way to shut off color constancy is a very important idea. If we really understand the underlying mechanisms, then we should be able to shut constancy off. There must be a set of circumstances in which identical *L, M, S* values in a complex image look the same in different illuminations.

27.9   References

Brown PK & Wald G (1964) Visual Pigments in Single Rods and Cones of the Human Retina, *Science*, **144**, 45.

CIE (1978) Recommendation on Uniform Color Spaces, Color Difference Equations, Bureau Central de la Paris: Psychometric Color Terms, Supplement 2 of CIE Publ. 15 (E-1.3.1).

Conway B (2010) *Neural Mechanisms of Color Vision: Double-Opponent Cells in the Visual Cortex*, Kluwer, Dordrecht.

Daw NW (1962) Why After-Images Are Not Seen In Normal Circumstances, *Nature*, **196**, 1143–45.

Daw NW (1984) The psychology and physiology of colour vision, *Trends in Neurosciences*, **7**(9), 330–5.

DeValois R & DeValois K (1986) *Spatial Vision*, Oxford University Press, New York.

Dowling JE (1987) *The Retina: An Approachable Part of the Brain*, Belknap Press.

Ebner M (2007). *Color Constancy*, John Wiley & Sons, Ltd, Chichester Chapter 7.

Foster DH & Nascimento SMC (1994) Relational colour constancy from invariant cone-excitation ratios, *Proc. R. Soc. London Ser. B*, **257**, 115–21.

Glasser G, McKinney A, Reilley C, and Schnelle P (1958) Cube-Root Color Coordinate System, *J Opt Soc Am*, **48**, 736–40.

Hering E (1872) *Outline of a Theory of Light Sense*, trans Hurvich LM & Jameson D, Harvard University Press, Cambridge, 1964.

Hunt RWG (2004) *The Reproduction of Colour*, IS&T John Wiley & Sons, Ltd, Chichester.

Khang BG & Zaidi Q (2004) Illuminant color perception of spectrally filtered spotlights. *J. Vision*, http://journalofvision.org/4/9/2/.

Land EH (1964) The Retinex, *Am. Scientist*, **52**, 247–64.

Land EH (1974a) The Retinex Theory of Colour Vision, *Proc. Roy Institution Gr Britain*, **47**, 23–58.

Land EH (1974b) Smitty Stevens' Test of Retinex Theory, in: *Sensation and Measurement, Papers in Honor of S. S. Stevens*, Moscwitz H et al. eds, Applied Science Publishers Ltd., London, 363–8.

Land EH (1977) The Retinex Theory of Color Vision, *Scientific American*, **237**(6), 108–28.

Land EH & McCann JJ (1971) Lightness and Retinex Theory, *J Opt Soc Am*, **61**, 1.

McCann JJ (1987) Local/Global Mechanisms for Color Constancy, *Die Farbre*, **34**, 275–83.

McCann JJ (1989) The Role of Nonlinear Operations in Modeling Human Color Sensations, Proc. SPIE 1077, 355–63.

McCann JJ (1990) Psychophysical measurements of Hannah color/distance effects, Proc. SPIE 1250, 203–11.

McCann JJ (1992) Color Constancy, Small Overall and Large Local Changes, *SPIE Proc.*, **1666**, 310–21.

McCann JJ (1997a) Color Mondrian Experiments Without Adaptation, Proc. AIC, Kyoto, 159–62.

McCann JJ (1997b) Magnitude of Color Shifts from Average-Quanta Catch Adaptations, *Proc. IS&T/SID Color Imaging Conference*, **5**, 215–20.

McCann JJ (2002) When Do We Assimilate in Color, *Proc. IS&T/SID Color Imaging Conference*, Scottsdale, **10**, 10–16.

McCann JJ (2004) Mechanism of Color Constancy, *Proc. IS&T/SID Color Imaging Conference, IS&T/SID*, Scottsdale, Arizona, **12**, 29–36.

McCann JJ (2005) Do Humans Discount the Illuminant? *Proc. SPIE*, **5666**, 9–16.

McCann JJ, McKee S & Taylor T (1976) Quantitative Studies in Retinex Theory, A Comparison Between Theoretical Predictions and Observer Responses to Color Mondrian Experiments, *Vision Res.*, **16**, 445–58.

Nayatani Y (1997) A Simple Estimation Method for Effective Adaptation Coefficient, *Color Res. Appl.* **20**, 259–74.

Pirenne MH (1962) Chapters 5–7, Davson H (ed.) in *The Eye, 2: The visual process*, Academic Press, New York.

Ripamonti C & Westland S (2003) Prediction of transparency perception based on cone-excitation ratios. *J Opt Soc Am A*, **20**, 1673–80.

Ripamonti C, Westland S & da Pos O (2004) Conditions for perceptual transparency, *J Electronic Imaging*, **13**, 29–35.

Shevell SK (2000) Chromatic Variation: A Fundamental Property of Images, *Proc. IST&SID Color Imaging Conference*, Scottsdale, **8**, 8–12.

Smithson H & Zaidi Q (2004) Colour constancy in context: Roles for local adaptation and levels of reference, *J Vision*, http://journalofvision.org/4/9/3/.

Westland, S & Ripamonti, C (2000) Invariant cone-excitation ratios may predict transparency, *J Opt Soc Am A*, **17**, 255–64.

Wyszecki G & Stiles WS (1982) *Color Science: Concepts and Methods, Quantitative Data and Formulae*, 2nd edn, John Wiley & Sons, Inc, New York, 429–51.

Zeki S (1980) The Representation of the Cerebral Cortex, *Nature*, **284**, 412–18.

Zeki S (1993) *A Vision of the Brain*, Blackwell Scientific Publications, Oxford.

28

Constancy's On/Off Switch

28.1 Topics

In real scenes the color of objects tends to be constant regardless of the color of the illuminant. Mechanisms using context recognition and adaptation of retinal sensitivities cannot explain Color Mondrian experimental results. This chapter presents experiments to test the hypothesis that colors are determined by the three independent channels of spatial relationship between the object of interest and all colors in the field of view. Further, these experiments test whether we can shut off color constancy.

28.2 Introduction

"Contrasts in Vision" was a fitting title for the meeting in September 1991 that honored Fergus Campbell, Cambridge University, for his many contributions to vision science. Instead of the usual Festschrift that featured a series of lectures, Fergus's colleagues organized a Festspiel. It was a kind of science fair in which each of the participants showed a demonstration of a new vision experiment. Although no prizes were awarded, Fergus made his way to each demonstration and discussed the experiment devised in his honor. John McCann had been a good friend of Fergus since the 4th Colloquium of the Pupil in 1964. He made the presentation of the Maximov's Shoe Box experiment, described below.

Color Contrast is the central idea of this demonstration of color constancy. The purpose of the demonstration was to highlight observations that large global changes in illumination produce very small changes in appearance. However, local changes in scene edges, particularly those associated with maxima, produce very large changes in appearances. This difference between global and local changes in receptor quanta catch is a basis for understanding color constancy.

28.3 Maximov's Shoe Boxes

Vadim Maximov (1989) wanted to test the effects of removing maxima from Mondrians. He devised the elegantly simple shoebox viewer (Figure 28.1, left). The hole cut in the top allows illumination to

The Art and Science of HDR Imaging, First Edition. John J. McCann, Alessandro Rizzi.
© 2012 John Wiley & Sons, Ltd. Published 2012 by John Wiley & Sons, Ltd.

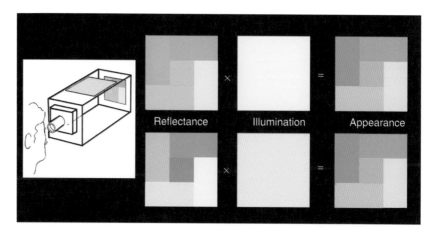

Figure 28.1 (left) Maximov's Shoe Box; (right) Color constancy shut off with identical stimuli. Tatami A (top) used a 40 CC cyan filter, and Tatami B (bottom) used a 40 CC red filter.

fall on the simplified Mondrian at the end of the box. Colored filters controlled the intensity and the spectral distribution of the light. Maximov's goal was to control the illuminants and the reflectances in the shoebox to shut off color constancy.

McCann (1992a) made a pair of Maximov Shoe Boxes. Each was a cardboard shoebox approximately $15 \times 12 \times 32$ cm, with two black 35 mm plastic film containers, plastic lenses, a 5×5 cm piece of black velour paper, a piece of diffuse drafting velum and a Kodak Wratten Color Correction filter, as illustrated in Figure 28.1 (left). A viewing tube with magnifying lens restricted the angle of view to a 3 degree circle. The simplified Mondrians with five or six papers are called Tatami, after Japanese floor mats.

In principle it is easy to do (Figure 28.1). Imagine two Maximov shoeboxes; one for the upper Tatami, and one for the lower. Select two filters that attenuate the color spectra, but do not reduce the light at any wavelength to zero. We used Wratten Color Correction filters: CC40R and CC40C. These filters are used by photographers to change color balance. The 40R filter has an optical density of 0.4 in the blue-green spectral region: and 40C has a density of 0.4 in the red. They are strong filters that reduce light to 40% of incident, but they do not block spectral regions entirely. These filters have different effects on appearances depending on how they are used. If we put them side by side on a lightbox, we see high chroma red and cyan areas surrounded by the white of the light box. They look like high chroma papers. Now, if we pick the 40R filter up and hold it close to one eye and look around the room, we see that the room has a pale pink cast. Replacing 40R with 40C makes the color cast cyan. The room colors are almost constant. When viewed side by side, they are highly colored, but in color constancy experiments they generate small changes in appearance.

We measured the difference in color between the two filters. The goal was to find pairs of colored papers that have exactly the same color changes. Both Maximov and David Stork (1976) found it very difficult to find existing papers because the constraints were too difficult.

Papers with such demanding specifications had to be manufactured to fit the measurements. Digital printers were not generally available in 1990. We used an early digital xerographic Canon CLC 500 printer (Canon, 1989) to make two Tatami with identical colorimetric shifts for all pairs of corresponding papers.

The color of each patch was created by specifying a digit between 0 and 100 for each of cyan, magenta, yellow and black toners, and using a Minolta ChromaMeter CR22 to measure their chromaticity, and L* value.

Table 28.1   Chromaticity data for Tatami A and Tatami B.

	Tatami A		Tatami B				
	x	y	x	y	Δx	Δy	ΔY
Top Right	0.38	0.33	0.30	0.32	0.08	0.02	10.9
Bottom Right	0.34	0.31	0.26	0.29	0.08	0.02	9.2
Bottom Left	0.35	0.39	0.28	0.35	0.07	0.04	16.3
Top Left	0.33	0.27	0.25	0.25	0.08	0.02	17.0
Center	0.30	0.32	0.22	0.30	0.08	0.02	14.4

A Wratten Color Correction 40R filter was selected for one shoebox and a 40C for the other. They have chromaticities of x = 0.35, y = 0.32 (40R) and x = 0.27, y = 0.30 (40C). The shift in chromaticity is Δx = 0.08, Δ y = 0.02.

Using the CLC 500 Laser Copier and the Minolta Colorimeter, we made two sets of five papers separated in color space by Δx = 0.08, y = Δy = 0.02. Table 28.1 lists the chromaticity of all five areas in Tatami A and Tatami B. This data shows that all of the colored papers in A are shifted by the same amount in CIEXYZ space. The amount of the shift is equal to the shift caused by changing from a Wratten 40R to a Wratten 40C.

28.3.1   Control – Change of Appearance from CC40R to CC40C Illumination

Observers were asked to hold, in turn, each filter close to their eye and to compare the appearance of the room with: no filters, 40R and 40C. They reported that the world looks neutral, warm, and cool as a result of the filters shifting the quanta catch of all receptors in the field of view by the same amount. The magnitude of the change in appearance is small. Objects do not change their color names; neutrals take on a warm, or a cool, cast. The general conclusion is that these color filters are relatively weak and do not change color appearance significantly.

28.3.2   Experiment – Change of Appearance in Shoebox

The demonstration experiment at the Fergus Festspiel was to compare the color appearances in the two Shoe Boxes. One (Figure 28.1, top) illustrates Tatami A with five colors that shifted away from red in CC40C illumination; the other (Figure 28.1, bottom) illustrates five colors shifted towards red in CC40R illumination. The papers were carefully manufactured to have the exact opposite shift in chromaticity as that caused by the filters.

Ordinarily illumination has little, or no noticeable effect. When we viewed the two Tatami side by side on a table in a room, there was very little change in appearance alternating the two filters.

When viewed in the Maximov Shoeboxes, the different sets of reflectances, in different illuminations, changed in appearance from looking different, to looking the same. Tatami A looked the same as Tatami B (Figure 28.1, right). The result was that we turned off the color constancy mechanism for complex images using this pair of Maximov Shoeboxes. Despite the fact that the reflectances were different, the color appearances were the same.

Why did Maximov's boxes turn off color constancy? The answer is that both Tatami have to look identical, because every pixel in their entire fields of view had identical cone quanta catches. The sets of papers were made to shift the entire image as much as the filters did. When viewed in isolation, the

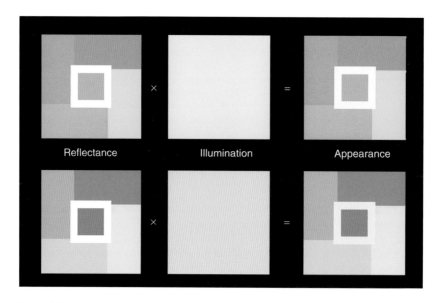

Figure 28.2   Illustrations of Tatami Aw (top) and Bw (bottom) that added white paper.

quanta catch for both were the same, everywhere in the field of view. Whenever two images have identical quanta catches everywhere, they look the same. We worked hard to find a set of papers that all shifted the same amount. The reward was that by shutting off color constancy, we succeeded in making the control experiment. Now, we could experiment to see what new papers in the field of view would turn constancy back on.

28.4   New Maxima Restores Constancy

Figure 28.2 introduced a white band around the central patch. If the white influences the appearance of all colors in the field of view, then the corresponding areas in the new *Tatami Aw* and *Bw* should no longer match in the Shoeboxes. If this is true, then it shows that color constancy is the result of spatial comparisons.

If the fundamental determinant of color appearance is the quanta catch at a pixel, then the small white frame should have a only small effect on appearance. Except for whites, every other pixel in the field of view is identical in *TatamiA* and *Aw*; as well as *B* and *Bw*. Compare the influence of the white on Tatami A and B (Figure 28.1– right) with *Aw* and *Bw* (Figure 28.2). Introducing a white reflectance in both Tatami revived color constancy.

Two careful observations are important here:

> First, the whites in Aw and Bw do not look exactly the same. Aw looks reddish in the CC40R box, and coolish in the CC40C box. The influence of the illuminant shift is visible.

> Second, the two sets of five original papers look almost the same as they do in the room. They still have a reddish, or coolish, cast depending on the illumination.

Nevertheless, the striking conclusion is that the introduction of white to both displays brought color constancy back to life. (McCann, 1992a; b).

These results support the early Retinex mechanisms using calculations that reset to the maxima in each waveband (Land & McCann, 1971). As well, observers noted the changes in color appearance of the white papers. That observation supports the hypothesis that small appearance changes are due to changes of overall quanta catches. (Hipparchus line, Chapter 21; McCann, McKee & Taylor, 1976: McCann, 1987; 1989).

The changes in color appearances are consistent with the colors expected by normalizing each receptor set independently to a maximum reference. In other words, the colors observed are consistent with the Retinex model.

In summary, the rules for experimental color constancy are:

- Exact color constancy is achieved by exactly equal quanta catches everywhere in the field of view.
- Local changes in maxima cone quanta catch cause large appearance changes.
- In a complex field of view:
  - Whites show subtle global changes
  - The rest of the image shows dramatic local changes
- Colors demonstrate independent spatial processing of each waveband

28.5   Independent L, M, S Spatial Processing

The Retinex theory is sometimes described as independent three-channel normalization. This is a good simplified description, as long as the properties of complex spatial interactions described in Sections B & D are not lost. Neural spatial comparisons of scene content perform the normalization process. Replacing that process with normalization to a single pixel with the highest radiance is not an accurate description of Retinex. We will use the term *spatial normalization* to differentiate Retinex usage from the usual arithmetic process.

Further experiments tested the Retinex hypothesis that humans spatially normalize to the maximum in each waveband (McCann, 1992b). If the information from long-wave cones is spatially normalized to the maximum L-cone response, then white and pure red, yellow and magenta will have the same effect. They all have reflectances that equal white for long-wave light.

The same principle holds for middle-wave cones spatially normalized to the maximum M cone response. Then white and pure yellow, green and cyan will have the same effect. For S-cone response, then white, and pure cyan, blue and magenta will have the same effect. The following experiments tested whether any new maxima in the L, M, or S channel could restore constancy.

28.5.1   Center-Surround Maxima Experiment

In an attempt to simplify the manufacture and viewing of different targets, we changed the experimental design from the Maximov Shoe Box to simultaneous viewing of two displays, using an overhead projector. The intent was to be able to change the local contrast components just as we did by changing the reflectance displays in the Shoe Box. The experimental design uses two transparencies: one uniform to represent the *Illumination Component* and the second to represent the *Reflectance Component*. These two transparencies multiply transmittances in the same manner that illumination and reflectance multiply. However, replacing two targets mounted on a single sheet of transparent media is more efficient than replacing two papers in two shoe boxes. Furthermore, this apparatus allowed observers to view two different targets at the same time in a dark room. This greatly simplified the comparison of colors between targets.

Figure 28.3 describes the transparencies on the top of the overhead projector. The bottom layer illustrates the glass on the top (image plane) of the projector. The second layer illustrates the left (pale-yellow white) and right (pale-blue white) uniform *Illumination Component*. The third layer shows the

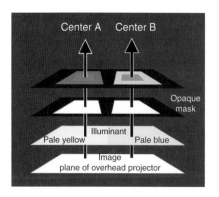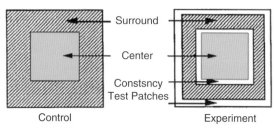

Figure 28.3 Diagrams of: (left) transparent layers used to make target; (right) spatial arrangement of Center, Surround and Constancy Test Patches.

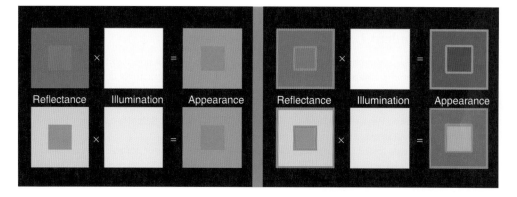

Figure 28.4 (left) Control target that shuts off constancy; (right) Constancy Test Patch that destroys the match.

opaque mask that restricts the field of view to just the areas of interest. The top layer shows the *Reflectance Component* that contains the local contrast information. The right of the diagram shows the relative size and location of Center, Surround and Constancy Test Patches.

In all the Center-Surround experiments the illumination and mask layers are fixed. Only the top layer representing the *Reflectance Component* is variable. Thus, changing the top transparency is equivalent to changing both Tatami in two Maximov boxes. The transparencies were made on the Canon CLC 500 laser printer under digital control. Since the dye sets are the same, it is possible to make convenient densitometer measurements while adjusting digits in the printer. All final measurements were confirmed using a colorimeter.

Two observers sat 3 meters from the projection screen. The control target was a Center square subtending 6 degrees in the middle of a Surround band of 12 degrees. The two Center Surround Targets were separated by 5 degrees.

28.5.2   Center-Surround Target – Control

The left side of Figure 28.4 illustrates Center-Surround A and B as they appear projected by themselves. On the far left are the *Reflectances Components*. The A Surround is gray; the A Center is a high-chroma blue (top). The B Surround is yellow; the B Center is gray. In the middle, the A Illumination is pale-yellow; the B Illumination is pale-blue (bottom). The right illustrates that the appearance of A and B are the same. They look the same because they are identical quanta catches everywhere in the field of view.

28.5.3   Center-Surround Target – Constancy Test Patches

The right side of Figure 28.4 illustrates the combination of the *Reflectance and Illumination Components*, this time with the addition of the *Constancy Test Patches*. This experiment adds new transmittances to the local *Reflectance Component* transparency. It changes the Local Contrast information. In this case, the addition of the new areas destroys the match, as the white did for the Tatami.

28.6   Model Predictions

The introduction of new radiances, or *Constancy Test Patches*, has different implications for different models of color vision. For example, let us compare the CIE models of colorimetry (CIE, 1931; Hunt, 2004; Moroney et al., 2002) and the Retinex model (Chapter 27). The fundamental difference in the models is input image data used in the calculation. Colorimetry makes measurements of single pixels in real-life complex scenes. Each pixel is independent of the other pixels. The Retinex model is a *spatial comparison* model.

28.6.1   Colorimetry Prediction

Colorimetry models predict the following results:

- Adding new values in other parts of the image does not enter into the calculation of X, Y, Z, or x, y.
- Pairs of pixels that match should continue to match, regardless of the presence of other areas.
- New patches will take on colors appropriate for their quanta catch at that pixel.

28.6.2   Retinex Prediction

The Retinex model predicts the following results:

- Adding new values in other parts of the image could change the appearance of all pixels in the image.
- Matches should continue to match, only if the new patch does not introduce a new maximum in any waveband.
- Matches should change, if the new patch introduces a new maximum.
- New patches will take on colors appropriate for their relationship to other pixels. The introduction a new maxima for any one of the three color channels will reset the color appearance of the entire field.

The pixel-based Colorimetry and spatial-based Retinex predictions are very different. Constancy Test Patches measure the effects of maxima on the entire image. Furthermore, they test whether the influence of new maxima operates in each color channel independently.

28.7   Center-Surround Target – Results

In the first part observers were shown the Control Center-Surround Target and asked if both left and right Centers and left and right Surrounds matched. They reported that they did. Next, the experimenter

Table 28.2 Effect of 16 constancy test patches added to the Reflectance component. The table identifies whether each test patch introduced a new maximum, and whether the observers reported that
constancy was restored.

	Constancy Test Patch	New L max	New M max	New S max	Restore Constancy
1	White	YES	YES	YES	YES
2	Gray	no	no	no	no
3	Black	no	no	no	no
4	Dark yellow	no	no	no	no
5	Bright Yellow	YES	YES	no	YES
6	Dark Magenta	no	no	no	no
7	Bright Magenta	YES	no	YES	YES
8	Dark Cyan	no	no	no	no
9	Bright Cyan	no	YES	YES	YES
10	Dark Red	no	no	no	no
11	Bright Red	YES	no	no	YES
12	Dark green	no	no	no	no
13	Bright Green	no	YES	no	YES
14	Dark Blue	no	no	no	no
15	Bright Blue	no	no	YES	YES
16	White & Black	YES	YES	YES	YES

replaced sequentially the control *Reflectances components* transparency with each of the 16 targets with
the *Constancy Test Patches*. The results are shown in Table 28.2 above. It lists the name of the
Constancy Test Patch in the left column. The next three columns report whether the Constancy Test
Patch introduced new maxima in any of the L, M or S reflectances. In other words, it reports whether
the new Test Patch has sufficient radiance to become the maximum radiance area in the field of view,
with the highest quanta catch of one of the L, M, S cones. The rightmost column lists the observers'
reports whether the new patches restored color constancy, destroying the matches of the two Centers
and the two Surrounds.

Sixteen Constancy Test Patches were used. Eight patches destroyed the color match; eight did not.
Whenever the match was destroyed for the Center, it also was destroyed for the surround. The colors
that destroyed the match caused a reset of all the colors in the field of view. All Constancy Test Patches
that destroyed the match introduced a new maximum in one of the color channels. That means that a
new maxima in any channel resets the color appearances and restored color constancy.

The combined White and Black patches were added at the suggestion of Andrew Moore in order
to test a *GrayWorld* assumption. Here, we have a set of white-black-white stripes that have the same

average radiance of a dark gray that fails to destroy the match. Do these white-black-white stripes have the same effect as a white, destroying the match? Or, does it behave as a dark gray, not destroying the match? As shown in Chapter 27, experiments designed to find *GrayWorld* effects fail to find them. The presence of local contrasts that reset the relationships controls color appearance. The white-black-white stripe behaves the same as the white Constancy Test Patch. (McCann 1992b).

The results are quite simple. If the Constancy Test Patches are not the highest quanta catch in any waveband, the color match is unchanged. Nothing happens. If the Constancy Test Patches are the highest quanta catch in any waveband, the color matches of both the center and the surround are destroyed. The bright red, green and blue Constancy Test Patches introduce a maximum for only one of the cone types. The yellow, magenta and cyan Constancy Test Patches introduce new maxima for two cone types. The white Constancy Test Patch introduces maxima for all three. In all of these cases, the color matches of the identical quanta catches are destroyed. The introduction of any new maximum causes a reset of color appearance. The introduction of any new maximum turns on the color constancy, or match destroying, mechanism. It follows that the mechanism controlling color constancy uses the individual maxima in each wave band to calculate color sensations. This is the result predicted by Retinex theory.

28.8   Summary

This is a two-part study of the human visual system's mechanism for normalization in color constancy. By combining the Tatami and the Constancy Test Patch experiments we can draw a number of important conclusions about the human color-constancy mechanism.

Exact color constancy is achieved by exactly equal quanta catches everywhere in the field of view. The introduction of global changes in quanta catch causes small changes. This is very different from local changes in quanta catch that cause large appearance changes.

The human color constancy mechanism spatially normalizes sensations to the maxima in the field of view. The addition of two white papers destroys the match of all of the areas in both Tatami. Each Tatami looks much closer to its appearance in the room. The introduction of the white area has destroyed the color match and has initiated the color constancy mechanism.

The conclusions in this chapter are very similar to those made in the study of lightness in Chapter 21. Maxima play a special role in appearance. Image segments that are darker than maxima behave in a different manner. The present observations about color follow all the same rules as found in Section D in each of the independent color channels.

Color constancy spatially normalizes sensations to the maxima in the field of view; it does this for each waveband separately (Retinex). Constancy Test Patches in Center-Surround targets showed that the introduction of any patch with a new maximum quanta catch for any cone causes a reset of color appearance. The introduction of any new maxima turns on the color-constancy mechanism. It follows that the mechanism controlling color constancy uses the individual maxima in each wave band to calculate color sensations.

A word of caution is needed here. We have used the term *spatially normalize* to describe the properties of a biological spatial-comparison – a neural mechanism. This usage of *spatially normalize* should not be confused with the mathematical usage of *normalize* to mean scaling all values in a set by the single pixel with maximum value. Mathematical normalization assumes that each member of the set will be treated exactly the same as all the others. This is where the term *normalize* as applied to human visual processing gets into trouble. As we have seen in Section D, maximal areas do not have exactly the same effect on every other pixel in the field of view. The normalization we observe here is responsive to many maxima and near maxima distributed throughout the scene. It is affected by distances between maxima and the area of interest.

28.9   References

Canon (1989) http://www.canon.com/about/history/04.html

CIE Proceedings (1931) Cambridge University Press, Cambridge.

Hunt RWG (2004) *The Reproduction of Colour*, IS&T, John Wiley & Sons, Ltd, Chichester.

Land E & McCann J (1971) Lightness and Retinex Theory, *J Opt Soc Am*, **61**, 1–11.

Maximov V (1989) Personal communication.

McCann JJ (1987) Local-Global Mechanisms for Color Constancy, *Die Farbe*, Florence: *Proc. AIC*, **15**, 275–83.

McCann JJ (1989) The Role of Nonlinear Operations in Modeling Human Color Sensations, SPIE Proc. 1077, 355–63.

McCann JJ (1992a) Rules for Colour Constancy, *Ophthal Physiol Opt*, **12**, 175–77.

McCann JJ (1992b) Color Constancy: Small Overall and Large Local Changes, Human Vision, Visual Processing, and Digital Display III, San Jose, California: *SPIE.*, **1666**, 310–21.

McCann JJ, McKee S & Taylor T (1976) Quantitative Studies in Retinex theory, A comparison between theoretical predictions and observer responses to Color Mondrian experiments, *Vision Res.*, **16**, 445–58.

Moroney N, Fairchild M, Hunt RWG, Li C, Luo M, Newman T (2002) The CIECAM02 Color Appearance Model, *Proc. IS&T/SID Color Imaging Conference*, **10**, 23–27.

Stork D (1976) MIT Undergraduate Thesis.

29

HDR and 3-D Mondrians

29.1 Topics

We studied color constancy of two 3-D Color Mondrian displays made of two identical sets of painted wooden shapes. We used only 6-chromatic, and 5-achromatic paints applied to 100+ block facets. The three-dimensional targets add shadows and multiple reflections not found in flat Mondrians. Observers viewed one set in nearly uniform illumination [Low-Dynamic-Range (LDR)]; the other in directional non-uniform illumination [High-Dynamic-Range (HDR)]. Both 3-D Mondrians were viewed simultaneously, side-by-side. We used two techniques to measure how well the appearances correlated with the objects' reflectances. First, observers made magnitude estimates of changes in the appearances of surfaces having identical reflectances. Second, a colleague painted a reproduction of the pair of Mondrians using watercolors. We measured the watercolor reflectances to quantify the changes in appearances. Both measurements give us important data on how reflectance and illumination affect color constancy. While universal constancy generalizations about illumination and reflectance hold for flat Mondrians, they do not for 3-D Mondrians. A constant paint does not exhibit perfect color constancy, but rather shows significant shifts in lightness, hue and chroma in response to nonuniform illumination. The results show that appearance depends on the spatial information in both the illuminations and reflectances of objects.

29.2 Color Constancy and Appearance

The psychophysics of color constancy has been studied for nearly 150 years. In fact, there are a number of distinct scientific problems incorporated in the field. These studies ask observers distinctly different questions, and get answers that superficially seem to be contradictory. The computational models of color constancy for colorimetry, sensation, and perception are good examples. The Optical Society of America used a pair of definitions for *sensation* and *perception* that followed along the ideas of the Scottish philosopher Thomas Reid. Sensation is "Mode of mental functioning that is directly associated with the stimulation of the organism" (OSA, 1953). Perception is more complex, and involves past experience. Perception includes recognition of the object. It is helpful to compare and contrast these terms in a single image to establish our vocabulary as we progress from 18th century philosophy to 21st century spatial image processing. Figure 29.1 is a photograph of a raft – a swimming float – in the middle of Mascoma Lake (McCann and Houston 1983a, McCann 2000). The photograph was taken

The Art and Science of HDR Imaging, First Edition. John J. McCann, Alessandro Rizzi.
© 2012 John Wiley & Sons, Ltd. Published 2012 by John Wiley & Sons, Ltd.

Figure 29.1. Photograph of raft.

in early morning: the sunlight fell on one face of the raft, while the skylight illuminated the other face. The sunlit side reflected about 10 times more 3000°K light than the 20 000°K skylight side. The two faces had very different radiances, and hence very different colorimetric values.

For *sensations*, observers selected colors they saw from a lexicon of color samples, such as the Munsell Book or the catalog of paint samples from a hardware store. The question for observers was to find the paint sample that a fine-arts painter would use to make a realistic rendition of the scene. Observers said that a bright white with a touch of yellow looked like the sunlit side of the raft, and a light gray with a touch of blue looked like the skylit side. The answer to the *sensation question* was that the two faces were similar, but different.

For the perceptions, observers selected the colors from the same catalog of paint samples, but with a different question. The perception question was to find the paint sample that a house painter would use to repaint the raft the same color. Observers selected white paint. They recognized that the paint on the two sides of the raft is the same despite different illuminations. The perception question rendered the two faces identical.

In summary, the raft faces are very different, or similar, or identical depending on whether the experimenter is measuring colorimetry, or sensation, or perception. Following discussions of the raft picture, subsequent experiments asked the same question, using a slightly different vocabulary (Arend & Goldstein, 1987. They found the same result. Namely, observer's responses depended on the observers' psychophysical task.

We need completely different kinds of image processing in order to model these three questions. Colorimetry models predict the receptor quanta catch (David Wright in Chapter 2); *sensation* models predict the color appearance that goes beyond light absorption; and *perception* models predict the observer's estimate of the object's surface using cognitive processes.

29.3   Color Constancy Models

Human color constancy involves the content of the scene. It depends on the reflectances of objects, the spectral content and the spatial distribution of the illumination, and the arrangement of the scene. There are a number of models of color constancy used to predict colors from the array of radiances coming to the eye, or the camera. They not only use a variety of image processing assumptions, they also have different sets of required information, and different goals for the model to calculate.

Table 29.1 lists four approaches to color constancy that have different assumptions and solve different problems. The interdisciplinary study in this chapter helps us to understand these model's strengths and weaknesses.

Table 29.1 Four classes of color constancy models. The columns list the names and the goals of the model's calculation; the information required to do the calculation; and references.

Color Constancy: 4 Types of Models

Model	Calculation Goal [Output]	Given Information [Model Input]	Reference
Retinex	appearance (sensation)	radiance array of entire scene	Land & McCann, 1971
Discount Illumination CIELAB and CIECAM	appearance (sensation)	pixel's radiance + pixel's irradiance	CIELAB, 1978 CIECAM, 2004
Computer Vision	reflectance	radiance array of entire scene	Ebner, 2007
Surface Perception	reflectance perception	radiance array + adaptation	Brainard & Maloney, 2004

- **Retinex** uses the entire scene as input and calculates color appearances by making spatial comparisons across the entire scene. Retinex models of human vision calculate appearance.
- **CIELAB and CIECAM** models predict single pixel appearances by discounting the illumination. It requires measurements of the light coming from the scene, and the light falling on the scene. These models calculate the *physical reflectance* of the object and scale it in relation to white. CIE models calculate *scaled physical reflectances*.
- **Computer Vision** models work to remove the illumination measurement limitation found in CIE colorimetric standards by calculating illumination from scene data. The image processing community has adopted this approach to derive the illumination from the array of all radiances coming to the camera, and discounting it. Computer vision calculates physical reflectance.
- **Surface Perception** algorithms study and model the observer's ability to recognize the surface of objects. Following Hering's (1905) concern that chalk should not be mistaken for coal, the objective is to predict human ability to recognize the paint on an object's surface. Surface perception calculates human estimates of reflectance, rather than physical reflectance. It does not calculate the surface's appearance.

There is an extensive literature on each of the above types of color constancy model. All are reviewed in detail in the full "3-D Mondrian CREATE Report" (2010).

All four models listed in Table 29.1 do well with their predictions in the flat, uniformly illuminated Color Mondrian. Under those special case restrictions, color appearance correlates with *Scaled Integrated Reflectance* (Chapter 27.3). In other words, "discounting the illuminant" models can predict appearances in flat, uniformly illuminated Mondrians. The experiments in this chapter measure whether such models are appropriate for complex 3-D scenes. Further, we study whether image processing algorithms should attempt to reproduce scenes using calculated physical reflectances.

29.4 Measuring Changes in Appearance from Changes in Illumination

The experiments in this Chapter are designed to study the interplay of reflectance, illumination and scenes' spatial content in human color appearance (sensation). Figure 29.2 shows photographs: the Low-Dynamic Range (LDR) Mondrian in nearly uniform illumination (left); the HDR version with two directional illuminants falling on identical wooden blocks (right).

We replaced the flat array of color papers used in Land's Mondrian (Land and McCann, 1971) with a collection of three-dimensional painted blocks. We replaced the 100 plus color papers used in Land's Mondrian with 11 reflectances: six chromatic and five achromatic. We replaced the spatially uniform illumination with a pair of different illuminant configurations: one as uniform as possible, and the other

Figure 29.2   (left) Low-dynamic-range (LDR) scene, (right) High-dynamic-range (HDR) scene.

Name	Munsell	Y	x	y
R	2.5R 4/14	43.2	.631	.335
Y	2.5Y 8/14	105.0	.553	.429
G	10GY 5/12	29.7	.370	.547
C	10BG 6/8	39.9	.350	.434
B	5PB 4/12	20.6	.351	.350
M	10P 6/10	65.1	.520	.359
W	N 10/	156	.464	.417
G7.5	N 7.5/	76.3	.456	.418
G6	N 6/	44.5	.456	.418
G4	N 4/	21.2	.452	.416
K	N l/	9.12	.462	.413

Figure 29.3   Flat painted test target; paint designations, Munsell designation and Y,x,y values.

highly directional and with different emission spectra. The uniform illumination has a low dynamic range (LDR), while the directional one has a high dynamic range (HDR) (Parraman et. al. 2009). While Land used many reflectances to make a complex array of radiances, we used the shadows and gradients created by the 3-D objects to generate complexity. The three-dimensional nature of these test targets adds shadows and multiple reflections. These properties enrich the targets and make them more like real scenes. Here we measure the effects of illumination on constant reflectances. Human vision models can take advantage of this data to assess how well their predictions match appearances.

29.4.1   Two Identical 3-D Mondrians

The experiment used two identical sets of objects in uniform and in non-uniform illumination in the same room at the same time. We painted each of the flat surfaces with one of 11 different paints [R, Y, G, C, B, M, W, NL (neutral gray G7.5), NM (neutral gray G6), ND (neutral gray G4), K) (Figure 29.3).

29.4.2   Characterization of LDR and HDR Illuminations

Above, in Figure 29.2 (left) we see a photograph of the LDR Mondrian in illumination that was as uniform as possible. The blocks were placed in an illumination cube. It had a white floor, translucent top and sides, and a black background. We directed eight halide spotlights on the sides and top of the illumination cube. The combination of multiple lamps with the same emission spectra, light-scattering cloth and highly reflective walls made the illumination nearly uniform. Departures from perfect uniformity came from shadows cast by the 3-D objects, and the open front of the cube for viewing.

Figure 29.2 (right) is a photograph of the HDR Color Mondrian illuminated by two different lights. One was a 150W tungsten spotlight placed to the right side of the 3-D Mondrian at the same elevation as the blocks. It was placed 2 meters from the center of the target. The second light was an array of WLEDs assembled in a flashlight. It stood vertically and was placed quite close (20 cm) on the left. Although both are considered variants of white light they have different emission spectra. The placement of these lamps produced highly non-uniform illumination and increased the dynamic range of the scene (McCann, et. al., 2009a, 2009b).

In the HDR 3-D Mondrian, the black back wall had a 10 cm circular hole cut in it. Behind the hole was a small chamber with a second black wall 10 cm behind the other. We placed the flat circular test target on the back wall of the chamber. The angle of the spotlight was selected so that no direct light fell on the circular target. That target was illuminated by light reflected from the dark walls of the chamber. The target in the chamber had significantly less illumination than the same paints on the wooden blocks. The target in the chamber significantly increased the range of the non-uniform display. However, human observers had no difficulty seeing the darker circular target.

One way of assessing the uniformity of illumination is to make a third set of blocks, all painted middle gray. We photographed this actual *GrayWorld* in the LDR lighting geometry (Figure 29.4, left). It shows that with 3-D objects uniform illumination is extremely difficult to achieve. Despite the use of eight light sources and light diffusers, the three-dimensional objects cast faint shadows on other block faces. Perfectly uniform illumination requires that the object be in the center of a perfect integrating sphere. The *GrayWorld* under HDR lighting geometry (Figure 29.4, right) shows a much wider range of luminances and apparent lightnesses.

Figure 29.4   (left) LDR GrayWorld 3-D Mondrian; (right) HDR. The same gray Mondrian blocks show that directional illumination can change a pixel's radiance and appearance. The camera digits for gray blocks in HDR illumination vary from 210 to 3.

29.5   Magnitude Estimation Appearance Measurements

Observers compared the HDR and LDR Mondrians (McCann, Parraman & Rizzi 2009a, 2009b). They were given a four-page form that identified a selection of 75 areas in the displays. The observers were shown the painted circular test target (Figure 29.2 left) placed on the floor of the display, in uniform light. This standard was explained to be the appearance of "ground truth". They were told that all the flat surfaces had the same paints as the standard.

Observers were asked if the selected areas had the same appearance as "ground truth". If not, they were asked to identify the direction and magnitude of the change in appearance. The observers recorded the estimates on the forms. Observers were asked to estimate hue changes starting from each of the six patches of colors [R, Y, G, C, B, M] (Figure 29.5, left). Participants were asked to consider the change in the hue as a percentage difference between the original hue, e.g. R, and the hue direction Y. For example, 50 % Y indicates a hue shift to a color halfway between R [Munsell 2.5R] and Y [Munsell 2.5Y] (Figure 29. 5, left). 50 % Y is Munsell 2.5YR. 100 % Y meant a complete shift of hue to Y.

Observers estimated lightness differences on a Munsell-like scale indicating either increments or decrements, for the apparent lightness value (Figure 29.5 center). Observers estimated chroma by assigning paint sample estimates relative to 100 % (Figure 29.5, right). In case the target patch appears more saturated than "ground truth", estimates can overtake 100 %.

We measured the Munsell Notation of chips of the 11 painted "ground truth" samples, by placing the Munsell chips on top of the paint samples in daylight. We know the direction and magnitude of changes in appearance from observer data. We made linear estimations to calculate the Munsell desig-nation of the matching Munsell chip for each area. We used the distance in the Munsell Book as described in the MLAB color space, (Marcu, 1998; McCann, 1999) as the measure of change in appear-ance. We assumed that the Munsell Book of Color is, as intended, equally spaced in color. MLAB converts the Munsell designations to a format similar to CIELAB, but avoids its large departures from uniform spacing (McCann, 1999). When the observer reports no change in appearance from illumina-tion MLAB distance is zero. A change as large as white to black (Munsell 10/ to Munsell 1/) is MLAB distance of 90.

The results (Figure 29.6) show the average of 11 observers' magnitude estimates of the selected areas in the pair of 3-D Mondrians. If appearance correlates with physical reflectance, then all facet data should fall on the bottoms of the graphs. The plots here show departures from this prediction for five facets in six colors of paint. In general, observed appearances in the LDR portion are closer to "reflect-ance ground truth" than those in the HDR. However, there are substantial departures from all paints with the same reflectance looking the same appearance.

(See 3-D Mondrian MagEst Data (2010) for all matching data.)

There are areas in the HDR scene that look like the ground truth standard colors. The change in appearance of individual areas depends on the illumination and the other areas in the scene. The sources

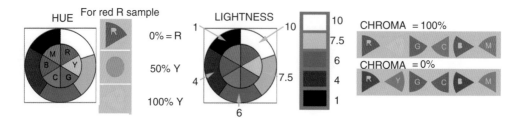

Figure 29.5   (left) shows the ground truth reflectance samples and illustrates the strategy for magni-tude estimation of hue shifts; (center) lightnesses; (right) chroma.

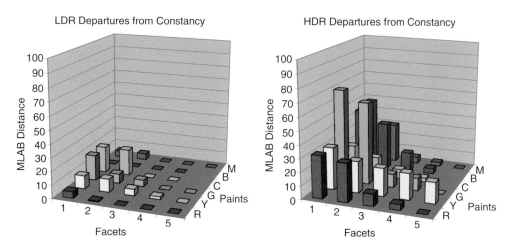

Figure 29.6   Bar graphs of sizes of departures from perfect constancy in LDR and HDR.

of illumination, the distribution of illumination, and inter-reflections of light from one facet to another, all play a part in generating appearance. One cannot generalize the influence of the surface property (reflectance) of the facet on appearance. Illumination, and all of its spatial properties, show significant influence on the hue, lightness and chroma of observed appearances.

29.6   Watercolor Rendition Measurements of Appearance

Fechner measured middle-gray lightness by asking an artist to paint it. After observers finished the Magnitude Estimations of Munsell designations, we left the pair of 3-D Mondrians in place. One of the paper's authors (Carinna Parraman) painted with watercolors on paper a rendition of both 3-D Mondrians (Parraman, et al., 2010). The painting took a considerable time. It reproduced as closely as possible the appearances in both displays.

Painters usually apply their particular "aesthetic rendering" that is a part of their personal style. In this case the painter worked to present on paper the most accurate reproduction of appearances possible. As with the magnitude estimation measurements, both LDR and HDR were viewed and painted together in the same room at the same time. Figure 29.7 is a photograph of the watercolor painting of the combined LDR/HDR scene. We made reflectance measurements of the watercolor with a Spectrolino® meter in the center of 101 areas. If the same paint in the scene appeared the same to the artist, then all paintings' spectra for this area should superimpose. They do not. The artist selected many different spectra to match the same paint on a number of blocks. The artist selected a narrower range of watercolor reflectances to reproduce the LDR scene. Many more paint colors are needed to reproduce the HDR scene. Nevertheless, some block facets appeared the same as ground truth, while others showed large departures.

We measured the reflectance spectra of both LDR and HDR paintings at each of the 101 locations using a Spectrolino® meter. The meter reads 36 spectral bands, 10 nm apart over the range of 380 to 730 nm. We calibrated the meter using a standard reflectance tile. The average reflectance for all wavebands for the all LDR samples is 39.9%; for HDR 32.2%. We scaled % luminance, and reflectance measurements using L* to correct for intaocular scatter and lightness appearance (Chapter 18).

Figure 29.7. Photograph of the painted watercolor of the LDR and HDR Mondrians. Reproduced by permission of Carinna Parraman

29.6.1   Watercolor Reflectances Measurements

We measured the watercolor spectra for all reproductions for all painted blocks in both the LDR and HDR scenes. The complete data on measurements of block reflectance, scene radiances, watercolor reflectances and observer magnitude estimates can be found with the complete report (see Wiley web site http://www.wiley.com/go/mccannhdr).

As before, we noted particular block facets where apparent watercolor reflectances were very similar to actual reflectances. That was true of all paints, achromatic and color. As well, there were many facets where watercolor reflectance was markedly different from actual reflectance. The HDR reproduction had more distinctly different measured reflectances, than the LDR painting. It is important to note that these differences are a result of the position of the blocks and their illumination, and are not related to the blocks' paint color. These results are in close agreement with the magnitude estimate data reported above in 29.5. Figures 29.8 and 29.9 provide some examples of observations about appearances of selected painted surfaces in the Mondrians.

29.6.2   Illumination Affects Lightness

Carinna Parraman's watercolor painting of the side-by-side LDR and HDR 3-D Mondrians, along with the information about the paint reflectances on the blocks, helps us to evaluate different computational models.

Figure 29.8 compares the LDR and HDR painting reflectances for five achromatic value blocks. We see a complex pattern of departures from perfect color constancy with significant departures caused by the specific illumination pattern.

First, by having an artist render the appearances of the LDR/HDR scene we represent appearance in the same easily measurable physical space as the paint on the blocks. The artist's rendition converts the high-dynamic range, caused by illumination, into the set of appearances expressed in the low-dynamic range of the watercolor. The conveniently measured watercolor reflectance is a measure of appearance. These measurements are ideal for evaluating computational algorithms.

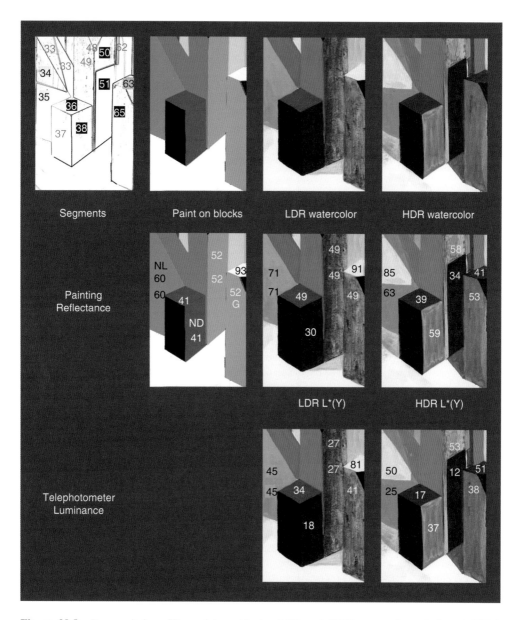

Figure 29.8 (top row) Area IDs; paint on blocks; LDR and HDR watercolor painting. (middle) Spectrolino® watercolor L*(Y); (bottom) telephotometer L*(Y) values.

Figure 29.8 shows central areas surrounding a dark gray and black block. The captured appearances of the LDR and HDR watercolor renderings have different values from the same paint on the blocks. Areas 36 and 38 have the same Neutral-Dark (ND) paint. The paint on the block for both top and side is CIE L*(Y) 41.4. These constant surface reflectances have different appearances in the LDR and HDR portions of the watercolor. In LDR area 36, the top is lighter [L*(Y) = 49] than the side [L*(Y) = 30]. In HDR, the top is darker [39], than the side [59].

In the HDR the order of the appearances changes in the different illuminations. Area 38 is the lightest of the block's faces in the HDR (36, 37, 38), and nearly tied for darkest in the LDR. These changes in appearance correlate with the changes in edges caused by the different illuminations. The bottom row of Figure 29.8 shows the telephotometer scaled luminances L*(Y). The LDR area 36, the top, is lighter [34], than the side [18]. In HDR, the top is darker [17], than the side [37].

The areas in Figure 29.8 illustrate that edges formed by illumination cause substantial change in the appearance of surfaces with identical reflectances. The direction of the changes in appearance is consistent with the direction of changes in illumination on the blocks. Edges in illumination cause substantial changes in appearance. The data do not show correlation of appearance with luminance, rather it demonstrates that change in appearance correlates with change in luminance across edges in illumination.

29.6.3   Illumination Affects Chroma

Figure 29.9 shows a different section, right of center, of the LDR and HDR scenes. These scene portions have a tall white block face that is influenced by shadows and multiple reflections. The white paint has constant reflectance values [L*(X) = 93, L*(Y) = 93, L*(Z) = 92)] from top to bottom. The LDR appearances show light-middle-gray, and dark-middle-gray shadows.

The HDR appearances show four different appearances. The watercolor shows: white at the top, a blue-gray shadow below it, a pinker reflection and a yellow reflection below that. In HDR, shadows and multiple reflections show larger changes in appearance caused by different illumination and reflections.

Figure 29.9 shows sections of the watercolor for LDR (left), and HDR (right) in three sections. The top section reports the L*(X), L*(Y), L*(Z) Spectrolino reflectance measurements of the painting. These measurements report appearance. The middle section reports on the telephotometer readings of the light coming from the blocks. The bottom section reports on the average magnitude estimates of observers in ML, Ma, Mb units.

The photographs of the LDR/HDR scene (Figure 29.2) and the watercolor painting show that the white block in LDR has achromatic shadows. The measurements shown in Figure 29.9 (left) of watercolor reflectances, light from the different parts of the block, and magnitude estimates show very similar achromatic shifts in appearance. The measurement on the right side of the figure are also similar to each other and indicate a chromatic shift from the two light sources (Area 83) and from multiple reflections (Areas 84, 85). The changes in illumination of the single white block caused relatively sharp edges in the light coming to the eye. These abrupt changes caused observers to report change in chroma as measured by the watercolor and the magnitude estimates. This data supports the observation that the changes in appearance correlate with the change in radiance at these edges.

Both Figures 29.8 and 29.9 show significant inconsistencies between appearance and object reflectance. These discrepancies are examples of how the illumination plays an important role in color constancy of complex scenes. Both sets of measurements give very similar results. Both sets of measurements show that appearance depends on the spatial properties of illumination, as well as reflectance. Edges in illumination cause large changes in appearance, as do edges in reflectance.

29.7   Review of 3-D Mondrian Psychophysical Measurements

This series of experiments studies a very simple question. Can illumination change the appearances of blocks with the same surface reflectance? We found a complicated answer. There is no universal generalization or result, rather a wide range of distinct individual observations. In the LDR, illumination changes appearance some of the time. In the HDR, illumination changes appearance most of the time. Appearance depends on the objects in the scene, their placement, and the spatial properties of the illuminations. In these experiments we found no evidence to support the idea that illumination has

Watercolor L*(X), L*(Y), L*(Z)

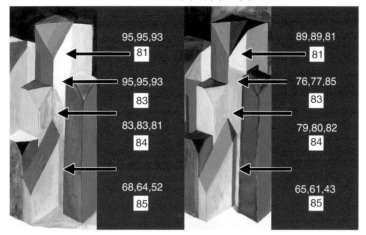

Telephotometer L*(X), L*(Y), L*(Z)

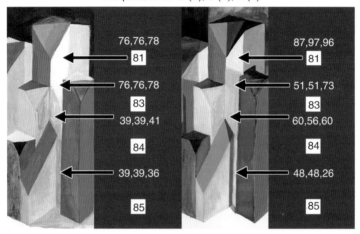

Magnitude Estimate (ML, Ma, Mb)

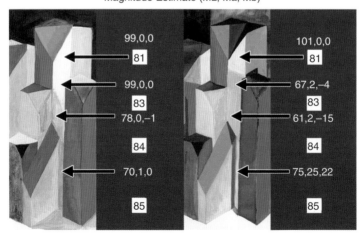

Figure 29.9 (left) LDR (right) HDR watercolor segment. Measurements of a single white block with Areas 81, 83, 84, 85. (top) watercolor reflectances [L*(X), L*(Y), L*(Z)]; (middle) photometer readings L*(X), L*(Y), L*(Z); (bottom) observed magnitude estimates [ML,Ma,Mb].

different properties from reflectance in forming appearances (*sensations*). Generalizations about illumination and reflectance do not fit the data.

29.7.1   Discount Illumination

Another simple question is whether observer data supports the *discount illumination* hypothesis. Hering (1905) observed that the process was approximate. The signature of the departures from perfect constancy provides important information about how human vision achieves constancy. The data here shows that in LDR/HDR Mondrians illumination alters the spatial information of the scene. Observer data correlated with spatial changes and not illumination measurements.

The present experiment varies the intensities of two similar white lights. Previous experiments described in Chapter 27 varied the amounts of long-, middle-, and short-wave illumination (27 different spectra) falling on a flat surface in uniform illumination. Those experiments measured the departures from perfect constancy. The results showed small changes in color appearances caused by illumination for highly colored papers, and no changes with achromatic papers. A "discounted illuminant" must have the same effect on all papers, but the signatures of the departures from perfect constancy do not support that hypothesis.

We did not find experimental evidence of "discounted illumination" when asking observers about appearances.

29.7.2   Real Paints and Lights

In the careful analysis of reflectance and illumination there is no room for errors and artifacts introduced by image capture and display technologies. In 1976 we began to study human vision using computer controlled complex image-displays (Frankle & McCann, 1983, McCann & Houston 1983a). Since then, we have been aware of the need for extensive calibration of electronic imaging devices (McCann & Houston, 1983b). For the experiments in this chapter, we chose to fabricate our test scene with real objects painted with exactly the same paints. We chose to use real light sources. We were able to measure the reflectance of each paint, the Y, x, y of the light coming from the surface, and the full spectra of the paints in the watercolor.

It is not a simple matter to verify the accuracy of an electronic display over its entire light-emitting surface, for all light levels, for its entire three dimensional 24-bit color space. The combination of reflectances (range = 100:1), and illuminations (range = 100:1) require great precision over a range of 10000:1. Rather than calculate the combined effects of reflectances and illumination for an image-dependent display device, we chose to use real lights and paints for this analysis (see McCann, et al., (2010) for a discussion of the problems in HDR display calibration).

29.8   Color Constancy Models

As one inspects the color appearances in the LDR and HDR watercolors, it is evident that these colors have a physical correlate. That correlate is not the XYZ values of a pixel. The correlate is the spatial relationship of XYZ values with different pixels. Darker regions of the same reflectance paint have edges created by the illumination. The appearances observed here are consistent with a model that builds colors from image structure.

If we return to the four computational models we discussed in the introduction, and add the properties of human vision, we can evaluate how they apply to these experiments, and their possible applications to image reproduction. These five approaches to color constancy have different assumptions and

Table 29.2   Five types of color constancy models. The columns list the names and goals of the model's calculation; the constancy mechanism; and whether the model <u>must</u> render physical reflectance as the output.

Color Constancy: 5 Types of Models

Model	Calculation Goal [Output]	Mechanism	[Output] = Reference
Human Vision	appearance (sensation)	visual pathway	depends on edges
Retinex	appearance (sensation)	build appearance from edges & gradients	depends on edges
Discount Illumination CIELAB & CIECAM	appearance (sensation)	measure reflectance stretch	always
Computer Vision	reflectance	estimate illumination to calculate surface	always
Surface Perception	reflectance perception	cues, local adaptation, Bayesian inference	depends on edges, adaptation & inference

solve different problems. Table 29.2 identifies the types of imaging problems that are appropriate for each type of model.

- **Retinex** uses the entire scene as input and calculates color appearances by making spatial comparisons across the entire scene. The McCann, McKee, and Taylor (1976) data show that flat Mondrian appearances correlated with Scaled Integrated Reflectances. This measured with human cone sensitivities, in uniform illumination, and calculated *integrated reflectances* by spatial comparisons. The intent was to show that it was possible to calculate those *integrated reflectances* without ever finding the illumination. Integrated reflectances are different from physical reflectances because of cone crosstalk.

  If one applies a spatial model to these 3-D Mondrians we would not calculate the paints' *physical reflectances*. Instead we would get a rendition of the scene that treated edges in illumination the same as edges in reflectance. The Retinex spatial model shows a correlation with *scaled integrated reflectance* sometimes, (in flat Mondrians), but not all the time (in 3-D Mondrians). Retinex models of human vision calculate appearance and can be applied to all images. Incorporating a model for vision in a reproduction system provides the much needed dynamic range compression of HDR scenes into LDR media (Chapter 32).

- **CIELAB and CIECAM** models predict single pixels' appearance by discounting the illumination. It requires measurements of the light coming from the scene, and the light falling on the scene. This model calculates the reflectance of the object and scales it in relation to white. CIELAB/CIECAM models measure the X, Y, Z reflectances of individual pixels and transform them into a new color space. There is nothing in the calculation that can generate different outputs from identical reflectance inputs, as is frequently observed in color appearances in 3-D scenes. These models predict the same color appearance for all blocks with the same reflectance. While useful in analyzing appearances of flat scenes such as printed test targets, it does not predict appearance with shadows and multiple reflections. CIE models calculate scaled reflectance and can be used to predict appearance only in uniform illumination. Incorporating CIE models in image reproduction systems can reduce metamerism problems, and other issues regarding the response of the cones (David Wright, Chapter 2). However, they cannot predict appearances in complex HDR scenes with nonuniform illumination.

- **Computer Vision** models work to remove the illumination measurement limitation found in CIE colorimetric standards by calculating illumination from scene data. The image processing community has adopted this approach to derive the illumination from the array of all radiances coming to the camera. Computer Vision has the specific goal of calculating the object's reflectance. The question here is whether such material recognition models have relevance to vision. If a computer vision algorithm calculated correctly cone reflectances of the flat Mondrians, then one might argue that such processes could happen in human vision (Ebner, 2007). Ebner cites McCann, McKee, and Taylor as the basis of his *appearance equals reflectance assumption* used throughout the book (page 2). Ebner says "Human color perception correlates with integrated reflectance (McCann et al., 1976)". The problem is that his interpretation of our work did not include spatial comparisons and crosstalk. Ebner and colleague's *physical integrated reflectance* (shown in Figure 27.6 left), is markedly different from McCann et al.'s *spatial integrated reflectance* (Figure 27.6 right) (Chapter 27).

  However, the 3-D Mondrians described above, and other experiments, show that illumination affects the observers' responses (Rutherford and Brainard, 2002; Yang & Shevell, 2003; McCann, 2004; 2005). If that same computer vision algorithm calculated 3-D Mondrian reflectances correctly, then these calculations are not modeling appearances. Computer Vision is a discipline distinct from human vision, with very different objectives. Computer vision calculates *physical reflectances*. Incorporating Computer Vision models in image reproduction systems can reduce dynamic range, but, if successful, they remove all traces of illumination. Most photographers believe that illumination is the most important component of aesthetic rendering.

- **Surface Perception** has the specific goal of calculating the observers' estimate of the object's reflectance. We did not ask the observers to guess the reflectance of the facets in these experiments. We told observers that the blocks had only 11 paints, identified in the color wheel. We asked them to estimate the appearances of the facets.

  If asked, observers are likely to get very high correlations of appearance with physical reflectance in the LDR because the 11 paint samples were so different from each other. However, in the HDR illumination, we would expect that there would be more confusion, as shown by the appearances in Figures 29.8 and 29.9.

  Modeling surface perception is a distinct field from measuring the appearances (sensations) in complex 3-D scenes. Since observers give different responses to the sensation and perception questions, surface perception models must have different properties from sensation (appearance) models (McCann & Houston, 1983a; Arend & Goldstein, 1987). Surface perception calculates human estimates of physical reflectance. A different set of experiments is necessary to measure human estimates of reflectances. Such models are not appropriate for this chapter's appearance data. We asked the observers and the painter to report on the colors they saw.

Humans exhibit color constancy using scene radiances as input. The appearances they see are influenced by the spatial information in both illumination and reflectance. Measuring, or calculating *physical reflectance*, is insufficient as a model of visual appearances of real complex scenes.

29.9 Conclusions

This chapter's experiments used two identical arrays of 3-D objects in uniform (LDR) and non-uniform (HDR) illumination. They were viewed in the same room at the same time. All flat facet objects have been painted with one out of a set of 11 paints. We used two different techniques to measure the appearances to observers of these constant reflectance paints. First, we recorded observer magnitude estimates of change in Munsell Notation; second, we measured an artist's watercolor rendition of both scenes. Both magnitude estimates and watercolor reflectances showed the same results. There is no general rule, based on illumination and reflectance, to describe the observations. Rather, we measured a great many individual departures from perfect constancy. In nearly uniform illumination more samples appear the same as *ground truth* than in complex HDR illumination. Even small departures from perfectly uniform illumination generate departures in appearances from physical reflectance. If an image-processing algorithm *discounted the illumination*, and succeeded in calculating accurately objects' *physical reflectances*, then that algorithm would not predict appearances of real-life scenes with complex non-uniform illumination. Edges and gradients in illumination behave the same as edges and gradients in reflectance.

29.10   References

3-D Mondrian CREATE Report (2010) http://web.me.com/mccanns/McCannImaging/3-D_Mondrians.html

3-D Mondrian MagEst Data (2010) http://web.me.com/mccanns/McCannImaging/3-D_Mondrians_files/MagEstApp.pdf

Arend L & Goldstein R (1987) Simultaneous Constancy, Lightness, and Brightness, *J Opt Soc Am A*, **4**(12), 2281–85.

Brainard DH & Maloney LT (2004) Perception of Color and Material Properties in Complex Scenes, *J. Vision*, **4**(9), i, ii–iv, http://journalofvision.org/4/9/i/.

CIE (Commission Internationale de l'Eclairage) (1978) Publication 15(E-1.31) 1972, Bureau Central de la CIE, Paris. Recommendations on Uniform Color Spaces, Color-Difference Equations, Psychometic Color Terms, Supplement No. 2.

CIE (Commission Internationale de l'Eclairage) (2004). Publication 159:2004, A colour appearance model for colour management.

Ebner M, (2007) *Color Constancy*, Wiley/IS&T, Chichester, Chapters 6 & 7.

Frankle J & McCann J (1983) Method and Apparatus of Lightness Imaging, US Patent 4,384,336, Issue date 5/17/1983.

Hering E (1905) *Outline of a Theory of Light Sense*. ed. Hurvich & Jameson, Trans., Harvard University Press, Cambridge.

Land E & McCann J (1971) Lightness and Retinex Theory. *Journal of the Optical Society of America*, **61**(1) 1–11.

Marcu G (1998) Gamut Mapping in Munsell Constant Hue Sections, Proc. 6th IS&T/SID Color Imaging Conference, Scottsdale, Arizona, 159–62.

McCann JJ (1999) Color Spaces for Color Mapping, *J Electronic Imaging*, **8**, 354–64.

McCann J (2000) *Simultaneous Contrast and Color Constancy: Signatures of Human Image Processing*, ed. Davis, Oxford University Press, USA, 87–101.

McCann JJ (2004) Mechanism of Color Constancy, *12th IS&T/SID Color Imaging Conference*, **12**, 29–36.

McCann JJ (2005) Do Humans Discount the Illuminant? Human Vision and Electronic Imaging X, *IS&T/SPIE*, San Jose, CA, *SPIE Proc*, **5666**, 9–16.

McCann JJ & Houston LK (1983a) *Color Sensation, Color Perception and Mathematical Models of Color Vision, Colour Vision*. ed. Mollon & Sharpe, Academic Press, London, 535–44.

McCann JJ & Houston K (1983b) Calculating Color Sensation from Arrays of Physical Stimuli, IEEE SMC-13, 1000–07.

McCann JJ, McKee S, & Taylor T (1976) Quantitative Studies in Retinex Theory: A Comparison Between Theoretical Predictions and Observer Responses to the Color Mondrian Experiments, *Vision Research*, **16**, 445–58.

McCann JJ, Parraman C & Rizzi A (2009a) Reflectance, Illumination, and Edges in 3-D Mondrian Colour Constancy Experiments, Proc.. of the 2009 Association Internationale de la Couleur 11th Congress, Sidney.

McCann JJ, Parraman C, & Rizzi A (2009b) Reflectance, Illumination and Edges, *Proc. IS&T/SID Color Imaging Conference*, CIC **17**, 2–7.

McCann JJ, Vonikakis V, Parraman C, & Rizzi A (2010) Analysis of Spatial Image Rendering, *Proc. IS&T/SID Color Imaging Conference*, **18**, 223–28.

OSA Committee on Colorimetry (1953) The Science of Color, *Optical Society of America*, Washington DC, 377–81.

Parraman C, Rizzi A, & McCann JJ (2009) Colour Appearance and Colour Rendering of HDR Scenes: An Experiment, Proc. IS&T/SPIE Electronic Imaging, San Jose, 7241–26.

Parraman C, McCann JJ, & Rizzi A (2010) Artist's Colour Rendering of HDR Scenes in 3-D Mondrian Colour-constancy Experiments, Proc.. IS&T/SPIE Electronic Imaging, San Jose, 7528–31.

Rutherford MD & Brainard DH (2002) Lightness Constancy: A Direct Test of the Illumination-Estimation Hypothesis, *Psychological Science*, **13**, 141–49.

Yang J & Shevell S (2003) Surface Color Perception Under Two Illuminants: The Second Illuminant Reduces Color Constancy. *Journal of Vision*, **3**(5)4, 369–79

3-D Mondrian CREATE website
http://web.me.com/mccanns/McCannImaging/3-D_Mondrians.html

Full report
http://web.me.com/mccanns/McCannImaging/3-D_Mondrians_files/RIA.pdf

CIC 10 webbsite "Analysis of Spatial Image Rendering"
http://sites.google.com/site/3dmondrians/

Album of 3-D Mondrian photos
http://web.me.com/mccanns/_CREATE_08/HDR_Albums/HDR_Albums.html

Measurements
http://web.me.com/mccanns/_CREATE_08/Paints.html
http://web.me.com/mccanns/McCannImaging/3-D_Mondrians_files/XYZmeasuprements.pdf
http://web.me.com/mccanns/McCannImaging/3-D_Mondrians_files/MagEstApp.pdf

30

Color Constancy is HDR

30.1 Topics

This chapter summarizes the convergence of HDR and color constancy experiments. Outside of the laboratory, and away from the beach, most scenes have complex nonuniform, multispectral (sun and skylight) illumination. All scenes that have nonuniform illumination introduce a degree of HDR imaging. Different spectral illuminants introduce color constancy. This chapter compares the similarities of these visual mechanisms. Further, the retinal rod cells add 3 log units of range to human HDR. We discuss the role of rods in color and in HDR appearance.

30.2 Introduction

At the start of this text, we described that HDR scenes were the result of illumination. We reviewed possible improvements in HDR technology. We measured limits in the dynamic range of light that falls on a camera's image plane. More important, we measured limits in the range of appearances because of intraocular glare Section C. That led us to the observation that post-retinal neural contrast mechanisms tended to cancel glare. Measurements of the effects of neural contrast showed the slow change in appearance for maxima, and the rapid changes with areas darker than the maxima (Section D).

In Section E, we found that color had these identical properties. Color constancy experiments vary the spectral content of the illumination, and observe small departures from perfect constancy. Measurements of these color appearances fit a model with three parallel color channels. Each channel exhibits the same slow change in appearance for maxima, and rapid changes with darker areas. Color appearance correlates with the triplet of apparent lightnesses generated by these parallel channels.

The appearances of HDR scenes, and those of Color Constancy experiments, are two sides of the same coin. Indifference to overall changes in illumination is the same for achromatic and colored lights. The same properties of appearance are found in HDR and Color Constancy experiments. (McCann, 2011a)

A number of times in this text we have discussed paradoxes in vision, in which we find major differences in vision response depending on how we ask a question.

If we experiment with spots of light, we see the visual mechanism from one perspective and are led to a number of seemingly logical conclusions.

Alternatively, if we change our point of view we can get a new set of experimental results that seem to contradict our previous conclusions. Again, the challenge is to dig a little deeper and search for a common understanding of both sets of experimental results.

The Art and Science of HDR Imaging, First Edition. John J. McCann, Alessandro Rizzi.
© 2012 John Wiley & Sons, Ltd. Published 2012 by John Wiley & Sons, Ltd.

30.3 Rod Receptors and HDR

Perhaps the greatest paradox in the study of human vision is the role of rod receptors in HDR and in color. We often read that the rods are an independent, colorless visual system that evolved for better night vision (*Duplicity Theory*). The rods play a major role in extending the range of vision because they are 1000 times more sensitive than the cones. It is pools of rod receptors that achieve the remarkable absolute sensitivity threshold of humans. Hecht, Shlaer, and Pirenne (1938) showed that we see the light from only a half-dozen photons caught by rods. That part of our human dynamic range, rod vision, requires that we sit in a darkroom for 10 to 30 minutes to regenerate rhodopsin, the rods' visual pigment. That absolute threshold part of our dynamic range is disabled by bright light, thus limiting our useful dynamic range in scenes.

Rod vision is a fully functional visual system. When we control the illumination to be below cone threshold we find the same spatial properties as with cone vision. Appearance responds to scene content. Luminance gradients are low in apparent contrast compared to edges. Simultaneous Contrast works well at light levels so low that only the rods are above visual threshold. In fact, White's Effect, the Checkerboard and Dungeon Illusions, and Benary's Cross behave the same using rod vision. Since these experiments are the result of spatial processes, it is possible that the different anatomy and physiology of rods and cones could limit the range of these effects. Remarkably, spatial effects at the lowest end of our visual HDR range are very similar to those at the top of the range in sunlight. (McCann, 2011b) Lightness does not correlate with rod quanta catch. Maxima follow the Hipparchus line, and areas with less luminance than the maxima get darker with smaller changes in light. The rods are a fully functioning Retinex. They use spatial comparisons to generate lightnesses.

The role of rods gets more interesting when we vary the shape of the illumination spectra. We can see color from the interactions of rods and long-wave cones. The best spectrum for these interactions is firelight (McCann, 2006). Experiments have shown that the range of colors from rod and L-cone interactions fall in the three dimensional appearance space of cone vision, and do not require a 4-D color space (McCann, 2007b; 2008). The rods are perfectly good color receptors when the spectral shape of the illumination is appropriate.

30.3.1 Duplicity Theory

Max Schultze (1866) proposed that humans have two independent visual systems; one for colorless night vision, and a second for high light-level color vision. Called *Duplicity Theory,* it states that the rods and cones operate as different systems. The many different properties of appearances produced by rods and cones have accumulated substantial support for the idea that the rod and cone processes are different. The rods and cones have different shapes and different distributions over the retina, (Max Schultze, 1866) different rates of dark adaptation, (Hecht, 1937) different spectral sensitivities, (Wald, 1945) different directional sensitivities, (Stiles & Crawford, 1933) different acuity properties (Shlaer, 1937) different flicker fusion properties,(Hecht & Verrijp, 1933) and independent desensitizing (Alpern, 1965) and sensitizing (Westheimer, 1970) retinal interactions. As well, there is substantial literature on the visual effect of spatial integration of retinal receptors. Threshold responses to spots of light vary with the size of spot and position on the retina (Pirenne, 1962).

30.3.2 Rod Retinex

The rods are the best example of a Retinex (McCann, 1973). Their extreme sensitivity to light means that by reducing the intensity of the light we can observe the properties of a single isolated Retinex. When the light is below absolute threshold for the three cones we see achromatic appearances. Because the rods have poorer spatial resolution and are absent from the fovea we observe a less-sharp world. Because the light level is very low, it falls on the lower end of the Hipparchus line and is less white (McCann, 2007a). Nevertheless, we see an image of achromatic lightnesses that are nearly constant over the three log-unit range of rod-only vision.

To study the range of human vision for scenes, rather than spots of light, we do the following:

Dark adapt in a lightless room for an hour.

Gradually increase narrow-band 546 nm illumination on a scene to absolute threshold.
- At first observers see just the presence of light with no sensation of shapes.

Increase the light to 10 times absolute threshold.
- Forms, namely light and dark regions, are visible (form threshold).

Keep increasing the light slowly and observe.
- Achromatic lightnesses are clearer, higher in contrast and slightly sharper with increased illumination.

Around 1000 times absolute threshold, or 100 times form threshold, appearances suddenly change with small increased illumination.
- The scene appears sharper and somewhat greenish. Experiments show that the light is now above cone threshold (McKee et al., 1977)

Having reached cone threshold the scene has a pale green wash; it is no longer achromatic. The scene is much sharper, and the contrast of the scene has increased. These appearances, as well as all the other rod characteristics (directional sensitivity or Stiles-Crawford, flicker, spectral sensitivity), allow us to measure with great accuracy the amount of light sufficient for the onset of cone responses. Between absolute threshold and the onset of cones, we can observe the visual properties of a single Retinex.

With the Rod Retinex we see the same Hipparchus line and neural contrast signature:

Overall increases in illumination cause small increases in the lightness of the maxima.

Lightnesses of the different areas in the scene are nearly constant with uniform changes in illumination.

Approaching cone threshold, the blacks appear slightly darker.

30.3.3 Rods are Color Receptors

Despite all this evidence that the rods and cones are independent systems, experiments have shown that color sensations can be generated by the interactions of rods and cones. Blackwell and Blackwell (1961) found that a rare type of color-blind observer reported different color names even though he had only rods and short-wave cones. Stabell (1967; 1998) reported color sensations from the interactions of afterimages (produced by lights above cone threshold) and stimuli below cone threshold.

McCann and Benton (1969) found that the rods and long-wave cones interact to produce color sensations. The technique was to determine the 546 nm flux necessary for a threshold response of the rods and the 656 nm flux necessary for a threshold response of the long-wave cones. In Figure 30.1 the shaded areas illustrate the technique of providing just enough 546 nm flux to excite the rods, and just enough 656 nm flux to excite the long-wave cones. Using those quantities of 546 nm and 656 nm flux, McCann and Benton produced color sensations even though the 546 nm flux was hundreds of times below cone threshold. They further showed that the color images produced by these interactions were nearly indistinguishable from images seen on two wavebands entirely above cone threshold.

McKee, McCann, and Benton (1977) asked observers to adjust a colored image for the best color. The observers were given a knob that changed the amount of illumination on the short-wave record. They adjusted the amount so the image was neither too warm, nor too cool. The colored image consisted of a long-wave record black and white transparency in 656 nm light, and a middle-wave record

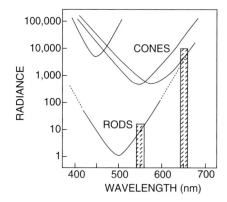

Figure 30.1 Radiance vs. wavelength necessary to excite a threshold response of the rods and the various types of cones in the retina. The shaded lines indicate the radiances necessary for exciting the rods with 546 nm light, and the long-wave cones with 656 nm light. The lower solid curve is the reciprocal of the scotopic luminosity curve.

in variable wavelengths of illumination. In the first part of the experiment, the 656 nm image was just above threshold. The experimenter selected a wavelength between 420 and 610 nm. By adjusting the illumination, the observers found the best color balance for that wavelength. Then, the experimenter changed the wavelength, and the observers found the best color balance for that wavelength. The plot of best-color short-wave radiance vs. wavelength fit the rod luminosity curve shown in Figure 30.1. The observers judged color appearance. Those judgements were made with the rods because their spectral plot fit the rod sensitivity curve. Benton and McCann (1977) reported color appearances from rod and Lcone separations identified by their fusion frequency.

McCann, Benton, and McKee (2004) compiled an annotated bibliography on rod-cone color.

30.3.4   Rod Retinex plus L-cone Color Appearance Demonstration

If you study rod- cone color with flat, equal energy spectra, or short-wave rich daylight spectra, you get the wrong impression about rods and color vision. The narrow-band illumination spectra in Figure 30.1 were chosen to selectively stimulate rods and *Lcones* separately, and then in combination. If we replace them with a horizontal line spectrum (equal energy at all wavelengths), and vary the amount of that light, we will see the following:

> At absolute threshold we will see a colorless scene.
>
> The scene remains colorless until the amount of light reaches cone threshold.
>
> The light spectrum reaches cone threshold for more than one cone type at the same intensity.

The illumination put the scene on the rods, and then on rods and several cones. In order to see the rod/Lcone interactions, we need to use light with much more long-wave light than short-waves. Studies have shown that firelight is the ideal stimulus for seeing such interactions (McCann 2006).

There is a simple demonstration using a candle that lets one compare rod/Lcone color with LMS cone color.

Figure 30.2   Rod/Lcone Color Tester. The letters C appear green, and I appears blue using L, M, S cone vision. These letters have the same reflectance for rod vision. When viewed with candle, at a distance, they appear the same cyan color with rod and Lcone interactions.

To do this experiment you need:

A print of the Rod/Lcone Tester: Figure 30.2, or one made from Wiley web site: http://www.wiley.com/go/mccannhdr
A candle
A light-free room (doing this experiment at night makes it easier).

First set up your candle

Find a candle, candle holder and a cigarette lighter, or cooking grill lighter.
Place your equipment on a tray that it is easy to find in total darkness.
Sit in the dark for at least 10 minutes. (To be fully dark-adapted, you need to sit in the dark for 30 minutes.)

Light the candle

After dark adaptation, light the candle, but try to avoid looking at the flame to avoid light adaptation.

Rod/L-cone colors – lower radiances

Move away from the flame. Hold the CIC print 6 to 10 feet away from the candle. Observe the colors.

Figure 30.3   Appearance from rod/L-cone interactions. The letters CIC has the same scotopic luminance, so letters [C] and [I] have the same lightness on the rods. The scene looks less sharp.

Note the similarity in color of letters **C** and the letter **I**.
When the firelight is far from the print, it is seen by rod/L-cones.

Cone colors – higher radiances

Hold the CIC print close to the flame.
Observe the colors. Note the difference in color between the letters **C** and the letter **I**.
When the firelight is near the print, it is seen by L-, M-, S-cones.

When the CIC letters are above cone threshold they have green and blue colors. When using
the candlelight at a distance, the reduced intensity results in seeing the letters with rod and L-cone
interactions. Hence the change in color. The letters C & I appear the same lightness on the rods (Figure
30.3).

30.3.5   Rod Retinex plus L-cone Retinex Color Space

Does having a fourth color receptor increase the range of color appearances? Are there new colors in
these conditions? Is color a four dimensional space? Experiments have shown that rod/L-cone colors
share the same 3-D color space (McCann, 2007b; 2008). The rod lightnesses share both the M- and
S- retinex color channels.

30.3.6   Rod Retinex plus L-cone Retinex

Performing experiments with only rods and long-wave cones provides a test of the hypothesis that color
depends on the lightnesses produced by independent systems. Certainly the rods are part of an independent system that forms an image in terms of lightness. We know this from the experiments supporting
the Duplicity Theory, and from observing the colorless rod image in very low illumination. In addition,
the image seen by the long-wave cones alone has no variety of color sensations, just an overall wash
of red covering the entire field of view. Here again, we have an image in terms of lightness, with a
reddish wash that must interact with the colorless, rod lightness image. The interaction of these two
images as already described is almost indistinguishable from an image seen on two wave bands entirely
above cone threshold. (McCann & Benton, 1969). This observation goes quite far in supporting the
idea that color sensations are generated by the comparison of lightnesses formed by independent
systems.

Can we press the hypothesis even further? Is it possible, in a complex image that excites only the rods, to produce two significantly different lightnesses from areas sending the same radiance simultaneously to the eye? Is it possible to combine such a rod image with an image seen by the long-wave cones, which has different lightnesses produced by areas having the same radiance? Do these areas produce different color sensations? Are the sensations consistent with the hypothesis that colors are determined by the comparison of lightnesses?

Land and Benton made a single scene in which one area appeared red and another area appeared green at the same time, even though the area's radiances were identical. The display, called the Street Scene, had a green awning on the left side and a red door on the right.

McCann (1972) repeated Land's experiment with only the rods and long-wave cones. Using two projectors with narrow-band interference filters (656, 546 nm) to illuminate the entire display. A pair of neutral density wedge transparencies was chosen so that the awning and the door had the same radiances in 656, and in 546 nm light. With 656 nm light alone the awning was dark and the door was light, with identical radiances. With 546 nm light alone the awning was light and the door was dark with identical radiances. When both projectors were turned on, the awning and the door were identical stimuli. The radiance of the 546 nm light was set below cone threshold, and that of the 656 nm light was slightly above long-wave cone threshold. The middle- and short-wave cones were below threshold response.

Just as in Land's experiments entirely above cone threshold, identical radiances produced very different color sensations: the awning was cyan, and the door was red. Color sensations from rod-cone interactions correlate with the lightnesses of each area in each illumination. This is an example of the mechanism proposed by Land's Retinex theory. Rods and cones act independently to form lightnesses using spatial comparisons. Lightnesses of the two physiologically different systems are compared to generate color sensations (McCann, 1973).

30.4   Assembling Appearance: Color Constancy, Rod Vision and HDR

Appearance has a number of ideas that keep recurring.

Vision synthesizes an LDR rendition of the HDR world, limited by glare and the dynamic range of neurons in the optic nerves.

Lightness scaling (cube-root function) happens before the retina, due to intraocular scatter.

Scatter plus logarithmic receptor response shapes lightness.

Scene dependent retinal image accounts for a wide variety of phenomena

As argued by David Wright, colorimetry is the domain of the retina receptors.

Uniform color space data requires a 3-D transform of cone receptor response.

Chroma has to be stretched relative to lightness. (Chapter 18)

As argued by Edwin Land, color constancy is best explained and modeled by three independent channels, or Retinexes.

The idea fits color constancy data for appearance

in Mondrians, Tatami, Assimilation, rod/Lcone interactions

Lightnesses in each color channel have the Hipparchus line/neural contrast signature behavior.

Experiments that were designed to measure human HDR vision have the same experimental results as experiments measuring color constancy. Experiments designed to study color from cones have the same results as those for rod/Lcone color.

Ever since von Kries became interested in matching afterimages on the retina to light entering the eye, the predominant hypothesis for color constancy has been that the eye *"adapts"* to the incoming light. The change in incoming light causes the receptors to change their sensitivity to render objects as constant perceptions. All neurons adapt. Although feedback signal processing has become a part of almost everything, from a camera's exposure control to ABS brakes in a car, there is no successful model of cone adaptation that can predict human color constancy from the array of scene radiances. Constancy predictions have to fit the crosstalk data.

30.5   Summary

We have seen in Section E that spatial processing is the only mechanism consistent with the many experiments presented here.

All of color constancy is HDR imaging. The color constancy experiment increases the scene's dynamic range by increasing the range of illumination. All the evidence compiled in Sections C and D correlate nicely with the color experiments in Section E. In the final section we will discuss the details of computational models of HDR imaging and color constancy.

30.6   References

Alpern M (1965) The Specificity of the Cone Interaction in the After-Flash Effect, *J Physiol*, **176**, 462–73.

Benton JL & McCann JJ (1977) Variegated color sensations from rod-cone interactions: Flicker-fusion experiments, *J Opt Soc Am*, **67**, 119–21.

Blackwell H & Blackwell O (1961) Rod and Cone Receptor Mechanisms in Typical and Atypical Congenital Achromatopsia, *Vis Res*, **1**, 62–107.

Hecht S (1937) Rods, Cones, and the Chemical Basis of Vision, *Phys Rev, Am Physiological Soc*, **17**, 239.

Hecht S, Shlaer S & Pirenne MH (1938) Energy, Quanta and Vision, *J Gen Physiol*, **25**, 819.

Hecht S & Verrijp C (1933) Influence of Intensity, Color and Retinal Location on the Fusion Frequency of Intermittent Illumination, *Proc Nat Acad Sci, US*, **19**, 522–35.

McCann JJ (1972) Rod-cone Interactions: Different Color Sensations from Identical Stimuli, *Science*, **176**, 1255–57.

McCann JJ (1973) Human Color Perception, in *Color Theory and Imaging Systems, Society of Photographic Scientists and Engineers*, Eynard R (ed.), Washington, 1–23.

McCann JJ (2006) Ideal Illuminants for Rod /L-Cone Color, Processing, Proc. SPIE 6058, 1–8.

McCann JJ (2007a) Aperture and Object Mode Appearances in Images, Proc. SPIE, 6292–26.

McCann JJ (2007b) Colors in Dim Illumination and Candlelight, Proc. IS&T/SID Color Imaging Conference, Albuquerque, 15.

McCann JJ (2008) Color Gamuts in Dim Illumination, Proc. SPIE, 6492, 680703–680709.

McCann JJ (2011a) HDR Imaging and Color Constancy: Two Sides of the Same Coin?, Proc. SPIE, 7866, 78660Q.

McCann JJ (2011b) Appearance at the Low-Radiance End of HDR Vision: Achromatic & Chromatic, Proc IS&T / SID Color Imaging Conference, San Jose, **19**, in press.

McCann J & Benton J (1969) Interactions of the Long-wave Cones and the Rods to Produce Color Sensations, *J Opt Soc Amer*, **59**, 103–6.

McCann J, Benton J & McKee S (2004) Red/White Projections and Rod-Lcone Color: An Annotated Bibliography, *Journal of Electronic Imaging*, **13**(01), 8–14.

McCann J & Rizzi A (2007) Veiling Glare: The Dynamic Range Limit of HDR Images, Proc. SPIE, 6492–541.

McKee S, McCann J & Benton J (1977) Color Vision from Rod and Long-Wave Cone Interactions: Conditions in which Rods contribute to Multicolored Images, *Vis Research*, **17**, 175–85.

Pirenne MH (1962) Chapters 5–7, Davson H (ed.) in *The Eye, 2: The visual process*, Academic Press, New York.

Schultze, M (1866) Zur Anatomie und Physiologie der Retina, *Atch mikr Anat*, **2**, 175–286.

Shlaer S (1937) The Relation Between Visual Acuity and Illumination, *J. Gen. Physiol.*, **21**, 165–88.

Stabell B (1967) Color Threshold Measurements in Scotopic Vision, *Scand J. Psycho I.*, **8**, 132.

Stabell U & Stabell B (1998) Chromatic rod-cone interaction during dark adaptation, *J Opt Soc Am A*, **15**, 2809–15.

Stiles W & Crawford B (1933) The Luminous Effect of Rays Entering the Eye Pupil at Different Directions, *Proc Roy Soc*, **1128**, 428.

Wald G (1945) Human Vision and the Spectrum, *Science*, **101**, 653–58.

Westheimer G (1970) Rod-cone Independence for Sensitizing Interaction in the Human Retina, *J Physiol*, **206**, 109–16.

Section F

HDR Image Processing

31

HDR Pixel and Spatial Algorithms

31.1   Topics

In Section F we will describe and analyze HDR algorithms that we have divided into a number of categories. We will emphasize the many variants of Retinex and ACE image processing and review analytical interpretations. We describe a number of different techniques to evaluate the success of different algorithms. Finally, we use what we have learned to discuss future research in HDR imaging.

31.2   Introduction – HDR Image Processing Algorithms

In Section A we described how the ideas of silver halide photography have influenced our thinking about images. We may never be sure if analyzing images one pixel at a time comes from Euclidean geometry, or the chemical mechanisms of silver halide photography, or the later influence of Thomas Young's tricolor hypothesis. It may be a combination of all three. Regardless, there are many useful systems that perform successful predictions with calculations that limit the input information to that of a single pixel. Silver halide film models and colorimetry are good examples. We can calculate the response of a film from the quanta catch of the silver-halide grains at a small local region of the film. The film's response to quanta catch is the same everywhere in the image. As well, we can predict whether two different spectra (wavelength vs. energy distributions) will match to humans. As we have seen many times in this text, calculations that are restricted to the information from a single pixel cannot predict human appearance, and are of limited value in HDR imaging.

Our analysis has the goal of evaluating an algorithm's ability to render all types of scenes. If the process mimics the human visual system, then it must render appropriately LDR, HDR, low-average and high-average scenes using the same algorithm parameters. Using this as a performance standard is very helpful in evaluating the great variety of algorithms.

We will review image processing algorithms that use light (quanta catch) from:

The Art and Science of HDR Imaging, First Edition. John J. McCann, Alessandro Rizzi.
© 2012 John Wiley & Sons, Ltd. Published 2012 by John Wiley & Sons, Ltd.

one pixel,

a local group of pixels,

all the pixels,

and all the pixels in a way that mimics vision.

31.3   One Pixel – Tone Scale Curves

It is easy and efficient to apply tone scale curves to digital images. While color negative-to-print tech-nology installed a fixed tone scale in the factory, digital photography can apply it using the individual's personal computer. Although the practice is completely different, the principles are the same. Jones and colleagues sought the best compromise for optimal rendition of all types of scenes. They measured the limits of veiling glare, and designed negatives that in single exposures captured all the dynamic range possible after glare (Section B). There is no fundamental difference between film and digital tone-scale rendering. Both cannot address the issue raised by Land's Black and White Mondrian, and "John at Yosemite" experiments. Nonuniform illumination, sun and shade, puts the same light on the camera's film plane from white areas in the shade, and black areas in the sun. HDR scene rendition that mimics human vision requires spatial image processing.

There is a great irony in using *Tone Scales* to improve digital imaging. *Tone scales* were built into the chemistry of film processing. A great deal of practical image-quality research on the scenes people photographed determined the optimal tone scale response for prints and transparencies (Mees, 1961; Mees & James, 1966). *Tone scales* for film photography were optimal for pixel based image processing of all scenes. Pixel-based processing had to be the best compromise. It cannot respond differently to scene content. However, digital imaging has made spatial processing possible. Unlike film, that restricted spatial interactions to a few microns, electronic imaging has freed us from the fixed compromise of a global tone scale.

31.3.1   One Pixel – Histogram

A popular tool among digital photographers is histogram equalization. Histograms that plot the number of pixels for all camera quantization levels (e.g., 0 to 255) are very helpful in evaluating camera expo-sures. Over- and under-exposure is easily recognized in histogram plots. Engineers have developed programs that group quantization levels into bins, and then redistribute the bins with the goal of having a more uniform distribution of digital values. Histogram equalization programs process the image using the pixel-value population. It assumes that the image will look better if all the bins have an equal count. Histograms equalization programs completely ignore the spatial information in the input, and signifi-cantly distort the spatial relationships in the output. Many scenes are harmed by histogram-equalization renditions because they change the spatial relationships of image content.

Figure 31.1 shows the effects of histogram equalization on a pair of test targets made up of gradients and uniform gray squares. The left and right targets have the same maximum, minimum and range. They have different gradients in the middle and more gray steps between max and min on the right. They differ in the image statistics; the number of pixels per digital value (0 to 255). In the gradients, pixels have very similar values to nearby pixels. Since histogram equalization just evaluates the statis-tics of the image, it does not preserve subtle local relationships. In the pair of test targets the gradient appears smooth and low in contrast. In the pair of *histogram-equalized* images below, the gradients are broken up into different bands of luminance, showing the algorithm's indifference to local relationships. Furthermore, the gray scales at top and bottom have been rendered with lower contrast near white and black, but with higher contrast for mid-grays.

Humans do not use image statistics in generating appearances. As we have seen in Sections D and E, the spatial relationships of the image content control appearances.

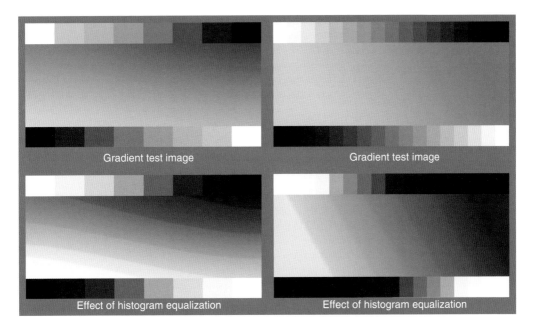

Figure 31.1 Banding created by histogram equalization. The top pair are different input images; the bottom pair are the outputs of the same histogram equalization algorithm.

31.3.2 One Pixel – LookUp Tables

There is a great advantage to single-pixel processing. It is fast, low-cost, and requires minimal hardware. HDR algorithms frequently use LookUp Tables (*LUTs*) because they require only a small number of memory locations per color to load the digital values of the output. The operation is to read the memory value out at the appropriate address. This operation is simpler and faster than multiplying two numbers.

Single pixel techniques are ideal for manipulating the appearance of individual images which have a particular problem. There are many HDR guide books that can help to solve particular problems for particular scenes using multiple digital exposures, *RAW* format cameras and *LUTs* (Bloch, 2007; Freeman, 2008; McCollough, 2008; Correll, 2009; Fraser & Schewe, 2009). Although they often use the less powerful three 1-D LUT approach, they do great things to individual images.

Three-dimensional LUTs are an extremely powerful tool in making accurate reproductions and controlling color profiles (Pugsley, 1975; Kotera et al., 1978 Abdulwahab et al., 1989, McCann, 1995; 1997). Recall the analogy of color reproduction as the act of moving from an old house to a new one with different size rooms. The 3-D LUT process lets one use a different strategy in each room. It is much more powerful than three 1-D LUTs. Such 3-D LUTs play a major role in the color calibration processes of digital printers, displays and color management systems (Green, 2010).

31.3.3 Using 1-D LookUp Tables for Rendering HDR Scene Captures

Kate Devlin (2002) and Carlo Gatta (2006) wrote extensive reviews of the many tone scale LUTs and their underlying principles. Reinhard et al. (2006, Chapter 6) review the digital versions of conventional film tone-scale techniques. Although Reinhard provides many color illustrations of the processes, it is difficult to compare these approaches because each of the examples uses a different test image.

Applying the same pixel processing LUT (global operators) to any, and all, scene reproductions is not realistic. Some algorithms use a two step process to incorporate models of human vision. First, they assume that multiple-exposures can measure accurate scene luminance. Second, they apply a global-tone-scale map derived from psychometric functions, measured using isolated spots of light, rather than using scenes.

Many algorithms use some kind of psychophysical data to convert scene radiance to visual appearance. As well, psychophysical data are sometimes used outside of their original context. Similarities with human vision are taken as an inspiration, rather than modeling calibrated scene input with precise implementation of the experimental data. They have been inspired by Steven's data on brightness perception (Tumblin & Rushmeier, 1993), Blackwell contrast sensitivity model (Ward, 1994), models of photopic and scotopic adaptation (Ferwerda, et al., 1996) differential nature of visual sensation (Larson et al., 1997), image statistics (Holm, 1996) adaptation and gaze (Scheel et al., 2000). Reinhard and Devlin (2004) used physiological data to modify uncalibrated cameras' scene capture. They used Hood's modification of Naka & Rushton's famous equation describing an electrode's extra-cellular post-receptor graded S-potential response of excised fish retinas.

There are many problems with such algorithms that prevent them from accurately predicting human vision. The captured scene data is an inaccurate record of both scene luminance and retinal contrast because of scene-dependent veiling glare (Section B). The psychometric functions do not apply to scenes, they model the appearance of spots of light (Sections C and D).

31.4   Some of the Pixels – Local Processing

As seen throughout this text, we need to spatially modify the captured image to make the best rendition of HDR scenes. It is only through spatial processing that we can avoid the need for individual tone scales for each picture. With digital imaging all the pixels in the image can influence the output of other pixels.

From Leonardo da Vinci and 19th century psychologists (Gilchrist, 2006), we know that the appearance of a test area changes, depending on the surround area around it. We can call this class of models Local Processing mechanisms. This is a non-linear transform, so that individual pixels with the same input value may not have the same output value. We discussed the use of spatial color opponent neurons for chroma enhancement in Chapter 18.5.2. We see local spatial processing in a wide variety of applications such as edge detection (Sobel operators) and edge sharpening. These applications use very small windows, or kernels. Many image processing and physiology models use difference of gaussians models with larger kernels (Pratt, 1991). In psychology surround models are also popular. Often the proposed mechanisms do not have enough specific information to be able to create an image processing model to verify them. Some perceptual framework oriented psychologists (See Gilchrist, 2006 for review) describe top-down models using the presence of a surround, rather than the specific mechanisms of interaction (size, shape, and distances of influence) needed in image processing techniques. A number of Siggraph papers proposed local-spatial-filtering techniques (see Reinhard et al., 2002). That paper described an automatic multi-resolution difference of gaussian model similar to physiological models. It also claimed that this process implemented Ansel Adams' Zone System with automatic dodging and burning. As we saw in Chapter 6, Adams' two main principles were:

> Adjusting the sensor response (film exposure and development) to control calibrated image capture,

> Selective scene-dependent treatment of image content to make each area have a specific Zone Value in the print.

Neither of these features are in this algorithm. It used uncalibrated camera digits for input, and scene independent criteria for multi-resolution local processing. In Figures 6.5 and 6.6 we saw the dramatic change in Adams' finished print and control. Such changes are not possible with this algorithm.

Bilateral filtering (Tomasi and Manduchi, 1998) is a powerful local operator able to modify a pixel according to its neighbors, without degrading the edges in the image. It has been studied to apply classic operators, like the gaussian filter, but in a way that depends on scene content. The problem they want to solve is to apply smoothing filters, or other averaging operators, to remove noise without losing edge information. The central idea of the bilateral filters is to define the filtering function not only in the space domain of the image (pixels coordinates), but also in the image content domain (pixel values) to weight the contribution of the pixels in the neighborhood. In this way, the process keeps the edge sharpness intact. This approach has been extended to several different applications, such as multispectral fusion, orientation smoothing, 3D fairing, image stylization and HDR tone mapping. The method for HDR contrast compression (Durand & Dorsey, 2002) does not attempt to imitate vision. It is based on the idea of multiscale decomposition, but only two layers are considered in the application: the base and the details. While the base range is compressed as desired by the user, details are kept unchanged. This reminds us of the ill-posed idea of separating physical reflectance from actual illuminant, as suggested in the paper. The same approach of base-details separation has been implemented in the iCAM06 model in which bilateral filtering replaced the gaussian filtering as the spatial component in their model (Khang et al., 2007).

Local operators can improve captured HDR data. The parameters require operator intervention depending on the source of uncalibrated scene radiance and scene content. Although local spatial operators, using parts of the images, are slower and require greater hardware processing power than LUTs, they are not a challenge to today's computer power. With appropriate, scene dependent, parameters they give good results.

31.5 All of the Pixels

Following World War II radar, optical and imaging research studied images in the frequency domain. We can call this class of model *spatial frequency* mechanisms. The practical use of spatial frequency filtering techniques was helped by the *Fast Fourier Transform (FFT)* invented by Cooley and Tukey (1965). Without it, the millions of pixels in today's images would make these techniques extremely slow. They are in wide use today with standard computer processing power.

A good example of such a process is: taking an image, converting to its Fourier transform, applying a spatial frequency filter, retransforming the filtered information back to the image domain. The retransformed image is the output of the model. This, too, is a non-linear transform so that all the pixels with the same input value need not have the same output value. As with tone-scale maps, spatial filtering can substantially improve the appearance of images. However, a given spatial frequency filter has a fixed effect on the image in the Fourier domain. Human vision behavior can be described with a given spatial-frequency filter for a particular image. However, a single filter cannot accurately describe the human appearances of all images. Human vision, in effect, uses scene dependent spatial filters (Chapter 21.5, McCann, 2006).

31.6 All Pixels and Scene Dependent – The Retinex Extended Family

The final type of spatial operations respond to the information in the individual input image. Whereas a spatial frequency filter will apply the same non-linear process to all images, this class of process applies a different modification to the image depending on the image content. We can call this class of model the *Retinex Extended Family*. It includes image processing algorithms that are modifications and extensions of the original Retinex, and other different algorithms that calculate the appearance of an image from the array of all scene radiances.

The *Retinex Extended Family* has three unique attributes:

It uses all the pixels in the image as input

It is influenced by the content of the image

It attempts to mimic human vision, or make the best rendition for humans

The best HDR imaging is done with processing that renders the scene the way that humans do; i.e., mimic human vision. When a scene has gray areas surrounded by white, those areas appear dark. When a scene has minimal white, the same gray areas appear lighter. The Retinex Extended Family alter the input values depending on the content of the scene. This type of spatial operation responds to the information in the individual input image. Whereas a spatial frequency filter will apply the same non-linear process to all images, the Retinex class of process applies a different modification to the image depending on the image content.

The first HDR computer and electronic implementation (Land & McCann, 1971) made the bold assumption that the model should use all the input pixels and mimic vision. It was a great computational challenge for computers in 1965. But, the goal in Land and McCann was to understand human spatial processes, and build their equivalent in special hardware. Frankle and McCann is a good example. First, it took on the challenge of having each pixel interact with each of the other pixels in the 512 by 512 pixel image. They invented multi-resolution techniques to calculate lightness. They used integer summations to replace floating point multiplication. These summations were performed for the whole image in one cycle time of the machine. Special purpose hardware using multi-resolution algorithms that mimic vision can make the difficult challenge of Retinex Extended Family approaches fast and low cost. A 512 by 512 image took less than a minute using 1975 hardware.

31.7   Retinex Algorithms

In Chapter 32 we review the many different variations of Retinex Lightness Models. We trace their history and compare their processing rules.

31.8   ACE Algorithms

In Chapter 33, we review the many different variations of Automatic Color Equalization (ACE), Random Automatic Color Equalization (RACE).

31.9   Analytical, Computational and Variational Algorithms

In Chapter 34, we review different types of mathematical calculations of the Retinex Extended Family using analytical, computational and variational formalizations. We summarize these computational techniques.

31.10   Techniques for Analyzing HDR Algorithms

In Chapter 35, we review some different measurement techniques for analyzing the effectiveness of spatial algorithms.

31.11   The HDR Story

In Chapter 36 we summarize the story of HDR imaging described in this text. It brings together the properties of spatial imaging in humans and machines.

Section F describes the many algorithms used to improve HDR images. They can be as simple as a lookup table, or as complicated as an array processor, or a variational calculation. This section points out the algorithms' similarities and differences.

31.12   References

Abdulwahab M, Burkhardt JL & McCann JJ (1989) Method of and Apparatus for Transforming Color Image Data on the Basis of an Isotropic and Uniform Colorimetric Space, US Patent 4,839,721 Filed Aug. 28, 1984: Issued Jun. 13, 1989.

Bloch C (2007) *The HDRI Handbook: High Dynamic Range Imaging for Photographers and CG Artists*, Rocky Nook, Santa Barbara.

Cooley J & Tukey J (1965) An Algorithm for the Machine Calculation of Complex Fourier Series, *Math Comput*, **19**, 297–01.

Correll R (2009) *High Dynamic Range Digital Photography For Dummies*, John Wiley & Sons.

Devlin K (2002) *A Review of Tone Reproduction Techniques*, Technical Report CSTR 02 005, Bristol.

Durand F & Dorsey J (2002) Fast Bilateral Filtering for the Display of High-Dynamic-Range Images, *ACM Trans on Graphics*, **21**(3), 257–66.

Fewerda JA, Pattanik S, Ferwerda J, Shirley P & Greenburg D (1996) A Model Visual Adaptation for Realistic Image Synthesis, *Proc. ACM SIGGRAPH 98*, 249–58.

Fraser B & Schewe J (2009) *Real World Camera Raw*, Peachpit Press, Berkeley.

Freeman M (2008) *Mastering HDR Photography*, Ampho Books, New York.

Gatta C (2006) Human Visual System Color Perception Models and Applications to Computer Graphics, PhD thesis, Università degli Studi di Milano, Dottorato di Ricerca in Informatica, XVIII ciclo.

Gilchrist, A (2006) *Seeing Black and White*, Oxford University Press, Oxford.

Green P ed (2010) *Color Management Understanding and Using ICC Profiles*, Wiley/IS&T, Chichester.

Holm, J (1996) Photographics tone and colour reproduction goals, In CIE Expert Symposium '96, 51–6.

Khang J, Johnson GM & Fairchild MD (2007) iCAM 06: A Refined Image Appearance Model for HDR Image rendering. *J Vis Comm & img Representation*, **18**(5), 406–14.

Kotera H, Hayami H, Tsuchiya H, Kan R, OIbida KY, Sbibata T & Tsuda Y (1978) Color Separating Methods and Apparatus using Statistical Method Techniques, US Patent 4,090,243, filed May 6, 1976: issued May 16, 1978.

Land E & McCann JJ (1971) Lightness and Retinex Theory, *Journal of the Optical Society of America*, **61**(1), 1–11.

Larson G, Rushmeier H & Piatko C (1997) Visibility Matching Tone Reproduction Opera for High Dynamic Range Scenes, *IEEE Trans Visualization & Computer Graphics*, **3**(4), 291–6.

McCann JJ (1995) Digital Color Transforms Applied to Fine Art Reproduction, *Proc. 2nd Int Conf Imag Sci & Hardcopy, Reprographic Scientists and Engineers Soc*, China Instrumentation Society, Guilin, **2**, 221.

McCann JJ (1997) High-resolution color photographic reproductions, *SPIE Proc*, **3025**, 53–9.

McCann JJ (2006) High-Dynamic-Range Scene Compression in Humans, *Proc. SPIE*, **6057**, 605707–711.

McCollough F (2008) *Complete Guide to High Dynamic Range Digital Photography*, Lark Books, North Carolina.

Mees CEK (1961) *From Dry Plates to Ektachrome Film: a Story of Photographic Research*, Ziff-Davis Pub, New York.

Mees CEK & James T (1966) *The Theory of the Photographic Process*, 3rd edn, Macmillan, New York.

Pratt WK (1991) *Digital Image Processing*, 2nd edn, John Wiley & Sons Inc, New York.

Pugsley PC (1975) Colour Correction Image Reproduction Methods and Apparatus, US Patent 3,893,166, filed May 6, 1974: issued July 1, 1975.

Reinhard E, & Devlin K, (2004) Dynamic Range Reduction Inspired by Photoreceptor Physiology, *IEEE Trans Vis & Comp Graphics*, **11**(1), 13–24.

Reinhard E, Stark M, Shirley P & Ferwerda (2002) Photographic Tone Reproduction for Digital Images, *ACM Transactions on Graphics*, **21**(3), 367–76.

Reinhard E, Ward G, Pattanaik S & Debevec P (2006) *High Dynamic Range Imaging Acquisition, Display and Image-Based Lighting*, Elsevier, Morgan Kaufmann, Amsterdam.

Scheel A, Stamminger M, & Seidel H (2000) Tone Reproduction for Interactive Walkthroughs, *Computer Graphics Forum*, **19**(3), 301–12.

Tomasi C and Manduchi R (1998) Bilateral Filtering for Gray and Color Images, Proc. of IEEE Int. Conf Computer Vision, 836–46.

Tumblin J & Rushmeier H (1993) Tone Reproduction for Realistic Images, *IEEE Computer Graphics and Application*, **42**, 13.

Ward G (1994) A Contrast-based scale factor for luminance display, in *Graphics Gems II* (ed P Heckbe), Academic Press, Boston, PP. 80–3.

32

Retinex Algorithms

32.1 Topics

This chapter describes the Retinex approach to High Dynamic Range imaging. It explains the early algorithms, their goals, assumptions and computational practices. Retinex processing changed dramatically with the advances in image processing and in digital photography. We recount the research on capturing real-life scenes, calculating appearances and rendering sensations on film, and other limited dynamic-range media. We describe: the first patents, the first computer simulations using 20 by 24 pixel arrays, psychophysical experiments and computational models of color constancy and dynamic range compression and the Frankle-McCann computationally efficient Retinex image processing of 512 by 512 images. It includes many modifications of the original approach including recent models of human vision and gamut-mapping applications. This chapter emphasizes the need for parallel studies of psychophysical measurements of human vision and computational algorithms used in commercial imaging systems. Further, it emphasizes that the spatial processing step is in the middle of the image processing chain.

32.2 Introduction

The word *Retinex* has come to describe a number of different ideas.

Strictly speaking, as coined by Land (1964), Retinex meant three independent color channels. It is a contraction of retina and cortex, and is the name of the mechanism that generates these independent spectral lightnesses. Over time, it has come to be used as the name of a variety of different spatial-image processing algorithms. This chapter will attempt to describe, review, differentiate, and suggest some consistent nomenclature for the different usages of the word.

32.2.1 Three Independent Spatial Retinexes Calculate Three Lightnesses – 1964

Land was working on color constancy. His key observation was that color appearance correlates with the lightness appearances in three wavebands of light. The triplet of long-, middle-, and short-wave

---

The Art and Science of HDR Imaging, First Edition. John J. McCann, Alessandro Rizzi.
© 2012 John Wiley & Sons, Ltd. Published 2012 by John Wiley & Sons, Ltd.

lightnesses correlates with color sensations for all kinds of color experiments (constancy, simultaneous contrast, color assimilation). The word Retinex was coined by Land (1964), because he needed a name for the three processes that generated the lightness triplets (Chapter 27).

Land and Smitty Stevens discussed the name for the output of each Retinex (Land, 1974). Land wrote:

> "Smitty felt that we were imposing on the word lightness a set of properties somewhat different from those it had had in the past. Our position would be better if we coined a new word that was defined in our own terms. He proceeded to coin a new word **grexity** and gave our definition of lightness to it. He derived it from a Greek word meaning gray woman and the German word to groan and grumble."

Smitty had written to Land:

> "**Grexity** is not dependent on illumination. Grexity is not reflectance. In order to see grexity one has to be only on the rods – or have the same grexity on all three retinexes. Grexity is an experience." [Stevens, personal communication to Land, 1967]

As mentioned in Chapter 2, the CIE (Wyszecki & Stiles, 1982) defines lightness and brightness as the attributes of a visual sensation as:

> Brightness: appears to emit more or less light

> Lightness: appears to emit more or less light in proportion to that emitted by a similarly illuminated area perceived as a "white" stimulus

LA Jones's OSA Committee (1953) had defined lightness as the "perception by which white objects are distinguished from gray, and light from dark colored objects". This is consistent with Gilchrist's (1994) and colleagues' usage.

Land decided to continue to use *lightness* as the name of the appearances generated by the *L-*, *M-*, *S-Retinexes*, although he did not accept the premise that lightness needed to include "past experience" as implied by Jones's description of lightness as a perception. Land did not believe that past experience plays a role in lightness appearances (sensations).

32.2.2   Ratio-Product

The three Retinex channels made it easy to predict the color appearances of objects in complex scenes. However, it emphasized a new version of an old problem. How can one calculate the lightness appearance?

Land believed that color constancy, simultaneous contrast, color assimilation, and his Red and White projection experiments showed that single-pixel information cannot predict appearances. As well, he believed that adaptation of retinal receptors cannot predict them (Chapter 27).

To calculate appearance one must use the spatial information in the scene. Hans Wallach (1948) had argued that edge ratios correlated with appearance in simple test targets. Wallach's idea was that lightness was a *local* process. We use *local* to mean the interaction between adjacent, or nearly adjacent pixels. The *Black & White Mondrian* with non-uniform illumination showed that this approach was very promising. It could establish the relationships of different Mondrian areas. However, local processes were insufficient to predict all the patches in the *Black &White Mondrian*.

Figure 32.1 (left) shows the reflectances of the papers in the Black and White Mondrian (Figure 7.1). The light paper at the top (75%) is 6.25 times higher reflectance than the dark paper at the bottom (12%). Figure 32.1 (right) shows luminance measurements of that display in non-uniform illumination. The low-reflectance paper at the bottom with higher illuminance measured 118 units. We adjusted the illuminance on the papers so that the high-reflectance paper at the top also had a reading of 118. Figure 32.1 (right) shows pairs of luminance readings at closely spaced points measured from the display in non-uniform illumination. As the position of luminance readings, on either side of an edge, approach that edge, then the ratio of those luminances approaches the ratio of the papers' reflectances.

The product of luminance edge ratios is equal to the product of the paper's reflectances. Land and McCann (1971a) combined the ratio of nearby pixel responses, (local processing) with the product of other ratios to propagate the calculated relationships across the image of the scene (more global processing). In photography, the quanta catch of the pixel is the fundamental measurement. They replaced the individual receptors with a bridge pair of receptors. The fundamental building block of the appearance image became the ratio of radiances, not the individual pixel's radiance. They demonstrated this idea by using the ratio of two luminances to modify the relationship of output values. They used an iterative process with an initial value. The OldProduct (initial value) multiplied by the ratio of radiances gave the NewProduct, or new relationship.

$$\mathrm{NewProduct} = \left[\mathrm{OldProduct} * (\mathrm{Ratio})\right] \qquad (32.1)$$

The ratio-product was the central element in the Retinex model for lightness. Land described the work leading to it in a paper presented as a Friday Evening Discourse, at the Royal Institution. He wrote:

"We have invented a strategy which our machine uses to determine the reflectance ratio. The key assumptions that make this strategy possible are: (1) changes in reflectance are discontinuous, are abrupt, and form edges; (2) changes in illumination are continuous,

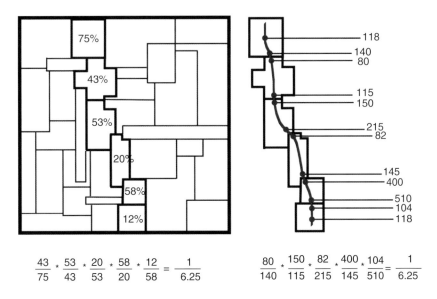

Figure 32.1 Ratios of paper reflectances (left) are equal to the ratio of luminances at points near to the papers' edges (right).

slow, and do not form edges, and (3) changes in illumination that are discontinuous and form edges may be regarded as changes in reflectance by both the human visual system and the machine. When two energy readings that form the bridge pair are made at positions closer and closer together, the influence of the illumination decreases. As the ratio of illuminations approach unity, the ratio of energies approach the ratio of reflectances.

. . . Our machine should have a way of comparing areas that are distant from one another without losing the properties of the bridges system. Some years ago, when puzzling about the next step, John McCann invented a new Gestalt, a way of looking at the reflectance relationships between widely separated areas.

. . . In other words, the product of the ratios across distant edges is the same as the ratio we would get if we placed the two remote areas next to each other. . . . Therefore by simply proceeding from one area to the next, by multiplying the bridge photometer readings across edges, we can establish a series of sequential products which yield the relative reflectance of any area relative to the first area without knowing or involving the illumination.

. . . Therefore if we had the biological ability for reading the ratio of energies across edges and for multiplying these ratios, we would arrive at a correct evaluation of reflectance of each area, all based on this program of sequential multiplication of ratios of energies across boundaries. Let us stress this again that this technique determines reflectance reliably, irrespective of the level of illumination and uniformity of the illumination." (Land, 1974)

In a subsequent paper McCann, Land, and Tatnall (1970) wrote:

"We set out to find a mathematical model that could take the information at these two areas (equal luminance white and black papers in B&W Mondrian), as well as the information in the rest of the scene, and compute a lightness value that agrees with what we see for each area. We have already seen that the luminance of an area need not correlate with lightness. While there is generally a strong correlation between lightness and reflectance, there are phenomena such as Mach bands which show this strong correlation does not always hold. Since the receptors in the retina respond to the luminous stimulus of objects, it would seem logical that the model should begin with luminance and then correct for departures from perfect correlation with lightness. We, however, took a different approach. Because in most situations there is a very strong correlation between lightness and reflectance, our model, although starting with luminance, will attempt to derive reflectances, and then make adjustments for imperfection in correlation between lightness and reflectance. . . . Having devised such as system, we began to modify that model to take into account the numerous situations in which lightness correlates less strongly with reflectance."

McCann, Land, and Tatnall (1970) described techniques for measuring the lightness appearances in a series of 12 different experiments to obtain quantitative data to be used to evaluate the accuracy of the lightness models.

In summary, there is a fine line between having a computational model that calculates reflectance, and one that calculates lightness. In the earliest days we set out to find techniques that could come close to calculating reflectance from the spatial array of scene radiances. As we began to have some success, we refined our goal to that of calculating lightness, with particular interest in cases in which lightness did not correlate with reflectance. After all, the origin of this goal of lightness was to do it three times for L, M, S radiances to calculate predicted color. We saw an example of constant reflectances with very different color appearances in Figure 27.12, called Color Assimilation. There, the triplet of lightnesses, not the triplet of reflectances, correlated with color.

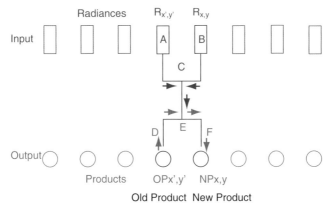

Figure 32.2   Ratio-Product process.

32.3   How to Calculate Lightness using Ratio-Products

Returning to the specific details of calculating lightness using Ratio-Product, we rewrite equation 32.1 using the arrays of input receptor and array of output pixels, as illustrated in Figure 32.2.

$$NP_{xy} = \left[log\, OP_{x',y'} + \left(log\, R_{x,y} - log\, R_{x',y'} \right) \right] \qquad (32.2)$$

Figure 32.2 shows two elements from two parallel arrays: input (receptors' responses to scene radiances) and output (product array values). Two scene radiances $R_{x',y}'$ and $R_{x,y}$ fall on two receptors (A, B) with logarithmic responses to light. The left receptor has negative output, while the right has a positive one. The sum of these outputs at C is the *ratio, (log[R_{x,y} / R_{x',y}'])*.

In order to calculate the product, we need a second array of elements that contain the values of the product for each receptor. There are a great many ways to initialize the calculation. In Land and McCann (1971a) we set all initial OldProduct pixels to equal 1.0. This is the equivalent to assuming that all output pixels have the value of white, or scene maximum. The process adds the OldProduct $(OP_{x',y}')$ at D (current output associated with location x',y') to the log ratio to calculate the *NewProduct* $(NP_{x,y})$ at E (new output associated with location x,y) and stored at F. The NewProduct replaces $OP_{x,y}$, or modifies the initial white value. The idea of the ratio-product was to provide the means to propagate information from one part of an image to another.

32.3.1   Normalization

As we discussed in Section A, human rods and cones respond to more than 10 log units of light. Again, we cannot respond to a scene with that range, because we can neither image it accurately on the retina (Section C), nor transmit that range to the visual cortex. What we see is normalized to maxima in the image (Chapter 21). If we want to model vision we need a normalization process. Ideally, that should mimic vision's normalization that is highly responsive to scene content (Section D).

The example of a path shown in Figure 32.1 takes the ratios at edges and calculates the products that relate the top white area to the bottom black area. In that example, we used only the receptors that fall across edges. What if, for every input sensor in the imaging array, there was an output element? Such a simple, iterative Ratio-Product model would calculate a NewProduct array proportional to the scene luminance array at completion. Ratio-product models were never intended to reach the

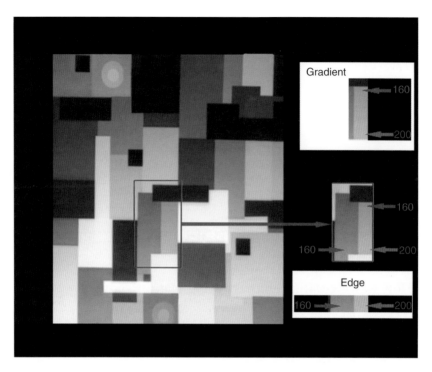

Figure 32.3 Two different appearances for luminance changes from 160 to 200. When abrupt, it is highly visible; when gradual, it is almost invisible.

steady-state of normalized input. Its goal was to calculate appearance (sensations). As discussed above, lightness does not always equal reflectance.

32.3.2 Models that Include the Limitations of Vision

The idea was that human vision cannot perform perfect image processing calculations with floating point precision across arrays with millions of receptors. The biological components that make up rods, cones, and neurons have real limits that can influence the *NewProduct* response. At first, we thought that the eye's ability to report accurately very small changes in luminance could act as a threshold. Thresholds emphasize edges and remove slow-changing gradients of light. Observations of the Black and White Mondrian show that effect. Figure 32.3 shows an example in which an edge ratio of 200:160 (adjacent pixels) generates a large change in appearance. However, a gradient of 200:160 (separated pixels) is barely above the detection threshold.

The idea behind Ratio-Product models was that the limitation caused by biological calculation shaped human vision's appearances. These limits are found in both the precision of calculation and the spatial interactions possible with neural pathways. We set out to study threshold of gradients, the role of local, instead of global, normalization, and sampling statistics on ratio-product calculations.

32.3.3 Gradient Threshold

Land and McCann (1971a) described the idea that there must be some finite limit to ratio detection of gradients. The Black and White Mondrian is made up of papers with sharp edges in reflectance. The illumination is a gradient generated from a single lamp near the bottom of the display. The small differences between adjacent receptors reported the small changes in illumination falling on the same reflectances. Rather than use an image segmentation approach to search for visually significant boundaries, we introduced the idea of a threshold. Receptors with foveal spacing are 1.0 minute of arc apart. There was data available on the smallest edges that people can detect. Blackwell (1946) had measured the threshold of edge detection to be three parts in a thousand at high luminance levels. Humans cannot detect edge ratios smaller than 1003/1000, or 1.003. If the Black and White Mondrian subtended 30 degrees of visual angle, and if individual foveal receptors subtend one minute of arc, then there are $30 \times 60 = 1800$ receptors in the image height of the Mondrian. If the threshold for each pair is 1.003, then the gradient it could remove is $(1.003)^{1800} = 219$. In other words, such a threshold applied to a 30 degree image can remove a gradient of over 200:1, while the gradient on the Mondrian is only 20:1. The threshold replaced ratios close to 1.0 with exactly 1.0. Using this threshold made the model responsive to edges and insensitive to gradients. This idea played an important role in the Ives Medal Address (Land, 1967), and the subsequent paper by Land and McCann (1971a).

In Figure 32.5 the blue square identifies the threshold described by Land and McCann. It changes any ratio value within 3 parts in 1000 to become a ratio of 1.0.

$$ if \left(0.997 < \left(\frac{R_{x,y}}{R_{x',y'}} \right) < 1.003 \right), then \left(\frac{R_{x,y}}{R_{x',y'}} \right) = 1.0 \qquad (32.3) $$

32.3.4   Reset

The idea for implementing the Ratio-Product model was to accumulate the results of a large number of paths across the array of receptor responses. The initialization to white meant that each path assumed that the OldProduct of the first pixel was white. Along the path all other edge ratios increased, or decreased, the values of other NewProducts. If a path happened to start in a white paper, then all the NewProducts would equal percent reflectance with white = 1.0. If the path started in a 5%, dark gray area, then all the values would be 20 times larger. If we look at the set of NewProduct values describing a single output pixel, we see the result for each path is scaled by its starting point. We adopted the idea that each path should *reset* itself, namely if the NewProduct output at a pixel was greater than white (1.0), then it should be set equal to 1.0. The *reset*, that normalizes the NewProducts to maxima, removes the dependence on the path's origin (equation 32.4).

$$ if \left(NP_{x,y} \right) > 1.0, then \left(NP_{x,y} \right) = 1.0 \qquad (32.4) $$

Reset is a very unusual computational non-linearity. We discovered its properties when working on the analog electronic demonstration for the Ives Medal Lecture (Figure 7.3). That device used logarithmic receptors so that any multiplication was performed by the summation of positive and negative voltages (excitatory and inhibitory responses). Using summations of logarithmic signals the *products* 1.0, or 100% maximum, is the log(1.0), or zero. We used an electronic system that was negative voltage-only, it *reset* any positive potentials to 0 voltage, thus limiting the *product* in analog electronics. The device performed the *reset* without logical (if statement) processing, just analog circuit design (Land et al. 1972).

The *reset* adds a self-normalization mechanism that can be easily implemented by a network of receptors. It is easy to compare a receptor's response to an average in an electrical circuit, but more difficult to compare to a maximum. Normalization in a computer program first has to poll all the receptor responses to find the maximum; second, it stores that value; third, it divides each receptor response by the stored maximum value. The process we implemented here is much simpler, and very efficient for a network. If the product exceeded 1.0 (whiter than white) that product was set to 1.0 and everything proceeded using the same rules. All previous contributions to NewProduct values were unchanged.

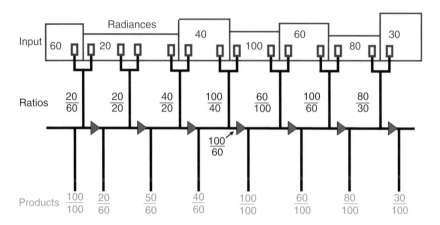

Figure 32.4   A specific example of how the reset mechanism operates. The pattern consists of a series of papers that reflect 60, 20, 40, 100, 60, 80, and 30 % in uniform illumination.

Using uniform illumination Figure 32.4 shows the *reset* mechanism's response at each receptor pixel. The initial state of 100/100 is on the left. The OldProduct is 1.0. assigning the first area a lightness equivalent to 100 % reflectance, The first pair of detectors read 60 and 20 (ratio 20/60 = 0.33). The NewProduct output for the second area is 33 %. These values do not correspond with the actual reflectances, but are proportionally correct. The next pair lies within the boundaries of an area and has a ratio of 20/20, or 1.0. Thus, multiplying by 1.0 transmits the edge-ratio signal across a uniform area. The next ratio on the path is 40/20, which when multiplied by 20/60 equals 40/60 or 0.67.

The operation of the next pair of receptors shows how the system finds the highest reflectance automatically and resets future products. The edge ratio from this pair is 100/40. We have already had fractions from edge ratios that are larger and smaller than 1.0. However, the NewProduct, in this case, 100/40 times the continued product 40/60 equals 100/60 and this is the first product that has a value greater than 1.0. Our procedure was that any value greater than 1.0 will be reset to equal 1.0. The remaining Old Products 60/100, 80/100, and 30/100 equal the relative reflectances of the papers in the display. Depending on location of the path and the image content, reset can occur more than once along the path. There is no attempt to modify responses prior to reset. As we will see these initial overestimates of reflectance can be used to model the discrepancies between lightness and reflectance.

Maxima play a special role in vision. We have seen this in all the measurements described in Section D, particularly in the Hipparchus line data (Chapter 21), and again in the FergusSpiel experiments (Chapter 28). If the objective is to mimic vision, then the normalization needs to be relative to the maxima, not relative to the average. In Figure 32.5 the red triangle indicates the *reset* operation.

32.3.5   Average Old and New Products

Although *reset* greatly reduces the variability of lightness predictions (NewProducts), it does not remove all of it. What do we do with the areas between the start and the first maximum? In the original Retinex, for simplicity and efficiency, we use all the information. The procedure assumes that the lightness prediction is the combination of many independent paths. The simplest way to combine different outputs from different paths is to average them. In Figure 32.5 the green circle identifies the average process. The NewProduct at x,y is the average of the OldProduct and NewProduct at x,y.

$$NewProduct_{x,y} = (OldProduct_{x,y} + NewProduct_{x,y})/2 \qquad (32.5)$$

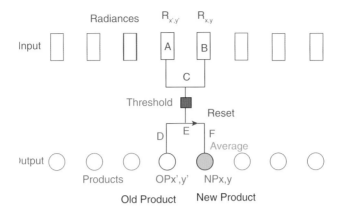

Figure 32.5   Threshold, Reset, and Average elements in the original Retinex model.

If we consider the example in Figure 32.4, the first path would have percentage outputs of [100, 33, 67, 100, 60, 80, and 30]. A path in the opposite direction would have [60, 20, 40, 100, 75, 100, and 100]. Thus averaging only two paths is closer to the papers reflectances than either of the separate paths.

32.3.6   Scene Dependent Output

The *reset* operation provided a means to normalize the array of receptor responses within the network. The maximum in the image, at some distant location did not need to be found and then stored, and retrieved, for each receptor. The other very interesting property was that this process introduced a strong dependence on scene content. If the scene had large, distributed areas of maximum radiance, then it appears high in contrast, modeled by few resets. If, however, the scene has only a small localized maximum radiance, then it appears low in contrast, modeled by many local resets.

32.3.7   Models Using Imposed Limits in Computation

The blue square in Figure 32.5 illustrates the use of threshold to remove light gradients from the new product output (Land et al., 1972). If the ratio near 1.0 was smaller than a certain value, then its ratio was set to 1.0.

The red triangle in Figure 32.5 illustrates the *reset* step. This acted as a local normalization. As stated above, if at onset the entire old product array is 0, or white, then the results of each path would be proportional to receptor response of the first receptor on the path. The sum of all paths on all pixels would reflect the average of the input scene. Again studies of vision have shown that maxima in scenes, not averages have the greatest influence on appearance. The *reset* adjusts the output to the maxima along each path.

The green circle in Figure 32.5 illustrates the average step. In a log implementation the average of the OldProduct and NewProduct at *x,y* is the geometric mean. Paths are made by continuing the same operations with the next pair of pixels.

32.4   A Variety of Processing Networks

Figure 32.6 shows examples from Land and McCann (1971a). Figure 32.6(a) shows a simplified conception of the scheme. Two opposed logarithmic receptors (A, B) first sum with each other (C) and

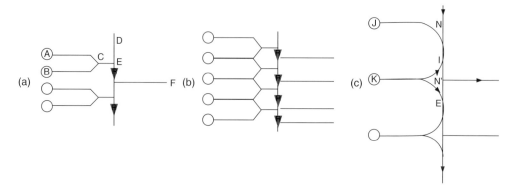

Figure 32.6   Equivalent ratio-products diagram models.

then sum with the OldProduct (D) at E. When the logarithm of the sequential product is greater than 0, it is reset to 0. This operation is indicated in the diagram by the red triangle below **E**. This total quantity is both the readout of the system (F), and the OldProduct that is combined with the output of the next receptor pair.

In 32.6(b), each photocell is the leading photocell for one bridge pair and the trailing photocell for the next bridge pair. In 32.6(c), shows a third synaptic model. The receptor K transmits its signal to its synapses I and E. Synapse I is an inhibitory synapse and adds to the sum of J and the product N. The new product is formed at N′ and is tapped off the chain between the two synapses I and E. Synapse E is excitatory and combines with this new sequential product N′ for the computation of the next product. This diagram 32.6(c) does not have a *reset*. Later, Land (1983; 1986) added a fourth, two-dimensional *Designator* model (see Chapter 32.7).

32.5   Image Content

The first computer model for accumulating the NewProduct array was written by Land's good friend Ed Purcell, the Harvard Physics professor who discovered nuclear magnetic resonance. We implemented it on the central research computer at Polaroid. To do image processing on 20 by 24 pixel images we needed to double the size of the computer's memory. Polaroid invested $18 000 to add 8 k of wound core computer memory for Polaroid Research's central computer. The study provided a wide range of background information on spatial interactions.

In the late 1960s, we constructed a series of test targets to measure the appearance of gray patches in white and black surrounds (McCann, et al., 1970). Observers matched the appearance of the gray areas to a standard target with nine equally spaced lightnesses: 9.0 equal to white; 1.0 equal to black.

We modeled the observer data with a wide variety of Ratio-Product parameters. We studied the effects of number of paths, length of paths, random vs. straight paths, the presence and absence of threshold, the average number of NewProducts per pixel. In McCann et al. (1976) we described a three channel model for Color Mondrians, based on our results. The input to the Retinex calculation was 480 luminances spaced regularly, in a 24 by 20 array. The computer model was not given the positions of boundaries of areas, just the luminances at all points. We chose to use unidirectional paths of length 200 for all three color channel calculations. The origin of the path and its direction in the 480-point array were determined by a random-number generator. The paths traveled straight ahead until they reached the perimeter of the target where they either reflected back into the target, or traveled along the perimeter. The direction of the reflection from the perimeter was also chosen by a random-number.

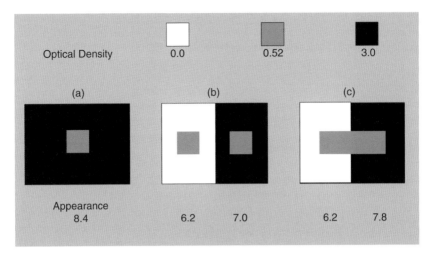

Figure 32.7   Three test targets with optical densities 0.0, 0.5, and 3.0: (a) density 0.5 looks light [8.4] in a completely black surround; (b) gray is darker [7.0] surrounded by white, and is lighter [6.2] surrounded by black; (c) grays at the same location are [7.8] and [6.2] with a gray bar.

At each point along the path we used the *Ratio-Product-Reset-Average* calculation (equations 32.2, 32.3, 32.4).

32.5.1   Normalization in Vision – Neither Local, Nor Global

The set of Simultaneous Contrast test targets in Figure 32.7 is a very sensitive test for vision models and their response to image content. It is a kind of litmus test for successful models of spatial interactions. First, the Gray in Black, Figure 32.7(a) with no other light in the room is an important control experiment. With the gray area (O.D. 0.52) as the highest luminance in the field of view, it appears near white [8.4].

If the model used only *local* comparisons, as suggested by Wallach (1948), then the gray square on black in 32.7 (b) should have the same appearance as in 32.7(a). It is the same local stimulus. It should appear nearly white. It does not. The gray in black matches a lightness value of 8.4, while in the presence of white in the scene it matches 7.0.

The other extreme would be if the model made accurate global comparisons, such as a computer program would do when finding the maximum value and normalizing all receptors to the maximum. Also, one could design a global Ratio-Product model with many long paths. The longer the path, the fewer the reports along the paths that have not reached their final reset. With accurate global comparisons, both gray patches in 32.7(b) must be equal, with a value of 6.2. They are not.

The observers' matching data (Figure 32.7 bottom row) reported that the gray in black (b) is about 13 % lighter [7.0 compared to 6.2] than the same gray in white. The spatial comparisons have neither local, nor global extent. They involve the entire scene in their normalization process, but it is an imperfect normalization. Image content plays an important role (Section D; McCann, 1987, 2000a).

The whites, grays and blacks in Figure 32.7 had constant transmittances, yet they were matched by different lightness values. Figure 32.7(c) shows the Albers (1962) experiment connecting the grays. The gray appears both uniform across its surface and slightly darker in white. This provides another important test for lightness models.

In 19th century psychological terms, the central gray squares in Figure 32.7 are influenced by their surrounds. In the context of 21st century image processing algorithms, assigning single values to large groups of white, gray and black pixels does not make sense. The observed lightnesses show content-dependent differences. If the gray square has sides of n pixels, with an area of n^2, then it would have $4n + 4$ adjacent pixels. A local spatial comparison could involve relatively few pixels. At the other extreme the entire test target is roughly $6n$ wide and $4n$ high, with an area of $862n^2$. *Global* algorithms compare every pixel, one at a time, with every other pixel. Global image processing would be slow and costly. It is difficult to imagine how the human neural anatomy could compare every pixel to every other pixel. The matches in Figure 32.7(b) show that vision does not. Instead, human vision involves large portions of the receptive field as we saw in Section D.

We need a term to describe the unique properties of vision. The gray square is not affected by just local influence, nor by universal global influence. Alessandro Rizzi gave the name *locality* for the neighborhood influence to describe vision's unique, intermediate properties – neither local, nor global. If the Retinex's model of lightness is going to mimic vision, then we have to measure and understand *locality* influence.

We studied a large number of different test targets including: high-average luminance, low-average luminance, gradients in illumination, edges and gradient targets (Chapter 35). At Polaroid, we varied the model's parameters of the paths using random, straight, long and short paths, along with varying the averaging techniques. Straight paths that reflected at the edge of the target were the most efficient. The length of the path determined the local vs. global extent of the comparison. Many test targets were relatively insensitive to the length of the path. Figure 32.7(b) and (c) were exceptions.

Figure 32.8 illustrates the role of resets along the length of the path. On the left side, many paths will start in the maximum white area and they will never reset. They produce a high-contrast rendition of the gray surrounded by white (see Path A). A smaller number of paths will start in the left gray square, but will finally reset to the white as soon as the path leaves the square.

Many paths will start in the black surround on the right (see Path B). These paths will have a local reset when they pass into the gray square on the right. They will not have the final reset until they enter

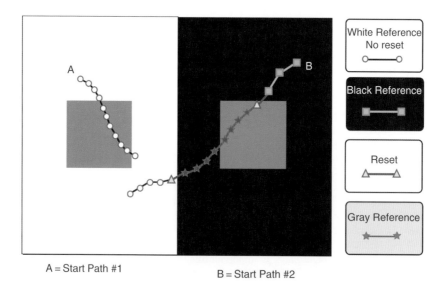

A = Start Path #1

B = Start Path #2

Figure 32.8 Path A starts in a maximum luminance white area: it never resets. Path B starts in a black and resets twice (blue lines), first to gray, and then to white.

the white surround on the left. These paths produce a low-contrast rendition of the gray surrounded by black because of the higher values reported before the final reset.

The effect of the length of paths is evident. Ratios and very short paths make local comparisons. Sparse sampling also contributes to emphasizing local relationships. Very long paths, that reflect at the edges of the array, approach, as a limit, relative reflectances of the papers. The longer the path, the more *global* the spatial comparison. The longer the path the smaller the *locality* influence.

32.5.2   Tools to Control a Response to Scene Content

As described above in Figure 32.4, the array of OldProducts *(OP)* is initialized to 1.0, or white. Each iteration at a pixel *x,y* modifies the uniform white output array, making that pixel darker. The very first ratio compares the value at *x,y* and compares it to x',y'(ratio); resets the intermediate NewProduct if greater than 1.0; and averages with the $OP_{x,y}$ value that in this case is 1.0. If the $R_{x,y}$ value is the maximum in the scene, it will have $NP_{x,y}$ equal to 1.0 for every iteration. If the input radiance value is less than the maximum, $NP_{x,y}$ will get darker until it reaches its asymptote. That asymptote is the input image of the scene divided by the maximum radiance (global normalization).

We can monitor the *NP* output array values as we add iterations. We start with a uniform white. Each iteration modifies a pixel in the output array, making it darker (lower digits). Pixels in close proximity to the maximum in the scene get darker faster, as we saw in Path A on the left side of Figure 32.8. There are no resets in that path. On Path B the initial pixel report was that the black paper was white. When the path reached the gray square, the output report was that the gray should be assigned the white value. When the path finally reached the white paper, the path had its final reset, and all further $NP_{x,y}$ reports were the same as those from Path A. The reports from paths prior to the final reset lighten the output image (lower the contrast) compared to the asymptotic normalized value. However, the lightening influence diminishes with increased length and number of paths.

This highly nonlinear operation has some interesting properties. It normalizes the image to the maximum in a peculiar way. We had no interest in the asymptotic case of arriving at radiance divided by the maximum. We were interested in the property that the *gray-near-white* areas get darker sooner than the same gray in a black surround. We also observed that a gray gets darker faster if the white area is larger, or surrounds the gray square. We also observed that local maxima, less than the maximum of the entire scene, have a strong local influence. All of these properties are needed to find a model that mimics spatial vision as measured in Sections D and E.

Reset is the antidote to glare. The greater the amount of white in the locality, the lower the contrast of the retinal image controlled by glare. The spatial properties of reset make pixels in a white locality have lower *NP* values predicting higher apparent contrast. This unusual operation has the same unusual properties as vision.

The HDR model that mimics vision is neither a local, nor a global process. It shows the influence of both. It is not a theoretically robust process that can be described by broad generalizations. It is an imperfect normalization process, which describes itself by the careful measurements of its imperfections. The study of color constancy is a good example. There the departures from perfect constancy give us the signature of the underlying process (Chapter 27). The measurements of appearances in lightness test targets tell us how to design the *locality influence*.

32.5.3   Edges and Gradients

Stanley Coren wrote a Letter to the Editor after reading Land and McCann (1971a). He argued that a threshold model that totally removed gradients could not accurately predict the appearances of the A and B targets in Figure 32.9. The edge ratios in all three are the same. We extended the experiment to include target C. he gradients connecting the edges either increase, remain constant, or decrease. The luminance that results is shown in Figure 32.9. We made three test targets A, B, C with white and

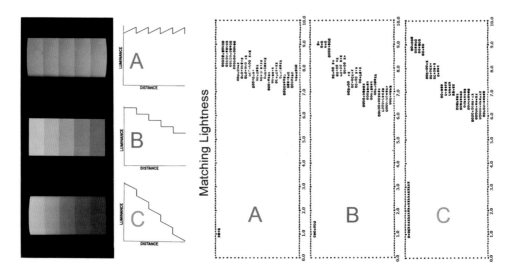

Figure 32.9 Constant edges with variable gradients: (left) photographs with plots of luminance vs. distance; (left center) plots of luminance; (right) observed lightness (O) vs. model predictions (X) (range ± 1 SDM). Note that the observed and predicted lightness for decreasing gradients show an increasing appearance slope in C.

black papers wrapped around a cylinder and spun it. The percentage of white and black papers around the circumference of the drum determined the % reflectance of the spinning drum. Figure 32.9 also shows photographs, plots of luminance, calculated lightness using the *Ratio-Product-Reset* model (X), and observer data (O) measuring the appearance along the face of the drum.

Coren's hypothesis is correct. If the edge ratios for these targets are equal, and if vision removed all gradients, then these three displays must appear identical.

The result of these experiments show:

1. Appearance does not correlate with luminance.
2. Appearances of A, B, C are similar, but they have important differences. The bands in all three get lighter left to right. The increasing-gradients target (A left) is lighter than flat reflectance (B middle), and it is lighter than the decreasing-gradient (C right).
3. All of the individual bands in all three targets appear lighter on the right side of the band, than on the left. Appearance follows luminance (left); shows a Mach Band with flat radiance (middle); and show increasing lightness with decreasing luminance gradients (right) (Ratliff, 1965).
4. We applied the *Ratio-Threshold-Product-Reset-Average* model to an array of luminance measured from these experiments. We used the parameters that were optimal for predicting the family of simultaneous contrast targets in Figure 32.7. The plots of calculated lightness vs. drum position superimposed on the plots of observed lightness vs. drum position. The model that included resets predicted appearance. The appearance of gradients is more complex than predicted by a threshold mechanism alone.

Coren presented a paradox. If vision removes gradients, then the three targets must appear identical. But, with further study we can understand it. Vision defies generalizations. Humans see gradients, but at much lower apparent contrast, than they see edges (Figure 32.3). The three targets look very similar, despite large differences in luminance profiles. The same properties of *reset* used to model Figure 32.7 were able to successfully model the measurements in Figure 32.9.

32.6   Real Images – 1975

In 1975, under the technical leadership of Jon Frankle, Polaroid purchased an I²S image processor (McCann and Houston 1983a, b; McCann, 2004). While the study of paths of *Ratio-Product-Reset-Average* model taught us about the nature of spatial interactions, it was hopelessly slow for processing real images. Influenced by the sampling techniques used by Stiehl et al. (1983), Frankle and McCann (1983) patented a multi-resolution algorithm that made it possible to process real images in real time. The Frankle-McCann multi-resolution process used the *Ratio-Threshold-Product-Reset-Average* model. The patent provides a complete description of the process. It provides FortranIV code for the I²S image processor and a description of the pre- and post-lookup tables (LUT). In today's world the 15 pages of the Fortran® code can be replaced by a half a page of MATLAB® code. The code and a discussion of the design of Pre- and Post-LUT parts of the model are included in the patent.

- First, this implementation of Retinex processed real images (512 by 512 pixels) in real time (seconds). This was a considerable challenge in 1975 considering the algorithm used the entire image in the calculation. The process combined special purpose hardware with an extremely efficient multi-resolution algorithm.

  The patent, "Method and apparatus of lightness imaging", was written by Hugo Liepmann and Bill Roberson in three conceptual levels with 86 claims, far more than the 17 claims in the first Retinex patent (Land and McCann, 1971b). The first level described the specific implementation of the multi-resolution *Ratio-Threshold-Product-Reset-Average* operations. The second level was a broader embodiment of ratio-product claimed as ". . . providing, for each pairing of segmental areas, a comparative measure of said radiance information at the paired areas,". The third level was an extremely broad restatement, that claimed ". . . A. receiving information responsive to the radiance values defining an image field, and B. deriving from said information a lightness field containing final lightness values for predetermined segmental areas of said image field . . . "

  This multilevel structure of claims was adopted because we realized the power of multi-resolution image processing. It was a timely publication because the Frankle and McCann patent was filed on August 28, 1980 and published on May 17, 1983. Well-known references in multi-resolution imaging introduced the term pyramid processing, e.g. "A Multiresolution Spline with Application to Image Mosaics" by Burt and Adelson was published in September (1983), and by Rosenfeld (1984).

- Second, it clarified the goals of the calculation, It specified that the algorithm calculated a lightness field, that always correlated with appearance, rather than saying that reflectance was the calculation's objective. As we saw in Chapter 29 reflectance correlates with appearance some, but not all, of the time. If an algorithm removed all traces of shadows, then the special light observed at sunrise and sunset would be removed. Such image processing would be counter-productive. It would make terrible pictures (Chapters 4–6).

- Third, it described that the lightness field algorithm resides in the middle of the image processing chain. Prior to the algorithm's spatial processing, care should be given to the array of digits used as input. If the 0 to 255 digits in the input image represent log radiance of the scene, then the difference in digits represents the ratio of scene radiances. A white-black edge in the sun could have 250 and 150 digits. The same papers in the shade could have 125 and 25 across their edge. The difference in digits (100) is constant because of the log luminance input images. If, instead the input image is a nonlinear transform of luminance, often used in photography, then the same papers in sun and shade will no longer have the same spatial relationships, i.e. ratio of radiances, or difference in digits. The property that such pairs of white and black papers have identical edge ratios in sun and in shade is very important for mimicking vision. The same objects need to have the same edge ratios in different illuminations. We need to take care that the image capture device has not introduced tone-scale transformations that alter the edge ratios. A change in H&D curve slope in sun and shade will be the same as changing the reflectances of the objects in the scene.

A similar problem occurs after the completion of the spatial processing. The digits in computer memory need an output device, printer or display, for viewing. These devices expect a particular color space that converts into reflectance on a print, and light on a display. There needs to be post-processing to shape the spatial output into the display device's expected color space. In the late 1970s, before digital cameras, device profiles and personal computers, electronic imaging hardware was designed and built as integrated systems from capture to display. The Frankle and McCann patent described in detail the steps required to calibrate input images so that 0–255 digits represented log scene luminance. As well, they described the post-processing steps to scale the calculated lightness fields to the color space that the display required for good looking images. This level of painstaking detail is necessary to compare different spatial imaging processes fairly. One cannot ignore the effects of input calibration and display response to digits (Chapter 35).

32.6.1   Real Images in Real Time

Frankle and McCann described the Retinex process using the three channel I^{2}S array processor in Figure 32.10. The original input image is read from computer memory to I^{2}S memory [28]. In step I, the original (digit = log radiance) image is assigned a negative value, and a second, positive, copy of the image is scrolled (x,y) into [36]. The values in 28 and 36 are summed to calculate log ratios for all pixels in the array in one cycle time of the device. In step II, the OldProduct stored in [44] is added to the ratio to calculate the IntermediateProduct. Step III resets pixel values greater than 100%. Step IV averages the OldProduct with the reset intermediateProduct and stores the array of NewProduct values in [44] for the next iteration.

The process did not use paths. The initial value of white for all pixels was modified by the reset NewProduct comparing the initial pixel with the scrolled pixel. That changed all the OldProduct values used in the next iteration. We used scrolls for eight directions by using plus, zero, and minus displacements of x and y. Starting with large x,y displacements, the first iterations compared distant pixels in one direction. The next iteration used the same displacement, but in a different direction. Different directions are needed to avoid artifacts in the image. With sufficient comparisons for that displacement,

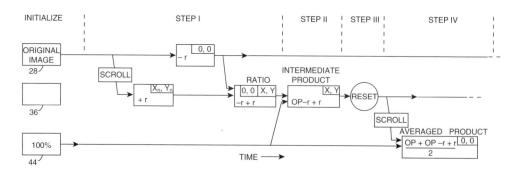

Figure 32.10   Diagram of the Frankle and McCann hardware image processing.

Figure 32.11 Photographs of a computer monitor; (left) shows scanned input images of independent records of the inside and outside scenes; (right) shows calibrated combination image used as input (top) and output of Retinex processing (bottom). The Retinex processed digital image renders the high-dynamic-range scene in a low-dynamic-range media.

the process was repeated with smaller displacements, until the final set of comparisons using one pixel displacements. The number and direction history of the displacements at each size determined the locality influence of the process. The device calculated Ratio, Product, Reset, Average for all 512 pixels in 56 seconds in 1978.

In the late 1970s it was difficult to digitize a photograph, and extremely difficult to digitize a high-dynamic range image for input to Retinex processing. We synthesized the test image shown in Figure 32.11. We took a photograph, using flash, of a model in front of a window frame in front of black paper. We selected a second outdoor scene photograph. We digitized both using a broadcast-quality video camera and a custom made scan converter. The two independent images are shown in Figure 32.11(left). We scaled the scanned images so that digits were proportional to log luminance. We scaled the outside scene to be dimmer than the indoor flash image. Figure 32.11(right) shows the combined input image (top) and the Retinex processed image (bottom). The prints were made by photographing a CRT display and were stored in a lab notebook.

In 1981, while at an Ansel Adams photographic workshop at Yosemite, CA, we made a number of photographs of John standing in the shade of a tree at noon on a sunny day. We used a Pentax

Figure 32.12   (left) Conventional prints with high and low exposures; (right) Retinex processed digital image capture on color negative film.

photographic spot photometer to measure the range of light between the white and black squares on the ColorChecker® test target. It was 5 stops, or 32:1 range. The white card in the shade was also 5 stops darker than the white square in the ColorChecker. The shade on that day was as dark as black paper. We had made measurements of shadows before, but not with such a clear sky. The range of this scene, from white in the sun to black in the shade, was (32*32), or 1024:1, or 3.0 log units.

This was a very exciting moment because it was the first time we measured and photographed a real-life scene outside the laboratory that had identical luminances coming from a white and a black paper at the same time. We made a number of photographs of the ColorChecker in sun and John with a white card in shade. We used some Polaroid Instant, and Kodak negative films. As we have seen in Chapter 11 the negative can capture 4 log units before glare, and more than 3.0 log units in a 35 mm camera. We used standard film development for the negative. We printed the negative using standard procedures with a variety of exposure times. The small top-left photograph shows a print made with a high exposure proper for John and the white card. The small bottom-left photograph shows a print made with a lower exposure proper for the ColorChecker. There was no intermediate exposure that made a satisfactory print (Figure 32.12 left).

We digitized the 35 mm negative along with a number of calibration test targets with an Itek graphic arts scanner. Using the calibration data, we converted the original digits into digits scaled by log radiance, so that digits 0–255 were proportional to log radiance over 3.0 log units. Figure 32.13 (left) plots the histogram of the input to Retinex. Figure 32.13 (right) plots the calibrated radiance range of the output of *Ratio-Product-Reset-Average* Retinex.

As a control we made a global compression of scene range. When we simply compressed the range of the scene from 3 to 2 log units using a Lookup Table, the scene looked as if John were in a fog. The contrast of everything in the image was 2/3 that of the original scene. The Retinex processing modified the entire captured image, preserved the edge information, and avoided fog in the rendition.

Figure 32.13 shows that the 3.0 log unit range of the scanned negative was converted to a 2.0 range of output that fit the range of the print material, and hence rendered both the sun and shade parts of the scene successfully (Figure 32.13 right). These images were described in a lecture at the Annual Meeting of the Society of Photographic Scientist and Engineers (now IS&T) in 1984.

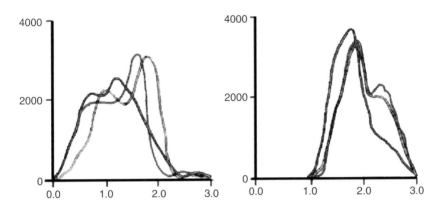

Figure 32.13   Histograms of: (left) calibrated input digits; (right) Retinex output digits vs. log luminance. The Retinex output was then scaled by Post-LUTs; printed in Figure 32.12 right.

32.6.2   Goal: Lightness Field Replaces Reflectance

In the Frankle and McCann (1983) patent they say:

"In the study of human vision one often finds in the literature the division of segments of visual images into two arbitrary categories: objects, and illumination. In addition, the literature contains numerous discussions of how human vision discounts illumination, so that information about objects in the field of view has greater emphasis. This arbitrary division of visual images has many exceptions. For example, shadows produce large changes in sensation, despite the fact that they are intensity variations in illumination. As another example, gradual reflectance changes across the surface of an object cause small changes in sensation, despite the fact that they represent changes in the properties of the object.

Instead of characterizing different portions of images as objects and as illumination, it is more useful to characterize them as radiance transitions that are abrupt, or as radiance changes that are gradual. Radiance transitions that are abrupt generate large changes in lightness, whereas radiance transitions that are gradual generate small changes in lightness. Signal processing systems that produce lightness fields produce quantities that correspond to lightness.

The foregoing lightness image processing of this invention realizes these properties of visual processing by calculating combination measures in such a way that abrupt changes in radiance are characterized by a set of reports all of which are the same. Furthermore, combination measures are calculated in such a way that gradients are de-emphasized by either a threshold or a technique using many comparison segmental areas with different spatial interaction histories, or both." (section 22 lines 33–64)

The formal language in the patent helped us to remove a considerable source of confusion in our descriptions. For making lightness appearances, gradients were less important to humans than edges, but are still important. In the Black and White Mondrians, edges were made by the papers' reflectances and the illumination was designed to be a gradient. By being more precise in our language, we shifted to distinguishing *abrupt* verses *gradual luminance changes*, instead of reflectance and illumination. By doing this we removed many unintended self-contradictions caused by the fact that appearance does not always correlate with reflectance. Again, our goal was to calculate appearance, and not reflectance.

32.6.3   Dynamic Range Compression from Threshold or Reset

The threshold was introduced to remove gradients, and *reset* was used to normalize output along a path. The practice of resetting and keeping the values prior to the final reset gave the desirable property of *locality influence*.

At first, we thought that threshold was the range compression mechanism. It stimulated our MIT neighbors' interest in the problem. Tom Stockham described homomorphic filters and Horn and Marr described Laplacian operators. These approaches applied mathematical functions to the analysis of gradients. Our research at Polaroid turned in a different direction. If the threshold mimicked our human visual system, our model should have exactly the same properties as vision. We needed to measure the rate of change on the human retina that was at the threshold of detection.

We undertook a major effort to understand the visibility of gradients. We felt we needed better data on the rate of change of radiance on the retina that was at detection threshold to improve our model. It took 10 years, but we learned that there is no universal rate of change at threshold. Vision does not work the way we suggested in Land and McCann (1971a, 1971b). We measured the magnitude of both continuous (McCann, et al., 1974) and sinusoidal gradients at threshold (threshold and supra-threshold matching: Savoy & McCann, 1975). To our surprise, there is no single threshold rate of change of luminance on the retina. All that matters is angular size and number of cycles of sinusoid, and the size of the surround (McCann & Hall, 1980; McCann, et al. 1978; McCann, 1978; Savoy, 1978). Although we had proposed the threshold mechanism, we could not find its correlate in the appearance of complex scenes. The Land and McCann iterative threshold, the Stockham spatial frequency filter, and the Marr and Horn Laplacian can improve some pictures, but do not have the same properties as vision.

About the time we were trying to understand our gradients' measurements, we found that the *reset* and short paths emphasized edges, compressed image range, and minimized gradients. We found that we could eliminate the threshold step, and still get good predictions of the Black and White Mondrian with gradient illumination (McCann, 1999).

We became fascinated with *reset* because it could model Simultaneous Contrast, in addition to auto-normalization. Even more important was the idea that *reset* provided a mechanism for calculating an output image that depended on the scene content. *Reset* was an extremely efficient *locality influence* generator.

Reset responded to scene content. That property is an important distinction from models that applied spatial-frequency filters to the input images (Campbell & Robson, 1968; Blakemore & Campbell, 1969; Stockham, 1972; Marr, 1974; Horn, 1974; Faugeras, 1979; Wilson and Bergen, 1979); as well as much later variations by Pattanik, et al. (1998) and Fairchild and Johnson (2002, 2004). They all looked to apply spatial filters to receptor images, but did not have a mechanism to independently "adapt" the filter coefficients to each object in each image.

We can study vision's response to scenes by analyzing the scene and its appearance in the spatial frequency domain. Take for example the eight gray test areas in Figure 21.2. We saw that these grays in a white surround matched the standard. The same gray luminances appeared lighter in the gray surround, and much lighter in a black surround. In order to predict these appearances using spatial frequency filters, the patches in a white surround need no filtering, patches in a gray surround need some filtering, and patches in a black surround need strong filtering. If we want to analyze sensations in the spatial frequency domain, human vision uses variable, scene dependent low-spatial-frequency filters (McCann, 2002).

32.6.4   Retinex and Lightness Field Hardware

In the early 1980s the cost of digital imaging electronics was very high. Although very desirable, consumer products, competitive in price with silver halide photography, were not possible. The approach we took was to find ways of incorporating the advantages of digital image processing without the prohibitive expense of full resolution electronics at the time.

Figure 32.14 Polaroid photographs of: left) conventional print; (right) low-slope film with out-of-focus illumination mask calculated using Retinex.

An example was a prototype device we made in the early 1980s that sampled a very low resolution electronic image of a positive 35 mm slide. We used Kodak slide duplicating film because it made a low-slope, high-dynamic-range capture of scene radiances (Chapter 5.8.6). At first, we scanned a 35 mm slide at 64 by 64 pixel resolution. We used the Ratio-Product-Rest-Average algorithm to calculate the desired low-resolution image. We then calculated the difference between the scanned image and the desired output to make a correction mask. We photographed the original scene using a conventional negative, and made a conventional print (Figure 32.14, left). We made a second photograph using low-slope slide duplicating film. We printed the positive transparency with the scanning fiber optics bundle. The amount of light was modulated by the Retinex calculation and device calibration (Figure 32.14, right). Jim Burkhardt and Karen Houston introduced a ratio limit in the Retinex calculation. The software that calculated the mask used the low-resolution scanned input data and applied the Ratio-Product-Reset algorithm. With a limit on the size of the input ratios, they found greater computational efficiency.

In the first prototypes, Jim Burkardt and Karen Houston built a scanner for image capture and a fiber optic bundle with modulated light output. A later prototype added the skills of Bill Wray, Ken Kiesel, and Dave Cronin and his research machine shop. They used a 64 by 64 pixel CCD for image capture and an LCD display provided by K Morizumi at Seiko. Figure 32.15 shows a picture of it and a diagram of the 1983 prototype.

The device shown in Figure 32.15 (left) had an on/off switch, a slot to insert a 35 mm slide, and a red button to start the process. The operator put the slide in the slot (top) and pressed the red button (front).

The device performed the following automatically:

Pressing the button tripped the strobe

The light passed through the transparent LCD display, the slide, reflected off the semi-silvered mirror and was imaged by a lens onto the low-resolution CCD array.

The digital color CCD image was transmitted to a microprocessor.

The Ratio-Product-Reset-Average algorithm calculated the low-resolution sensation of the scene.

System calibration data of the scene capture film, the CCD sensor, the LCD display, and Polaroid print film were used to calculate a negative mask image to correct the information in the slide.

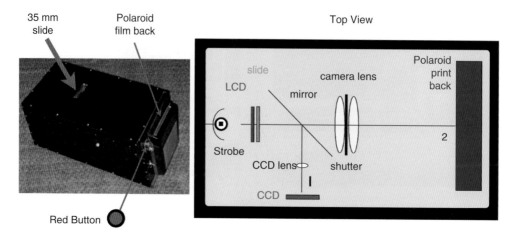

Figure 32.15   (left) Photograph of working prototype; (right) Top view diagram.

The calibrated negative mask image was sent to the Seiko LCD screen behind the slide, and displayed as an out-of-sharp-focus mask.

The microprocessor opened the print camera shutter.

The strobe flashed a second time to expose the masked image on the film.

The Polaroid back processed and ejected the print.

The prototype worked very well. It was designed for a future Polaroid instant positive transparency film that would have more than 3.0 log units' dynamic range for scene capture. At the time we completed the prototype, interest in Instant Polaroid 35 mm film began to decline.

Another example of low-resolution spatial processing is the patent by Kiesel and Wray (1987) "Reconstitution of Images". Here a full resolution radiance field is captured and averaged to form a coarse image field comprising a small fraction of the number of the full resolution pixels. The coarse field image is processed using small affordable image processing hardware to produce an improved coarse image. The improvement was isolated by comparing the coarse input image with the coarse improved image. The improvement is interpolated to full resolution and applied to the full resolution input. This could be a second scan of the input image corrected by the scaled improvement. Such techniques required very small digital storage and small processors, but provided significant improvements to images by adjusting their low-spatial frequency components. This process is the electronic equivalent of silver halide's unsharp masking techniques. In this case the mask is both unsharp, to preserve the edges, and spatially modified to compress the dynamic range. This use of applying digital unsharp masks to images is very effective, with minimal hardware cost, and speedy processing. It is used frequently today.

Another example is the application of the zoom principle described by Frankle and McCann (1983). Wray (1988) processed image pixels with operations on a relatively coarse representation of the full image field, and with operations on a high-resolution representation using only a portion of the full field. The process enables a computer with comparatively small memory capacity to execute relatively complex image processing tasks.

The experience we gained from these prototypes was twofold:

> The spatial information needed to improve the high-dynamic-range images worked best at low-spatial resolutions. Obviously, the final result has to be high resolution. But, the improvements in images are most efficiently processed at low-spatial resolution. Investing in full-resolution image processing time and processing hardware does not improve the results of the process. It only adds burdens of time and cost.

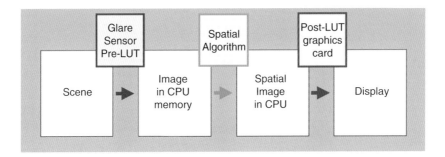

Figure 32.16   Image processing chain.

The issues described above concerning device calibration, are more time consuming, and often more demanding than designing the image processing algorithm. While the spatial processing works well over a wide range of processing parameters, the calibration of the image sent to the output device is much more critical. In the late 1970s we had a slogan in the Polaroid lab: "Whoever controls the last lookup-table will win any beauty contest between algorithms". In other words, the spatial processor output needs to fit the color space that the output devices expect.

32.6.5   Spatial Processing is in the Middle of the Imaging Chain

Figure 32.16 illustrates that the Retinex spatial image processing is in the middle of the image processing chain. Its input is modified by its capture history (Pre-LUT) and the device that displays the digits (Post-LUT).

In a sense, what we are calling Post-LUT processing is the same as the Tone-Scale mapping described decades later by Tumblin, Ward, and Reinhard (Chapter 31). Pixel-based processing leaves out the spatial-processing step that makes a general solution possible. Pixel processing just tunes the captured image using post-LUTs. It works well for individual images, and some groups of similar images, but not for all images.

32.6.6   Zoom Processing

Retinex Zoom processing reads in a full resolution image. First, it makes and stores a family of smaller images by averaging four adjacent pixels. If the input image has 2048 by 1024 pixels, then the set of smaller zoom levels will have 1024 by 512, 512 by 256, 256 by 128, . . . 4 by 2, and 2 by 1 pixels. We end up with a pair of left and right pixels that are the average values of half the image.

In order to build the output image we begin with the coarsest 2 by 1 array. If we assign the OldProduct to 1, or white, then the ratio of the left pixel to the right will darken the NewProduct of the pixel with the lower average. The amount it darkens depends on the ratio of the averages, the number of iterations, and the average technique assigned to this zoom level. The coarsest zoom level establishes the initial relationship of the left half of the output image to the right half.

We use smooth interpolation to make the OldProduct array for the next zoom level. The NewProduct of the 2 by 1 pixel array is interpolated to size 4 by 2 and used as the OldProduct for spatial processing at this second level. We use the *Ratio-Reset-Product-Average* process on all adjacent pixels to make the 4 by 2 pixel NewProduct array (Figure 32.17). There are six horizontal edge ratios (top plus bottom) of adjacent pixels. There are four vertical edge ratios, and six diagonal pixel ratios. The NewProduct output establishes the relationships of the eight image sectors, by modifying the OldProduct input. In this Zoom process we apply the Ratio-Product-Rest-Average process to adjacent pixels only at each level. Again there are no paths.

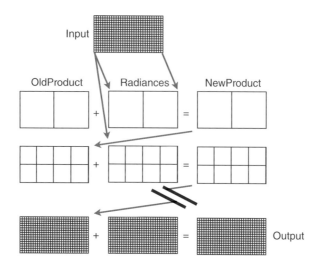

Figure 32.17 Zoom process first stores an average of adjacent pixels for all zoom levels. It takes ratios of adjacent horizontal, vertical and diagonal pixels to modify the OldProduct. It starts with coarsest level and interpolates the output NewProduct to be the OldProduct at the next less coarse level, until it reaches full resolution.

Figure 32.18 Input and output photographs of a zoom Retinex process.

In the next zoom level, the NewProduct of the 4 by 2 pixel array is interpolated to size 8 by 4 and used as the *OldProduct* for spatial processing at this level. There are 28 horizontal edge ratios, 24 vertical edge ratios, and 42 diagonal pixel ratios. The NewProduct output establishes the relationships of the 32 image sectors, by modifying the OldProduct input.

The process continues until it incorporates the full resolution image with 2,096,128 horizontal edge ratios, 2,095,104 vertical edge ratios, and 4,188,162 diagonal pixel ratios to make the final output image of 2,097,152, or n pixels. The process has incorporated comparisons of all pixels with all other pixels, or $[n*(n-1)]$ comparisons $(4.4*10^{12})$, but only used $[5n]$ comparisons $(1.1*10^{7})$ of adjacent pixels.

Figure 32.18 (left) shows a photograph of two identical JOBO test targets, taken on a clear September day in Massachusetts. Most summer days in New England have shadows that are 10 to 20 times darker than the sunlight. In the fall, with a clear blue sky, the shadow cast on the picnic table was 32 times darker, so that the white in the shade had the same luminance as the black in the sun.

Figure 32.19   Equal linear luminance digits (119) for white-in-shade, and black-in-sun. As well, it shows constant edge ratios in sun and shade.

Figure 32.19 illustrates the point made earlier that edge ratios in sun are equal to edge ratios in shade. Local spatial comparisons can be used to build an output lightness image. As before, the number of iterations tunes the locality influence of the output image. One tunes the process depending on characteristics of the input image, and the degree of desired locality influence. That includes tuning the number of iterations for each zoom level separately.

Figure 32.18 (right) shows the output of three parallel RGB lightness calculations for the JOBO image. The MatLab software for a Zoom, or pyramid, process is found in Funt et al. (2000). This paper provides both the Matlab® code and some guidance for pre-LUT and post-LUT processing. The desired input image is one in which digit is linearly proportional to log R, log G and log B radiances. This requires calibration of digital cameras so as to remove their built-in nonlinearities. Figure 32.20 shows the relationship of spatial processing parameters and post-LUT calibration.

Figure 32.20 shows how the number of iterations and Post-LUTs work together. We see the input image and its histogram at the top. Here, 0–255 maps log luminance over 3.5 log units. In this scene the contents in the sun are nearly the same as those in the shade. We see the sunlit part from 128 to 255 in the histogram. The shade part is between 0 and 128.

Although this calibration preserves scene content, it will make very poor prints and displayed images. To render the scene for viewing we need to: first compress the range of scene content, and then fit the compressed image to the color space that the output device expects. That means that the compressed image must be transformed into the color space of the output device profile. Here, we will just discuss the examples of linear LUTS for simplicity, even though we understand the problem of fitting it to the device profile is much more complicated.

Below the input image, we see three extreme examples of different numbers of iterations: one, four, or 128 iterations for all direction of edge ratios, at all levels. These processed image and their histograms are shown on the left side of Figure 32.20.

Using just one iteration severely over-compresses the dynamic range. In order to send this to a display or printer we have to stretch the lowest Retinex digit to 0, or black. For one iteration that is a four time stretch or slope 4. The histogram of the Retinex output plus slope 4.0 post-LUT expands the output range for display. This combination of over-compression and over-stretching with the Post-LUTs shows compression, but with a number of undesirable artifacts.

The second example that uses four sets of iterations has considerable range compression, and requires only half the Post-LUT stretching. It demonstrates considerable locality influence, and a lack of sensitivity to PostLut artifacts.

The example of 128 sets of iteration takes much longer to calculate and makes poorer pictures. It shows much less compression, less Post-LUT stretch, and an improved, but duller, rendition. The additional processing has moved the output image closer to the input image as seen in the final histogram.

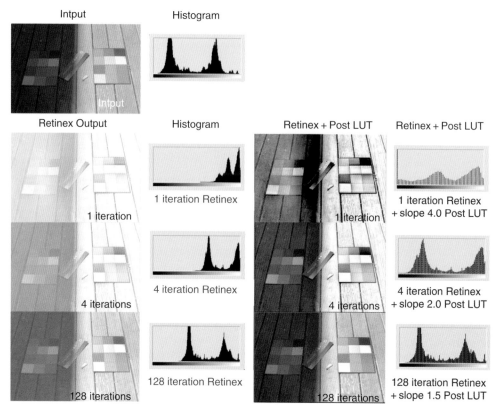

Figure 32.20 Role of the number of iterations and Post-LUTs. The input image is at the top. The columns show: the calculated (Ratio-Product-Reset-Average) images; the histogram of the processed images; the final images; the histogram after Post-LUT stretch.

In Figure 32.20 we used oversimplified linear post-LUTs to illustrate how calibration, number of iterations and post-LUT work together. To optimize the image these post-LUTs should be shaped so as to take into account the profile of the output device and the tone reproduction curve desired (see Appendix II & III of Frankle and McCann, 1983 for details). Using pictures from different cameras and uncalibrated post processing will reduce the quality of the results because of unwanted nonlinearities.

Although the Funt et al. (2000) paper provides the Matlab code of the spatial processing algorithms, it should not be regarded as a cookbook recipe for evaluating Retinex processing. As we have seen, the source of the scene capture, the scaling of the input data, and the expected color space of the output device are all independent variables in the process. Applying arbitrary parameter values to images of unknown calibration weakens our ability to analyze the results. We can only make conclusions about individual images because general conclusions about the spatial algorithm are unreliable.

Whether using paths as implemented by Purcell and Taylor, or using multi-resolution ratios as implemented by Frankle, or zoom as implemented by Wray and McCann, the reset is the most powerful tool in implementing locality influence. Keeping all the reports before the final reset increases the efficiency of the computation. All of these algorithms developed in the 1970s and early 80s shared the same process that created scene dependent lightness calculations. The measurements of human vision (Sections C & D & E) were used to optimize model parameters.

Recall the results of monitoring the New Product (*NP*) values as we add iterations. The *gray-near-white* areas get darker (lower *NP* values) sooner than the same gray in a black surround. Reset is the antidote to glare. The greater the amount of white in the locality, the lower the contrast of the retinal image controlled by glare. The spatial properties of reset have the opposite effect. They make pixels in a white locality have lower *NP* values predicting higher apparent contrast. This unusual operation has the same unusual properties as vision. The reset operation mimics the *neural contrast* process measured in Sections C and D. If we were modeling the physiology of vision we would need to take the measured scene radiances, convolve them with the Glare Spread Function to calculate the retinal image, and then apply the model of *neural contrast*. Since the retinal image has very low contrast, then *neural contrast* must be very powerful. It has to make the lowest contrast retinal images (with white) appear to have the highest contrast.

32.7   The Extended Family of Retinex Models

Ironically, over the years the term Retinex has drifted away from its original intent, namely, three independent spatial processes to predict color. It has become associated with computational models for lightness. It has also become the name of a great many different computational models, that have different characteristics. The family of Retinex models has diverse properties and conflicting goals. Below, we will summarize the different branches of the family and discuss their goals and results (Table 32.1). The term *Retinex*, in some circles, has strayed so far as to be the trademark name of a retinol cosmetic compound used in shampoo, skin creams and lip enhancers.

32.7.1   Land's Designator – 1983

In 1980 Edwin Land retired as President of Polaroid. In 1983 he retired as Polaroid's Chairman of the Board to spend his time doing research at the Roland Foundation in Cambridge, MA that he established and funded. There were now two independent labs working on Retinex: Land's on theory and McCann's on theory and application.

Land (1983) introduced the first departure from the original Retinex lightness model. It considerably simplified the algorithm, and lost some of its most powerful properties. In this paper Land describes

Table 32.1   Family of Retinex Algorithms.

Retinex	Idea	Author
Retinex	3 independent, spatial color channels	Land, 1964
Retinex Lightness Model	Spatial comparisons calculates lightness [Ratio, Threshold, Product, Reset, Average]	Land & McCann, 1971
Multi-resolution Retinex	Fast array processor, coarse first-then fine [Ratio, Product, Reset, Average]	Frankle, McCann, 1983
Spatial Designator	Spatial comparisons calculates lightness [ratio of center and surround]	Land, 1983, 1986
NASA Retinex	Multi-Scale "Land's Designator" + "Color Restoration"	Jobson, et al, 1996 Kotera & Fujita, 2002
Gamut Mapping Retinex	Multi-resolution optimization of spatial color reproduction	McCann, 2000b
Brownian Path/ Random Spray Retinex	Brownian paths/Random spray aray to compute lightness	Marini & Rizzi, 2000 Provenzi, et al., 2008
Sobol Retinex	Retinex used in HP cameras	Sobol, 2002, 2004

two versions: the first was the same as Land and McCann (1971a) without the reset; the second, calculates a Designator, the average relative reflectance. The first used a series of paths that began in the surround and traveled to a central pixel of interest. Since each path did not reset, it used the average of relative reflectance as a way of normalizing the output to the maxima in each waveband, or Retinex. Thus, the calculation became a *GrayWorld* comparison, while the original Ratio-Product-Reset normalized to the maxima using locality influence. The paper discussed calculating color appearances found in the flat Color Mondrian.

As discussed in Section E, appearance in flat Mondrians in uniform illumination correlates with the Scaled Integrated Reflectances of the papers in the Mondrian. That is true as long as the reflectance calculation is a spatial ratio. Land's Designators performed well in predicting colors in the Color Mondrian, and were used to emphasize the three independent color channels in the Retinex theory.

In the second version, Land (1986) calculates the Designator as the ratio of a central area to an extended surround. This meant that the threshold operation was replaced by averaging over a large area, and the *Grayworld* average in 1983 was replaced with local-average dependence in 1986.

Land's emphasis in the Designator papers was still the three-channel Retinex model of color. Work by others (Hurlbert, 1986; 1989; Brainard & Wandell, 1986; Hurlbert & Poggio, 1988) analyzed the properties of the Designator model of lightness. While the reset non-linearity of the original ratio-product models decoupled any dependence on global and local averages, Land's Designator algorithms linked the output results to global and local averages.

Brainard and Wandell (1986) criticized Designator Retinex models for being too sensitive to changes in the color of nearby objects. They also analyzed the original reset Retinex, but only considered the case of "infinitely long paths" because it could be fit by an analytic expression. That expression lost the *locality influence* of scene content. As we have seen above, their analysis used a special case of the algorithm that was never intended by its authors. Karen Houston calculated for David Brainard the appearance using the Ratio-Product-Reset-Average model, and showed that it predicted the appearances observed by Brainard and Wandell. The original algorithm, used as intended, predicts their observations (McCann, 1989). In that paper McCann wrote the following about Brainard's analysis:

"The goal of calculating sensation is clearly different from the goal of finding reflectance of objects. (McCann & Houston, 1983a) If one chooses the goal of calculating physical reflectance there is not enough information to arrive at a proper solution (Hurlbert & Poggio, 1988). If one states that the goal is to choose the appearance of an object, then there is adequate information in the two-dimensional array of radiances. The sensation problem is simpler. One no longer has to differentiate gradients in illumination from gradients in reflectance or edges in illumination from edges in reflectance; they are treated the same. Both illumination and reflectances are incorporated in color matches. There is sufficient information to calculate a color match (color sensation)."

Moore et al. (1991) developed a real time neural net of Land's (1986) *Designators*, using resistive grid technology. It extended the approaches of Horn (1974), Marr (1974), and Hurlbert (1986) that depended on gradient removal. Moore points out that Land (1986) is a variation on *GrayWorld* average comparisons. The resistive grid approach allowed them to simplify the electronic comparison process. Instead of connecting each pixel to all other pixels they used the grid iteration process to make the spatial comparisons. They reproduced Hurlbert's one-dimensional results, and extended the study to color constancy images.

Moore concluded that the Land (1986) model was too simple. First, it did poorly on Simultaneous Contrast, and second it "suffers from overreliance on the gray world assumption". To lessen the induction effects and the reliance on the gray world assumption, they modified the surround weight from point to point by adding an edginess modification operation.

32.7.2   NASA Retinex – 1996

Jobson, Rahman and Woodell, (1996; 1997a; 1997b; Rahman et al., 1996a, b; 2004) used Land's Designators with a post processing color adjustment. They made Single-Scale Retinex (SSR), and then Multiple-Scale Retinex (MSR) images using Land's Designator and Frankle and McCann's multi-resolution approach. They used the SSR and MSR images to compress the range of the input image. They applied this process to a wide variety of images from different sources. They found better results with a gaussian surround, rather than Land's inverse-square distribution. They reported that this process had excellent color constancy, but had poor color fidelity. They added a color restoration step based on their experience from a number of input images. Their goal was to demonstrate range compression, and maintain color constancy with the multi-scale Designators.

There are many Retinex versions based on the Designator and the NASA Retinex: Barnard and Funt (1997), Herscovitz et al. (2003, 2004), Meylan and Süsstrunk (2004, 2006), Cherifi and Beghdadi (2010).

Reinhard et al (2006, page 281), described Retinex as an expansion of Chiu's gaussian blur because of a reported similarity between NASA Retinex and a 1993 paper.

Kotera and colleagues made a number of extensions of Designator Retinex. They recommended the linear processing camera images. (Yoda & Kotera, 2004) Camera images are already approximately logarithmic in the mid-tone regions, so taking the log of a log is undesirable. They studied: adaptive gain (Kotera & Fujita, 2002); integrated surround in which they calculated an unsharp mask to apply to the input image (Wang et al., 2006, 2007); time sequential processes (Horiuchi et al., 2006) and Weber-Fraction contrast (Kotera et al., 2006). Bai and Miyake also studied Retinex processes with sLab scaling (Bai et al., 2006).

Ramponi and colleagues modified NASA Retinex (Ramponi, et al., 2003) using nonlinear scaled Laplacian operators (Carrato et al., 2003), or low pass edge-preserving recursive rational filters (Saponara et al., 2007) to process the image. They worked mainly on luminance, not dealing with color constancy and color correction, and optimized the algorithm for hardware implementation to process video sequences in real time.

As with Land's Designators, the *reset* feature was not included in NASA Retinex. The absence of the *reset* makes it harder for the process to mimic vision (Section D) and makes the analysis quite similar to that of the Blakemore & Campbell (1969), Stockham (1972), Faugeras (1979) spatial processing mechanisms. The presence of multi-resolution elements differentiates them from the above, except for Campbell.

The NASA Retinex work in the late 1990s was able to take advantage of the remarkable progress in digital imaging hardware since the mid 1970s. The Frankle-McCann (1983) patent made three important contributions.

1. Used multi-resolution processing for real-time speed and image quality
2. Refined the definition of the goal of the calculation
   [replaced illuminations and reflectance, with abrupt and gradual luminance changes]
3. Described the pre- and post-processing that surrounded the spatial-imaging core of the process

The NASA Retinex, and the work that derived from it, used only one of three of these principles: it adopted the multi-resolution approach.

It did not follow Frankle and McCann's definition of the problem. It took a turn in the opposite direction from Frankle/McCann and from Land. The spatial transform of the input luminances got the name "illumination", even though it had no correlation with the physical definition of illumination.

Finally, the NASA Retinex group did not work with calibrated images. Frankle and McCann described how they had to synthesize HDR images in 1975. By 1995 there were innumerable images on the internet that came from any number of sources. Originally, Frankle and McCann and Land made their own scenes to log luminance specifications. NASA Retinex took the log of images that were transformed by many different tone scale curves, probably pseudo-logarithmic in the mid-tones. Kotera's use of linear camera digits made things closer to the original specifications of log scene luminances. Jobson regarded the use of uncalibrated images as a feature. He felt that the process worked well on all kinds of images.

32.7.3 Color Gamut Retinex – 1999

By the Retinex theory, color in humans is generated by a spatial comparison process. Can color gamut calculations, using spatial comparisons, make displays with different color gamuts appear much more similar to each other? We applied spatial comparisons to the mismatch of different media. This approach minimizes the spatial errors introduced by limited color gamut and employs human color constancy mechanisms, so as to reduce the color appearance differences caused by limited color gamut (McCann, 1999; 2000b; 2002; McCann & Hubel, 2000, 2001, 2003).

McCann described a Reset-Retinex approach to Color Gamut Mapping. Its idea was to preserve the edge ratios from the original in the smaller color gamut reproduction. In this analysis we call the original with large color gamut the Goal image; and the limited color possible in the smaller color gamut medium the Best image. McCann used *Ratio-Product-Reset-Average* to calculate the in-gamut image with better color appearance. Here he calculated the NewProduct (*NP*) for the output pixel *x,y*. We begin at the starting pixel x',y' using its OldProduct (OPx',y'). All *OP*'s are initialized with the value in the Best image [B] for that waveband. The NewProduct multiplies the Goal's ratios times the OldProduct. It is reset twice: if greater than the [*Bx,y*] because the NP separation value is lighter than the gamut allows; and if less than 0 because the *NP* value is darker. The NP is then averaged with the previous NewProduct.

Figure 32.21 illustrates the color gamut Retinex calculation. For gamut mapping we begin with two input images, instead of one. We have the *Goalin* image that has the large, original gamut, and the *Bestin* image that represents the limited gamut of the reproduction medium (Figure 32.22 top). Again, we begin by averaging down each of the R, G, B separations to a small number of pixels for both the *Goalin* and the *Bestin* image. We take the OldProduct initialized to *Bestin* and multiply by the Goal ratios from the *Goalin*. This NewProduct is reset to the *Bestin*. This process is repeated and the New Product values from this resolution are interpolated up to the next resolution. The process is repeated for R, G, B.

The gamut retinex process takes the spatial comparisons from the *Goalin* and limits the products by the *Bestin*. The iterative process keeps reinforcing the ratios found in the *Goalin*, while the reset forces the NewProduct to migrate toward an image with all the same ratios, regardless of the absolute input values of the *Goalin* image. The resulting image *NewProductOut* shows a substantial improvement in appearance compared to the *Bestin* (Figure 32.22, bottom).

Many familiar gamut mapping processes evaluate the colorimetric values of each pixel (Morovic & Luo, 1999). If the pixel is in-gamut, it is unchanged. If out of gamut, it is replaced with the nearest in-gamut color. This process distorts color appearance. Take two areas next to each other. Let us assume that one area is in-gamut and the other is not. If we leave the in-gamut pixel value unchanged, while changing the out-of-gamut pixel, we have replaced the ratio of these two areas with a new ratio, and a new color relationship.

The superior reproduction is the one that preserves the most spatial relationships (McCann, 2000b; 2001; 2002). Conventional color gamut maps make global substitutions treating all pixels with the same R, G, B values, regardless of where it is in the image. The Gamut Retinex lets pixel values change locally, and attempts to preserve the ratio values between pixels. We want to recall that this method differs from the "perceptual" rendering intent of the ICC gamut mapping technique since the ratio

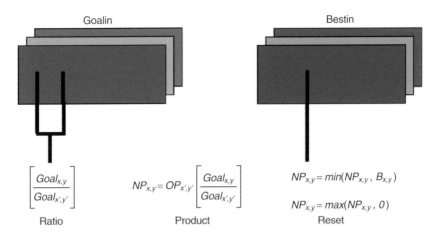

$$\begin{bmatrix} \dfrac{Goal_{x,y}}{Goal_{x',y'}} \end{bmatrix} \qquad NP_{x,y} = OP_{x',y'} \begin{bmatrix} \dfrac{Goal_{x,y}}{Goal_{x',y'}} \end{bmatrix} \qquad \begin{array}{c} NP_{x,y} = min(NP_{x,y},\, B_{x,y}) \\[2mm] NP_{x,y} = max(NP_{x,y},\, 0) \end{array}$$

Ratio   Product   Reset

Figure 32.21   Gamut mapping Retinex compares ratios from the large gamut original (Goalin), but resets the NewProducts to the small gamut medium limits (Bestin).

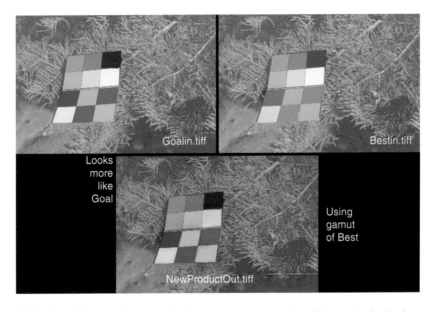

Figure 32.22   Spatial comparisons applied to color gamut mapping. Edge ratios in the large gamut Goalin are preserved in the much smaller gamut medium NewProduct.

preservation acts locally according to the scene content. In other words, it is image dependent, and cannot be computed simply on the digital values in color space. It has to be recomputed for each individual image. That is not a problem because the multi-resolution process is very efficient.

Retinex calculations extended to the problem of gamut-limited reproductions show promise. It solves for global and regional shifts in color, similar to those found in color constancy. It preserves edge ratios. It produces much smaller changes in appearance than local individual pixel-based color shifts. Further, we argue that color-gamut transformations using spatial comparisons can generate in-gamut reproductions that look more like the original, because it employs the benefits of human color-constancy processing. These reproductions have greater average colorimetric differences between original and reproduction, but look better. Human color constancy uses spatial comparisons between different parts of the image. The relationships among neighboring pixels are far more important than the absolute differences between the colorimetric values of an original and its gamut-limited reproduction. If all the pixels in an image have a reproduction error in the same direction (red, green, blue, lightness, hue, chroma), then our color constancy mechanism helps to make large colorimetric errors appear small. However, if all the errors are randomly distributed, then small colorimetric errors appear large (Chapter 35, McCann, 2000b; 2001; 2002).

32.7.4   Brownian Path Retinex

In 1993 Marini and Rizzi modified the Land and McCann (1971a) Retinex formulation (Marini & Rizzi, 1993; Marini & Rizzi, 2000; Rizzi et al., 2002, 2003; Marini et al., 2004). They studied algorithms using straight paths between pixels, Brownian Motion paths, and Random Spray Retinex (RSR) sampling. We will call this group the MI-Retinexes, after the Milan group that developed them.

The interesting point of this approach is that the sampling geometry affects locality i.e. the balance between global and local influence. In other words, the path's, or spray's, sampling geometry can vary, and consequently varies the local behavior of the algorithm. The study of Brownian paths and RSR helps understand Retinex locality influence.

Paths are used in Brownian Retinex to sample the image similar to the paths in 32.23 (left). The algorithm computes the output image calculating the NewProduct of each pixel, one by one, independently.

As with other versions of Retinex, input values need to be calibrated, if one aims at modeling human vision. Alternatively, one can just use digital input values, if the goal is image enhancement; and that was the main goal of MI-Retinexes. We will refer to the input value as I. We recall that is always possible to use all the following algorithms with calibrated radiance inputs to model, and investigate our visual system.

For each pixel, the Brownian Retinex selects a series of paths that start at randomly selected locations, and move across the image, one pixel at a time, towards the pixel of interest. Along each path, the algorithm computes the ratio of input values of the second pixel to the first, and multiplies it by the initial value 1.0, or OldProduct, to form the NewProduct. Instead of modifying the NewProduct array (as described in Chapter 32.3), this algorithm just records the current NewProduct value as it travels along the path, continuing the sequence of ratio-products until the path reaches the final point, namely the pixel of interest. Along the path, computing the ratio-product sequence, if the NewProduct gets greater that 1.0, they applied the Retinex reset (equation 32.4) and restarted the ratio-product sequence from the unitary value. The final product at the pixel of interest ($NewProduct_{x,y}$) is the only output from each path. The final output pixel value is the average of the result values from all paths.

The MI-Retinexes (Brownian Path and RSR) compare the pixel of interest to the maximum value found in the selected set of pixels in the path or spray. The use of selected pixels, instead of all pixels, introduces the *locality influence*.

The Milano group tested two versions of Brownian-path Retinex: with and without threshold (equation 32.3). The results on the images tested were similar. If no threshold is used, the ratio-product computation along the path is equivalent to searching for the maximum value along the path. Thus, the $NewProduct_{x,y}$ output can be simplified with the ratio of $I_{x,y}$ to maximum input value I_{maxN} found along the path (Provenzi et al., 2005).

Figure 32.23 (left) Brownian paths start randomly (gray dots), and proceed to the pixel of interest (black dot);(right) Brownian Retinex processing.

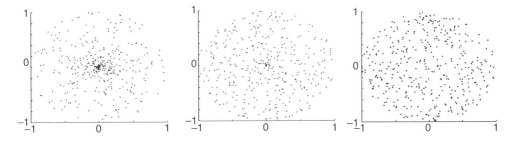

Figure 32.24 Examples of sprays calculated by different functions.

In the case when threshold is used, with visual configurations containing gradients without edges, *NewProduct$_{maxN}$* value can be different from the I$_{maxN}$ along the path. Tests made by the Milano group on natural images found that this difference is limited. Moreover, the value of the single path will be averaged with other paths to obtain the final pixel output value, making the introduced difference even smaller (Provenzi et al., 2005).

At the beginning of the study of locality influence, Rizzi used paths that were just a straight line from a random point in the image to the pixel of interest. These very short paths gave results which were very noisy with a small number of paths. They also had unwanted "ghosts". It was clear that short, direct straight line paths between a random point and the pixel of interest was under-sampling the radiance information; moreover straight lines introduced a correlation between the sampled quadrant and the area travelled by the path. Brownian paths (Figure 32.23, left) do a much better job calculating images because it increases the area of the image sampled and removes the correlation between sampled quadrant and sampled portion of the image. Brownian paths have been computed by the technique of mid-point displacement: the mid-point between the random and the target pixels is displaced by a random distance along both axes, this generates two segments from the original one; on these two segments the same mid-point displacement is applied recursively a certain number of times that is one of the parameters of the algorithm (Marini & Rizzi, 2000). Figure 32.23 (right) shows an example of Brownian Retinex processing with 200 paths per pixel.

32.7.5   Random Spray Retinex

Provenzi et al. (2007) described the Random Spray Retinex (RSR). Similar to Brownian Retinex, it uses a different sampling technique. It samples a spray around the pixel x,y that is a calculated set of random points, or a function generated statistical sample. In RSR these are N different sprays, each one composed by n points.

It differs from the Land (1986) designator because RSR keeps the *reset* and is not a *Gray World* average. Land took the average of randomly selected pixels in the pixel of interest's locality. The designator value was the ratio of input pixel values $I_{x,y}$ to the average of all pixels selected in the sample. In RSR, the highest input value found within the spray I_{maxN} is used as a local maximum, or reference white, and the final NewProduct$_{x,y}$ value is the average of N independent random sprays. The advantage of RSR compared to Brownian Retinex is a smaller computational cost and that the spray can be controlled by a statistical function, testing in this way different localities (Provenzi et al., 2007). Figure 32.24 shows examples of different sprays made using different functions and consequently realizing different image sampling.

The RSR Retinex output is the following:

$$NewProduct_{x,y} = \frac{1}{N}\sum_{k=1}^{N}\left(\frac{I_{x,y}}{I_{maxN_k}}\right) \tag{32.6}$$

where N is the number of individual spray, $I_{x,y}$ is the input value of the pixel, I_{maxNn} is the maximum value found among the n spray pixels of the Nk-th spray. Different spray densities have different sampling and cause different local changes in the image (Provenzi et al., 2007).

RSR can substitute the paths with the random sprays, since it is derives from the Brownian Retinex without threshold. Thus the ratio of $I_{x,y}$ over I_{maxNn} and the ratio-product sequence value at the end of the path are the same. Choosing the max point is equivalent to having traveled across all the points in the spray and having restarted the virtual path from I_{maxNn}. Figure 32.25 shows two examples of RSR processing.

32.7.6   Different Properties of Locality

The original and MI Retinex models for computing lightness are very similar, but have subtly different spatial properties. All of these Retinex models interpret the information in the scene that is neither local, nor global. All use the maximum and local maxima to normalize the scene, and modify the darker lightnesses away from the value predicted by a global normalization. All these models emphasize the edges and de-emphasize the gradients.

The models differ with regard to:

1. How they sample the radiance input values (paths, arrays, sprays, averages)
2. How they normalize the scene (L&M *reset* vs. Sample *reset*)
3. What they average to form the *NewProduct x,y*

The departures from global scene normalization have different causes:

Land and McCann (1971) paths: Every computation at every pixel modified the *NewProduct*. Each iteration, at every pixel is the average of the OldProduct and Reset-NewProduct. The initial values of all *OP* is white. The output before the first iteration is uniform white. The asymptote for very large numbers of paths, and pixels per path, is the input radiance divided by the maximum radiance in the scene (global normalization). The

Figure 32.25   (left) Control; (right) RSR processing with 150 sprays of 20 points per pixel. Reproduced by permission of Cristian Bonanomi.

early iterations report lighter *NP* than normalized radiance for most pixels. When the path reached a pixel where the *NP* was greater that 1.0, then it was set to 1.0.

Land and McCann kept all the *NPs* larger than the global normalization value. This had the beneficial effects of slowing the convergence toward global normalization, and introduced a dependence of local scene content. The *NPs* before the last reset introduced the *locality influence*. As the number of iterations increases because of the increased number paths through that pixel, the number of resets decreases. With very long paths the *NP* approaches global normalization. Thus, the length of the path was a key factor in the relative local vs. global contributions. The other key factor was the distribution of maximum and local maxima in the scene.

In summary, Land & McCann's *locality influence* mechanism has the following properties:

The initial condition sets all output pixels to white.

The maximum radiance remains white through all iterations.

Other output values get darker with increased iterations.

Output values get darker at different rates.

Spatial content of the scene determines the darkening rate (size and location of the maximum and local-maxima).

Path length, pixel selection, and number of paths control the departures from global normalization.

The combination of these factors made it possible to tune the model's *locality* properties to find the best match for vision's appearances.

MI Retinex algorithms   They used the path, or the spray, to select a set of pixels used to evaluate the pixel of interest (x,y). The operation searched the set of radiances to find the maximum value. As seen in equation (32.6) the output *NPx,y* is the ratio of the radiance at that pixel to the maximum found in the selected set of pixels. We will call this *sample reset* to differentiate it from the *Land & McCann reset*. In the Marini and Rizzi (1993) case, with increased numbers of paths, or sprays, and with more pixels per path/spray, the output approaches the global normalization asymptote.

Here the selection of pixels and its interaction with the spatial content of the scene determines the *locality influence*.

Frankle and McCann Multiscale Retinex – 1983   In the early 1970s there were severe hardware limitations to image processing. The goal of the two approaches described in Frankle and McCann was to significantly reduce the amount, and the time of computation. They used special purpose hardware that processed images that were 512 by 512 pixels. Both approaches divided the input image into different layers. In the so called *Frankle and McCann* algorithm the first layer took the ratio of pixels that were separated by 128 pixels. Each successive layer reduced by half the separation of receptors used to calculate ratios, until the distance was 1 pixel.

In the so called "McCann 99" version (based on Frankle & McCann Zoom; Wray, 1988), the first processing layer of a 512 by 512 images has four pixels: averages of the four quarters of the image. The processing of the first layer calculated *Ratio-Product-Reset-Average* values of these four pixels. In the following layers, there were 4^2, 8^2 . . . 256^2. 512^2 pixels. Both *Frankle and McCann*, and *McCann99* initialized the output to uniform white. Starting with coarsest representation, two by two pixels (averages of 128 by 128 radiances) the number of iterations, and the direction of the ratios influenced the amount of contrast between the maximum average radiance and the other pixels. The *NP* array was interpolated to twice the horizontal and vertical sizes and used as the *OP* for the next layer.

There were no paths in either algorithm. In each succeeding layer in *Frankle and McCann* only single ratios of pixels separated by 64, 32, 16, 8, 4, 2, 1 were used. In *McCann 99*, ratios of adjacent pixels that calculated averages of averaged 64^2, 32^2, 16^2, 8^2, 4^2, 2^2 pixels were used.

By selecting the number of iterations at each layer of the calculation you can tune the local-global contrast, or *locality influence*, of that layer. The selection of the orientation of ratios and their order can prevent overly systematic sampling that generates artifacts, such as halos.

In these algorithms, the tools of path length are not available, but the number of iterations at each layer provides a similar tuning mechanism. Although *Land and McCann, Frankle and McCann, and McCann 99* generate slightly different final results, they all have local *resets*, that introduce very efficient *locality influence*.

Land Designator: The Land (1983, 1986) Designator does not specify *reset*. As a consequence, it normalizes its output to a global average of the input image. (Hurlbert, 1986; Moore et al., 1991) This is often called a *GrayWorld* algorithm. Although, as Land pointed out, it has some very interesting analogies to the spatial measurements of neurons, it has the unfortunate consequence of modeling measurements of human spatial vision poorly (Moore, 1991). Land's Designator, NASA Retinex, Kotera's Retinex and Bilateral filtering and many others achieve locality from comparison of the pixel of interest to the neighborhood.

32.7.7 Sobol's Digital Flash Retinex – 2002

Bob Sobol led the software spatial processing development for a series of HP digital cameras. His spatial processing used a color-balanced image as input so that it did not need to calculate color constancy. It used a combined RGB signal as the input Retinex. It scaled that output to apply it to the RGB color space.

In the context of making a fast camera algorithm Sobol (2002, 2004, 2006) improved the *Ratio-Product-Reset-Average* algorithm and described it as:

$$NP_{x,y} = \left[ResetToWhite\left(RMO\left(log R_{x,y} - log R_{x',y'} \right) + log OP_{x',y'} \right) \right]/2 \tag{32.7}$$

where, $R_{x,y}$ refers to a segmental area of the input image at location x,y. NPx,y refers to a segmental area in a shifted version of the output image where the image has been displaced by some distance d_x and d_y relative to the original image. The *Ratio Modification Operator (RMO)* function limits the range of the ratios. The *ResetToWhite* function constrains the results of the product of ratios. The process initializes the OldProduct to the input images, rather than white. These modifications resulted in a stable output image using fewer iterations. The process was specifically designed to produce an output color space for known digital printers.

This formulation has two properties that significantly increase the speed of the calculation. First, it applies a threshold to size of the ratio that reduces the number of iterations needed for the desired image compression. Second, it adds a mask to the captured camera image. Forming the output image *NP* iteratively from the shifted input image at the full resolution of the input and output images, is a computationally intensive process. The process made a scalable mask that improves the *NP* with less computation along with using Frankle and McCann (1983). This combination of methods reduced the number of calculations required to produce an improved output image (Figure 32.26).

Figure 32.26 Bob Sobol's Retinex processing. Reproduced by permission of Hewett Packard Company.

Figure 32.27 Six pairs on normal control image and constant Retinex processing. The top left pair shows changes that are dramatic, while those in the bottom-right are minimal. The changes in rendition are due to scene content, not changes in the constant Retinex algorithm.

Hewlett Packard made and sold a series of cameras using Sobol's algorithm. It worked extremely well. The marketing name for the algorithm was "Digital Flash", implying that it performed the digital equivalent of adding automatic fill-flash to the scene to equalize the illumination.

Figure 32.27 shows six different pairs of pictures taken with an HP 945 camera. The left pictures in each pair were made using the camera's normal settings. The right pictured used full "Digital Flash", or Retinex processing. All Retinex pictures were made using the same algorithm and camera settings. The upper left pair show dramatic changes, while the bottom right has almost none. Since the image processing was constant, the individual scene content generated the sometimes large, and sometimes minimal changes, in the image. The rooster images show very subtle decrease in contrast and improvement in chroma. As discussed above, the goal of the Retinex calculation is to capture all scenes, high-

and low-dynamic range, and render them in the range of the display device. The goal is to write appearances on the media, not record the radiances accurately.

HDR imaging should compress the range of scenes with markedly uneven illumination. HDR imaging must not compress LDR illumination because it would render such scenes as if they were in a fog. The Figure 32.27 pictures provide an excellent example of HDR renditions that compress some scenes and not others. Bob Sobol pointed out that there is a complication in these examples. The HP-945 firmware pipeline uses an adaptive post-LUT called preferred picture rendition (Holm, 1995). This was designed to improve the rendition of Control images using image statistics, rather than scene content. That part of the pipeline operates on both Control and Retinex intermediate output. Their image statistics are different so the adaptive post-LUTs are different; large changes with Control and much smaller changes with Retinex. So, the examples in Figure 32.28 are the comparison of PPR with combined result of Retinex and PPR. Despite this complication, the pictures provide an example of an essential property of HDR, response to scene content. (Many more examples of Retinex processing are available on the Wiley web site, http://www.wiley.com/go/mccannhdr)

If this commercial camera actually mimics vision then we can use the camera to render some scenes that illustrate the spatial processing in vision. The HP 945 camera has a menu setting that disables Retinex processing. The image on the left in Figure 32.28 (left) shows the conventional photograph of Simultaneous Contrast on a computer monitor. Figure 32.28 (right) is the HP 945 Retinex picture of the same computer screen.

The conventional picture on the left had equal camera digits in the gray squares, so it accurately reproduced the equal values on the display. The Retinex picture had the same gray value for the white surround, but had a significantly higher value for gray in a black surround. Figure 32.29 plots both pictures' digital values across the middle of the target.

Figure 32.29 shows range compression in that it lowered the white, and increased the black digital values. It successfully mimicked human vision. It wrote sensations of the gray squares in the output image.

Another example uses the appearance of a white square in the center of a black cross on a gray background (Figure 32.30, top left). If we replace the abrupt white black edges with gradients, the whitesquare glows (Figure 32.30, bottom left). As with Mach bands, the end of the gradient looks lighter. We made a pair of photographs with and without Retinex spatial processing. The digits in the square were equal in the conventional image (control), while Retinex picture of the apparent glow was 13 digits higher. Again, the camera mimicked vision.

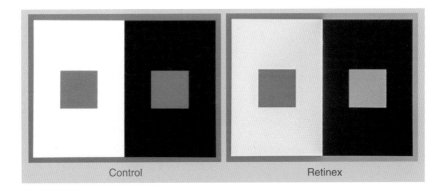

Figure 32.28   Shows two HP 945 photographs without and with Retinex processing.

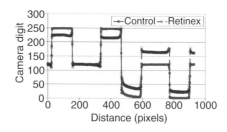

Figure 32.29 Plots of average camera digits vs. horizontal distance through gray squares. In the Control photograph (blue) the gray square have the same value (121) in both white and black surrounds. In the Retinex picture (red) the gray in black has a much higher value (163).

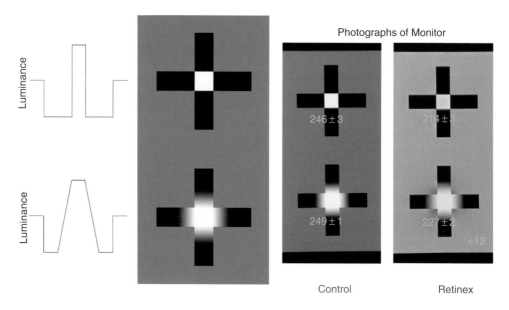

Figure 32.30 Change in appearance of Central Square. (left column) horizontal luminance profile of adjacent Cross with White Center; (left center) image on the computer screen; (right) pair of HP945 photographs of a computer monitor (Control and Retinex).

Other examples include Adelson's Tower version of the Black and White Mondrian. Gilchrist (2006) describes the tower experiment as evidence of a top-down perceptual framework mechanisms in vision. This is quite ironic because the Black and White Mondrian, which has two identical stimuli and similar edges and gradients, was first used to illustrate a bottom-up spatial mechanism. In the original computer display, the checkerboard squares A and B have equal digits. Figure 32.31(left) is the control HP 945 photograph that render the scene with A = B. In the Retinex photograph rendition, square A is 28 digits darker than square B, which correlates with their appearance.

Pawan Sinha (2011) designed the lightness experiment photographed in Figure 32.32 (left). On a uniform light gray background there is a highly visible outer gradient that is light on the left. In that gradient, there is a thin rectangle that appears nearly the same gray.

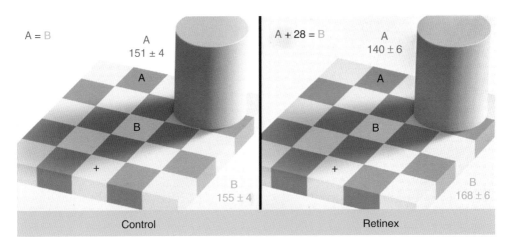

Figure 32.31   HP 945 photographs of Adelson's Checkerboard and Tower. The Retinex photograph renders square A 28 digits darker than B.

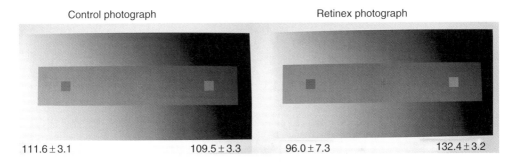

Figure 32.32   Photographs of Pawan Sinha's experiment.

In fact, it is a gradient with a slower rate of luminance change with distance than the one around it. It starts out darker than the outside gradient on its left side, and ends up lighter than it on the right side. The luminance edge ratio of inner gradient to square is less than 1.0 on the left, and more than 1.0 on the right.

The combination of the outer and inner gradients makes the inner one look nearly uniform. The experimenter asked the observers to compare the lightnesses of the two small squares. They reported that the right square was lighter, despite the fact that they had equal display digits, and hence equal luminances. The edge ratios of the gray square to its surround correlate with the changes in lightness.

Figure 32.32 shows a pair of HP 945 photographs. The left shows the control in which the camera digits for the two squares are equal. The camera in control mode reproduced the equal scene luminances. The right shows that the Retinex spatial image processing rendered the gray on the left as 96, and the one on the right 132. That scene rendering is consistent with the visual appearance of the scene. Sinha designed these experiments to show that the appearances of the squares could not be predicted by the apparently uniform area around it. The photographs of Sinha's experiments show that spatial processing can mimic vision.

The examples in Figures 32.28 through 32.32 demonstrate again an important observation. The data show that the process has significantly changed the image digits in converting Control to Retinex. But, when we study the pictures, we have to look very carefully to see any changes. We cannot look at a picture of a scene to evaluate what has happened to the digits that make up the scene. Looking at pictures turns on the entire, very-complex visual system that adds its own processing to that of our algorithm. If we want to evaluate our image processing, we have to look at the digits and not look at the print (Chapter 35).

There is a lot going on in a commercial camera's digital image processing firmware. While these photographs of vision experiments are interesting, and provide proof of the things spatial processing can do, they are not the same as more controlled measurements and models of vision. The HP 945 is not a substitute for such models, but does provide an interesting benchmark for spatial processing. The fact that algorithms, designed to make better HDR renditions of the world, respond the same as humans to time honored vision problems goes well beyond coincidence. There are no cognitive frameworks in the camera's firmware. The spatial processing of scene data is central to vision and HDR.

32.8   Algorithm's Goal

As we have seen there are many Retinex algorithms. They all address the fundamental issue of HDR imaging. They all use spatial information, from arrays of sensor responses, to calculate scene appearance.

As illustrated in Figure 32.33 the judgement of scene rendition is heavily influenced by the calibration and firmware/hardware imaging chain. The time to process and hardware cost also play a big role in evaluating the merits of the different processes. The goal of the calculation varies as well. One goal is to model vision. A different, but related goal is to make an imaging device that mimics vision. If the camera acts like vision, it will make a better picture. That was the mixed goal that Land and McCann set out to achieve in the early 1960s; namely, to study both humans and cameras in parallel and learn

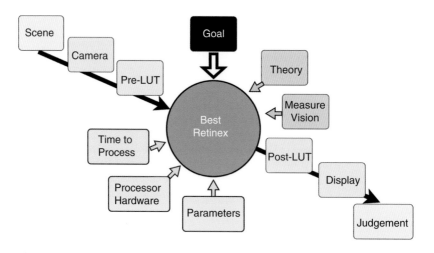

Figure 32.33   Factors that play important roles in Retinex evaluation.

things from each that can benefit the other. Since there is no universal goal for a Retinex calculation, then there cannot be a singular best calculation.

32.8.1   Reflectance and Illumination

As well, over the years there has been a dispersion of the goals of the Retinex calculation. Recall that the last sentence in the Land & McCann "Lightness and Retinex Theory" paper (1971a) reads:

> "Whereas the function of colorimetry is to classify reflectances into categories with similar visual properties, the function of retinex theory is to tell how the eye can ascertain reflectance in a field in which the illumination is unknowable and the reflectance is unknown."

Recall, as well, that paper studied flat, uniform reflectance surfaces, in uniform, or smooth gradients of illumination. Although it was an enormous increment in scene complexity at that time, these test targets did not approach that of real natural scenes, and 3-D Mondrians. It was not possible then to apply image processing algorithms to real scenes. It is always a bit of a shock when a colleague states unequivocally that the goal of Retinex is to use spatial information to find the illumination, so as to be able to discount it, so as to find the objects' reflectances. The original intent of Retinex was to ignore the illumination, just the way that human vision does. The goal of mimicking human vision requires that the model departs from calculating reflectance when appearances differ from matching the physical measurements of reflectance. The physical dichotomy of reflectance and illumination generates the light on the retina. However, it is the dichotomy of edges and gradients, in that retinal image, that generates HDR appearances and color constancy.

By describing how vision calculates appearances, we can formulate image calculations that render captured scenes mimicking vision. The idea of finding the reflectance of objects goes back to the 19th century, and as we saw in the 3-D Mondrian experiments, reflectance does a poor job of predicting appearance in complex illumination. Further, the act of removing shadows for scenes is a curious anomaly. It does not mimic vision, in which shadows are clearly visible, and shadows are important for calculating perceptions of objects. As well, removing shadows makes for very dull photographs and "flat" paintings.

The components of the original Land and McCann model for lightness were *Ratio, Threshold, Product, Reset and Average*. The initial image (OldProduct) and the array of input radiances were processed to make the output (NewProduct).

The Ratio Product is the fundamental component that gets beyond the limitations of colorimetry. It assumes that human vision is a spatial mechanism that responds to the entire array of input radiances. Dynamic range compression, color constancy, and most of the phenomena we study in vision, are easy to understand as spatial processing.

Very early in our studies we concluded that human spatial processing was neither local, nor global. It had mixed properties of both.

In HDR imaging, Tone Scale processes apply a global mapping that is helpful to many individual images. In the study of spatial HDR processing, the vast majority of work has studied algorithms that make better pictures by preserving edges and manipulating less visible gradients. Compression of gradients can be the result of thresholds, resets, local averages, local sampling and ratio limits. They all work: they just require different parameters for the same amount of dynamic range compression. Some processes are more powerful than others.

Some models use computers to render the images, while others use special hardware for increased efficiency. Some labs used carefully calibrated images; others just grabbed stuff off the web. The selection of a favorite Retinex process depends on all of the above.

All of these algorithms sample the input radiances in such a way as to have the locality of a pixel influence it. Locality goes beyond the local surround, and is less than the entire image. This locality

Table 32.2 Major Retinex models along with their properties and initial pixel values.

Retinex	Properties	Initialization
Land & McCann (1971)	Ratio, Threshold, Product, Reset, Average	White
Frankle & McCann (1983)	Multi-resolution (Pixel separation & Zoom pyramid) Ratio Threshold Product Reset Average	White
Land Designator (1986)	Center – Surround (Sampled Pixels) No reset	None
NASA Jobson et al. (1999) Kotera (2002)	Multi-scale Designator No reset	None
Gamut Retinex McCann (1999)	Ratio, Product, Reset, Average	Best Small Gamut
Milano Brownian Path (1993) & Random Spray (2005)	Average ratio (pixel/maximum of selected pixels) Reset	None
Sobol (2002)	Ratio, Clip, Product, Reset, Average	Scene input
ACE & RACE (Ch33 – not Retinex)	Distance weighted ratio *Grayworld* scaling	None

sampling is the key feature that makes human vision and its models have a scene dependent response. We have a vast number of algorithms that apply locality influence and produce better pictures. However, we do not have a clear, data driven understanding of human spatial processing. We have a great many approaches, but we do not have a way to integrate them into a general solution.

The issue going forward should not be who has the best looking, fastest, lowest cost, or mathematically elegant algorithm for a favorite scene. The important issue is that we understand the general solution to rendering all scenes that include HDR, normal and LDR inputs. Vision has the unusual property that it does not compress LDR scenes, while compressing the dark regions of HDR scenes. Applying the same Retinex algorithm to both LDR and HDR scenes is a good test of a model's effectiveness.

Table 32.2 lists the major variations in lightness models that have developed in the past 40 years. It lists the techniques for spatial comparisons and the initialization used in the calculation.

We have assumed that the general solution for rendering all scenes is that the Retinex process should mimic vision. Unfortunately, we still do not know enough about its spatial mechanisms. We have evidence that the maximum radiances normalize the scene, but local maxima play an important role. We know that the distance from, and enclosure by the maxima are important (Section D). We know that the size of the maxima relative to the size of the test hints at separate spatial frequency, or size based channels. Experiments using Mondrians and assimilation targets also show the importance of the maxima in the independent spectral channels. We know the results of many different spatial experiments, but the general theory that unites them all still remains a challenge.

In the original Land and McCann Retinex the iterative calculations start at white and every pixel except the maximum radiance gets darker. The amount darker depended on the image content, the length of the path, or the number of iterations, i.e, the spatial locality influence. The goal was to find the best description of locality to mimic vision. The idea of iterating to completion was never an option. As we learned that this process simulates vision, we put forth the hypothesis that HDR rendition should calculate sensation, and write that to film (McCann, 1988a; b). That proposal remains relevant today.

32.9 References

Albers (1962) *The Interaction of Colors*, Yale University Press, New Haven, CT.

Bai J, Nakaguchi T, Tsumura N & Miyake Y (2006) Evaluation of image corrected by Retinex method based on S-CIELAB and gazing information, *Ieice Trans Fundamentals Electronics Comm & Comp Sci*, **E89A**(11), 2955–61.

Barnard K & Funt B (1997) Analysis and Improvement of Multi-Scale Retinex, *Proc. IS&T/SID Color Imaging Conference*, CIC5, Scottsdale, 221–26.

Blackwell H (1946) Contrast Thresholds of the Human Eye, *J Opt Soc Am*, **36**, 624–43.

Blakemore C & Campbell F (1969) On the Existence of Neurons in the Human Visual System Selectively Sensitive to the Orientation and Size of Retinal Images, *J Physiol*, **203**, 237–60.

Brainard D & Wandell B (1986) Analysis of Retinex Theory of Color Vision, *J Opt Soc Am A*, **3**, 1651–61.

Burt P & Adelson T (1983) A Multi-resolution Spline with Application to Image Mosaics, *ACM Trans Graphics*, **2-2**, 17–36.

Campbell F & Robson J (1968) Application of Fourier Analysis to the Visibility of Gratings, *J Phyiol*, **197**, 551–66.

Cherifi D & Beghdadi A (2010) Color Contrast Enhancement Method Using Steerable Pyramid Transform, Signal, *Image and Video Processing*, **4**, 247–62.

Carrato S, Marsi S, Ramponi G, & Crespi B (2003). A nonlinear pseudo-Retinex for dynamic range compression in CMOS imaging systems. IEEE-EURASIP Nonlinear Signal and Image Processing, Grado, Italy, NSIP-03, June.

Fairchild M & Johnson G (2002) Meet iCAM: A Next Generation Color Appearance Model, *Proc. 10th IS&T/SID Color Imaging Conference*, Scottsdale, **10**, 33–8.

Fairchild M & Johnson G (2004) iCAM Framework for Image Appearance Differences and Quality, *J Electronic Imaging*, **13**, 126–38.

Faugeras O (1979) Digital Color Image Processing Within the Framework of a Human Visual Model, *IEEE Transactions on Acoustics, Speech and Signal Processing*, **27**, 380–93.

Frankle J & McCann JJ (1983) Method and Apparatus of Lightness Imaging, US Patent 4384336, filed 8/29/1980, issued 5/17/1983.

Funt B, Ciurea F & McCann JJ (2000) Retinex in Matlab, *Proc. 8th IS&T/SID Color Imaging Conference*, Scottsdale, Arizona, **8**, 112–21.

Gilchrist A (1994) *Lightness, Brightness and Transparency*, Lawrence ErlbaumAssociates, Hillsdale.

Gilchrist A (2006) *Seeing Black and White*, Oxford University Press, Oxford.

Herscovitz M, Artemov E, & Yadid-Pecht Y (2003) Improving the Global Impression of Brightness of the Multiscale Retinex Algorithm for Wide Dynamic Range Pictures, *Proc. SPIE*, **5017**, 393–405.

Herscovitz M & Yadid-Pecht Y (2004) A Modified Multi Scale Retinex Algorithm with an Improved Global Impression of Brightness for Wide Dynamic Range Pictures, ed. Mach. *Vision Appl*, **15**, 220–28.

Holm J (1995) A Strategy for Pictorial Digital Image Processing (PDIP), *Proc. IS&T/SID Color Imaging Conf*, **4**, 194–201.

Horiuchi T, Kotera H & Wang L (2006) Serial Retinex Algorithm for Time-Sequential Processing, *CGIV*, 87–91.

Horn B (1974) Determining Lightness from an Image, *Comp Gr Img Proc*, **3**, 277–99.

Hurlbert AC (1986) Formal Connections Between Lightness Algorithms, *J Opt Soc Am*, **A3**, 1684.

Hurlbert AC (1989) The computation of color, Ph.D. dissertation, Mass. Inst Technol, Cambridge.

Hurlbert AC & Poggio T (1988) Synthesizing a color algorithm from examples, *Science*, **239**, 482–5.

Jobson D, Rahman Z & Woodell G (1996) Retinex Image Processing: Improved Fidelity To Direct Visual Observation, Proc. IS&T/SID Color Imaging Conference: CIC4, Scottsdale, 124–6.

Jobson D, Rahman Z & Woodell G (1997a) Properties and Performance of a Center/Surround Retinex, *IEEE Trans Image Processing*, **6**, 451–62.

Jobson D, Rahman Z & Woodell G (1997b) A Multi-scale Retinex for Bridging the Gap Between Color Images and the Human Observation of Scenes. *IEEE Trans. Image Processing: Special Issue on Color Processing*, **6**, 965–76.

Kiesel K & Wray W (1987) Reconstitution of Images, US Patent 4,649,568, filed 10/22/1984, 3/10/1987.

Kotera H & Fujita M (2002) Appearance Improvement of Color Image by Adaptive Scale-Gain Retinex Model, Proc. 10th IS&T/SID Color Imaging Conference, 166–71.

Kotera H, Horiuchi T, Saito R, Yamashita H & Monobe Y (2006) Visual Contrast Mapping Based on Weber's Fraction, Proc. 14th Color Imaging Conference, 257–62.

Land EH (1964) The Retinex, *Am Scientist*, **52**, 247–64.

Land EH (1967) Lightness and Retinex Theory, *J Opt Soc Am*, **57**, 1428A, 1967; For citation see (1968) *J. Opt. Soc. Am.*, **58**, 567.

Land EH (1974). The Retinex Theory of Colour Vision, *Proc. Roy Institution Gr Britain*, **47**, 23–58,

Land EH (1983) Recent Advances in Retinex Theory and Some Implications for Cortical Computations: Color Vision and the Natural Image, Proc. Natl Acad Sci USA 80, 5163–9.

Land EH (1986) An Alternative Technique for the Computation of the Designator in the Retinex Theory of Color Vision, *Proc Nat Acad Sci*, **83**, 3078–80.

Land EH, Ferrari LA, Kagen S & McCann, JJ (1972) Image Processing system which detects subject by sensing intensity ratios, filed Jan, 22,1970, U.S. Patent 3,651,252, Mar. 21, 1972.

Land EH & McCann JJ (1971a) Lightness and Retinex Theory, *J Opt Soc Am*, **61**, 1–11.

Land EH & McCann JJ (1971b) Method and System for Reproduction Based on Significant Visual Boundaries of Original Subject, US Patent 3,553,360, filed 2/8/1967, issued 6/5/1971.

Marini D & Rizzi A (1993) Colour Constancy and Optical Illusions: a Computer Simulation with Retinex Theory, ICIAP93, 7th Int Conf Image Analysis & Processing, Monopoli, Italy.

Marini D & Rizzi A (2000) A Computational Approach to Color Adaptation Effects, *Image and Vision Computing*, **18**(13), 1005–14.

Marini D, Rizzi A & Rossi M (2004) Post-Filtering for Color Appearance in Synthetic Images Visualization, *Journal of Electronic Imaging*, **13**(1), 111–19.

Marr D (1974) The Computation of Lightness by the Primate Retina, *Vision Res*, **14**, 1377–88.

Meylan L & Süsstrunk S (2004) Color Image Enhancement Using a Retinex-Based Adaptive Filter, Color Graphics in Imaging and Vision, Aachen, *Proc. IS&T CGIV*, **2**, 359–63.

Meylan L & Süsstrunk S (2006) High Dynamic Range Image Rendering Using a Retinex-Based Adaptive Filter, *IEEE Transactions on Image Processing*, **15**, 2820–30.

McCann JJ (1978) Visibility of Gradients and Low-spatial Frequency Sinusoids: Evidence for a Distance Constancy Mechanism, *J Photogr Sci Eng*, **22**, 64–8.

McCann JJ (1987) Local/Global Mechanisms for Color Constancy, *Die Farbre*, **34**, 275–83.

McCann JJ (1988a) Calculated Color Sensations applied to Color Image Reproduction, *Proc. SPIE*, **901**, 205–14.

McCann JJ, (1988b) The Application of Color Vision Models to Color and Tone Reproductions, *Proc. Japan Hardcopy '88*, 196–9.

McCann JJ (1989) The Role of Nonlinear Operations in Modeling Human Color Sensations, *Proc. SPIE* 1077, 355–63.

McCann JJ (1999) Lessons Learned from Mondrians Applied to Real Images and Color Gamuts. *Proc. IS&T/SID Color Imaging Conference*, Scottsdale, 7, 1–8.

McCann JJ (2000a) *Simultaneous Contrast and Color Constancy: Signatures of Human Image Processing*, ed. Davis, Oxford University Press, USA, 87–101.

McCann JJ (2000b) Using Color Constancy to Advantage in Color Gamut Calculations, *Proc. IS&T PICS Conference*, **3**, 169–76.

McCann JJ (2001) Color Gamut Mapping Using Spatial Comparisons, *Proc. SPIE*, **4300**, 126–30.

McCann JJ (2002) *A Spatial Color-Gamut Calculation to Optimize Color Appearance, in Colour Image Science: Exploiting Digital Media*, ed. MacDonald & Luo, John Wiley & Sons, Ltd, Chichester, 213–33.

McCann JJ (2004) Capturing a Black Cat in Shade: Past and Present of Retinex Color Appearance Models, *J Electronic Imaging*, **13**, 36–47.

McCann JJ (2006) High-Dynamic-Range Scene Compression in Mumans, *Proc. SPIE*, **6057**, 605707–11.

McCann JJ & Hall JA (1980) Effects of Average-luminance Surrounds on the Visibility of Sine-wave Gratings, *J Opt Soc Am*, **70**, 212–19.

McCann JJ & Houston K (1983a) *Color Sensation, Color Perception and Mathematical Models of Color Vision, Colour Vision*, ed. Mollon & Sharpe, Academic Press, London, 535–44.

McCann JJ & Houston K (1983b) Calculating Color Sensation from Arrays of Physical Stimuli, *IEEE SMC-13*, 1000–07.

McCann JJ & Hubel P (2000) Image Processing, European Patent, PN EP1049324A, filed, 4/27/2000, 11/2/2000.

McCann JJ & Hubel P (2001) In-gamut Image Reproduction Method Using Spatial Comparison, Japan Patent, PN JP 2000350050, 4/1/2000, 12/15/2000.

McCann JJ & Hubel P (2003) In-gamut Image Reproduction Method Using Spatial Comparison, US Patent 6,516,089, filed 4/30/1999, issued 2/4/2003.

McCann JJ, Land EH & Tatnall S (1970) A Technique for Comparing Human Visual Responses with a Mathematical Model for Lightness, *Am J Optometry and Archives of Am Acad Optometry*, **47**(11), 845–55.

McCann JJ, McKee S & Taylor T (1976) Quantitative Studies in Retinex Theory: A Comparison Between Theoretical Predictions and Observer Responses to "Color Mondrian" Experiments, *Vision Res*, **16**, 445–58.

McCann JJ, Savoy R & Hall J (1978) Visibility of Low Frequency Sine-Wave Targets, Dependence on Number of Cycles and Surround Parameters, *Vis Res*, **18**, 891–4.

McCann JJ, Savoy R, Hall J & Scarpetti J (1974) Visibility of Continuous Luminance Gradients, *Vis Res*, **14**, 917–27.

Moore A, Allman J & Goodman R (1991) A Real-Time Neural System for Color Constancy, *IEEE Transactions on Neural Networks*, **2**, 237–47.

Morovic J & Luo R (1999) *Developing Algorithms for Universal Color Gamut Mapping, Colour Imaging: Vision and Technology*, ed. MacDonald & Luo, John Wiley & Sons, Ltd, Chichester, 253–83.

OSA Committee on Colorimetry (1953) *The Science of Color*, Washington, DC: Optical Society of America, 373.

Pattanik S, Ferwerda J, Fairchild M & Greenburg D (1998) A Multiscale Model of Adaptation and Spatial Vision for Image Display, *Proc. SIGGRAPH 98*, 287–98.

Provenzi E, Rizzi A, De Carli L & Marini D (2005) Mathematical Definition and Analysis of the Retinex Algorithm, *J Opt Soc Am A*, 22.

Provenzi E, Fierro M, Rizzi A, De Carli L, Gadia D & Marini D (2007) Random Spray Retinex: A New Retinex Implementation to Investigate the Local Properties of the Model, *IEEE Transactions on Image Processing*, **16**(1), 162–71.

Provenzi E, Gatta C, Fierro M & Rizzi A (2008) A Spatially Variant White Patch and Gray World Method for Color Image Enhancement Driven by Local Contrast, *IEEE Transactions on Pattern Analysis and Machine Intelligence*, **30**(10), 1757–70.

Rahman Z, Jobson D & Woodell G (1996a) A Multiscale Retinex for Color Rendition and Dynamic Range Compression, *Proc. SPIE*, **2847**, 183–91.

Rahman Z, Jobson D & Woodell G (1996b) Multiscale Retinex for Color Image Enhancement, *Proc. IEEE International Conference on Image Processing*, 965–76.

Rahman Z, Jobson DJ & Woodell G (2004) Retinex Processing for Automatic Image Enhancement, *J. Electronic Imaging*, **13**, 100–10.

Ramponi G, Tenze L, Carrato S & Marsi S (2003) Nonlinear contrast enhancement based on the Retinex approach, *Proc. SPIE*, **5014**, 169.

Ratliff F (1965) *Mach Bands: Quantative Studies on Neural Networks in the Retina*, Holden-Day, San Francisco.

Reinhard E, Ward G, Pattanaik S & Debevec P (2006) *High Dynamic Range Imaging Acquisition, Display and Image-Based Lighting*, Elsevier, Morgan Kaufmann, Amsterdam.

Rizzi A, Marini D & De Carli L (2002) LUT and Multilevel Brownian Retinex Colour Correction, *Machine Graphics and Vision*, **11**(2/3), 153–68.

Rizzi A, Rovati L, Marini D & Docchio F (2003) Unsupervised Corrections of Unknown Chromatic Dominants using a Brownian Paths Based Retinex Algorithm, *Journal of Electronic Imaging*, **12**(3), 431–41.

Rosenfeld A ed (1984) *Multiresolution Image Processing and Analysis*, Springer-Verlag, Berlin.

Saponara S, Fanucci L, Marsi S, Ramponi G, Kammler D & Witte EM (2007) Application-Specific Instruction-Set Processor for Retinex-Like Image and Video Processing, *IEEE Trans Circuits & Systems II*, **54**(7), 596–600.

Savoy R (1978) Low Spatial Frequencies and Low Number of Cycles at Low Luminances, *J Photogr Sci Eng*, **22**, 76–9.

Savoy R & McCann JJ (1975) Visibility of Low Spatial-Frequency Sine-wave Targets: Dependence on Number of Cycles, *J Opt Soc Am*, **65**, 343–50.

Sinha P (2011) Brightness Induction via Unseen Causes. *Perception*, in press.

Sobol RE (2002) Improving the Retinex algorithm for rendering wide dynamic range photographs, *Proc. SPIE*, **4662-41**, 341–8.

Sobol R (2004) Improving the Retinex Algorithm for Rendering Wide Dynamic Range Photographs, *Journal of Electronic Imaging*, **13**(01), 65–74.

Sobol R (2006) Method for Variable Contrast mapping of Digital Images, US Patent, 7,046,858, filed, 3/15/2004, issued 5/16/2006.

Stiehl W, McCann JJ & Savoy R (1983) Influence of Intraocular Scattered Light on Lightness Scaling Experiments, *J Opt Soc Am*, **73**, 1143.

Stockham TP (1972) Image Processing in the Context of a Visual Model, *Proc. IEEE*, **60**, 828–84.

Wallach H (1948) Brightness Constancy and the Nature of Achromatic Colors, *J. Exptl. Psychol*, **38**, 310–24.

Wilson HR & Bergen JR (1979) A four mechanism models for threshold spatial vision, *Vision Res*, **26**, 19–32.

Wray W (1988) Method and Apparatus for Image Processing with Field Portions, US Patent 4,649,568, filed 8/14/1986, issued 6/7/1988.

Wyszecki G & Stiles WS (1982) *Colour Science: Concepts and Methods Quantitative Data and Formulae*, 2nd ed, John Wiley & Sons, Inc, New York, 486–513.

Wang L, Horiuchi T & Kotera H (2006) HDR Image Compression by Integrated Surround Retinex Model, Proc. NIP22, 339–42.

Wang L, Horiuchi T & Kotera H (2007) High Dynamic Range Image Compression by Fast Integrated Surround Retinex Model, *J Imaging Sci Tech*, **51**, 34–43.

Yoda M & Kotera H (2004) Appearance Improvement of Color Image by Adaptive Linear Retinex Model, *Proc. IS&T NIP20*, 660–3.

33

ACE Algorithms

33.1 Topics

There are several algorithms in the Retinex Extended Family that share the idea of computing each single pixel from a comparison among all the other pixels in the image and mimicking vision. This chapter presents Automatic Color Equalization (ACE), RACE (Retinex + ACE) and mentions other methods as representatives of the extended family of algorithms.

33.2 Introduction

The basic property of the extended Retinex family is to mimic the behavior of the human vision system (HVS), by calculating a new image influenced by local-global content for the scene. After many experiments with Brownian Retinex, Rizzi et al. (2003; 2004; Gatta et al., 2007) described an alternative method of computing spatial pixel relationships. They proposed a series of simple operations called Automatic Color Equalization (ACE).

33.3 ACE Algorithm

The ACE algorithm was inspired by a set of mechanisms related to the modeling of low-level features of the human visual system like independent computation of chromatic channels (McCann, 1999), lateral inhibition mechanisms (Hartline et al., 1956), *global/local* mechanisms (McCann, 1987), global normalization to white (von Kries, 1970) and centering of the histogram's center of mass. ACE incorporated these mechanisms in the way described below. This resulted in some color and contrast equalization properties in line with the behavior of our vision system.

The algorithm ACE uses two functions: d to implement the spatial influence, (global/local effect), and r that affects the image contrast as described below. A diagram of the ACE algorithm is shown in Figure 33.1.

The Art and Science of HDR Imaging, First Edition. John J. McCann, Alessandro Rizzi.
© 2012 John Wiley & Sons, Ltd. Published 2012 by John Wiley & Sons, Ltd.

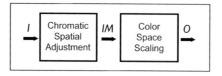

Figure 33.1   ACE structure.

The first stage of ACE compares the value of the *pixel of interest x,y* to each of the other pixels in the scene and weights the difference by the distance between them. Adding a weighting factor makes the output of the comparison local, like human vision (see Section E)

In Figure 33.1, *I* is the *input* digital image, *IM* is an *intermediate matrix* and *O* is the *output* image.

33.3.1   ACE Stage 1

The first stage, called chromatic/spatial adjustment, calculates an output matrix called *IM*. The input image *I* can be an RGB camera image, a calibrated radiance image, an LMS calculated cone response image or a black and white achromatic image. Color channels are treated separately.

For each camera digit in *(x,y)* in the original input image, we compute a value for the corresponding intermediate matrix *IM*:

$$IM_{x,y} = \sum_{\substack{x',y' \in Im \\ x',y' \neq x,y}} \frac{r(I_{x,y} - I_{x',y'})}{d_{x,y:x',y'}} \tag{33.1}$$

where $I_{x,y} - I_{x',y'}$ is a pixel difference, *d* is a distance function which weights the amount of local or global contribution, and *r* is a function that amplifies and distorts the pixel difference, acting as a contrast tuning.

Following equation 33.1, we calculate the *intermediate value $IM_{x,y}$* by:

Selecting a pair of pixels: x,y as the pixel of interest; and x',y' all other pixels in the image evaluated one at a time.

Computing an input difference for these pixels

Amplifying the difference by the contrast tuning function *r*.

Calculating d, using a weighting function of the distance between them

Taking the ratio of the amplified difference and the weighted distance

Summing all the ratios for the pixel of interest.

Each contribution is weighted by the relative distance. Near pixels count more, but far pixels are many, thus wide distant areas count as well.

Figure 33.2 shows examples of *r* contrast tuning functions. The *r* function slope variation acts as a contrast tuner: the higher the slope, the higher the contrast. The *r* function slope parameter can vary from the unitary value to infinite (Signum). The function *r* is defined in the following way:

$$r(\rho) = \begin{cases} -1 & if\ \rho \leq -1/slope \\ \rho \cdot slope & if\ -1/slope < \rho < 1/slope \\ 1 & if\ \rho \geq 1/slope \end{cases} \tag{33.2}$$

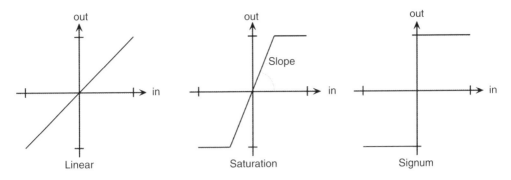

Figure 33.2   The r function set.

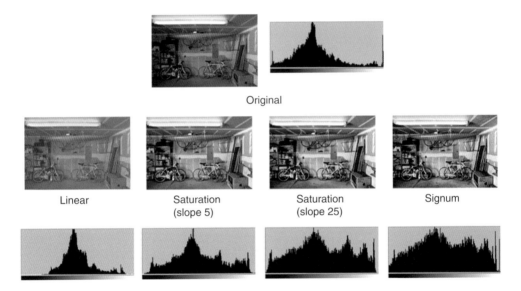

Figure 33.3   Sample image with different r functions and Euclidean distance as d.

When the r contrast tuning function is set to the maximum, the pixel difference degenerates to the signum function (Figure 33.2). This gives back values +1 if pixel difference is positive and −1 if negative. Before being summed, each +/−1 is weighted by its weighted distance.

Figure 33.3 shows an example of the effect of r slope on contrast tuning and its effects on the histogram of the grayscale channel.

The function d is a distance measure. When we started to study ACE we set d equal to the Euclidean distance between the pixels in the input image. Then, we tested a series of alternative distance functions. Our experimental results showed that distance formula is not a critical parameter, as long it is a quadratic distance (L2), similar to euclidean distance (Rizzi & Chambah, 2010).

In equation 33.1, no compensation is computed for the distance of the pixel from the picture margins. We recall that the value of each output pixel derives from the sum of all the other pixel contributions weighted by the relative distance. If the pixel to compute is near the margin, the number of pixels close to itself decreases significantly on the edge side. These neighboring pixels give an important contribution, since the inverse of d has high values only around the pixel. Consequently, we have modified the equation (33.1) with a normalization coefficient to avoid vignetting near the borders of the image, in the following way:

$$IV_{x,y} = \frac{\displaystyle\sum_{\substack{x',y' \in \text{Im}, \\ x',y' \neq x,y}} \frac{r(I_{x,y} - I_{x',y'})}{d_{x,y:x',y'}}}{\displaystyle\sum_{\substack{x',y' \in \text{Im}, \\ x',y' \neq x,y}} \frac{r_{max}}{d_{x,y:x',y'}}} \tag{33.3}$$

where r_{max} is the maximum value of function r.

The result of the first stage is a *chromatic spatial* adjustment of each pixel according to all other pixels in the image, resulting in a *global/local* contrast and color correction.

33.3.2   ACE Stage 2

ACE's first stage takes into account neighboring and dimensional relationships. At the end of the first stage, ACE gives back an intermediate matrix IM, composed of positive and negative floating point values, more or less centered around the zero value, that is not ready to be displayed. The second stage fills the image color space range.

The intermediate matrix IM, which are real numbers, can range from *min* to *max*, where *min* and *max* are arbitrary values which depend on the content of the picture, thus, for each chromatic channel, we have to map this range into the integer interval [0, 255] (or higher if there are more bits available) of the final output image O, choosing reference values to translate and scale this interval. The proposed method scales linearly the values of IM for each chromatic channel, into the available device color space range $[0, D_{Max}]$ with the following formula:

$$O_{x,y} = round(D_{mid} + S \cdot IM_{x,y}) \tag{33.4}$$

where D_{Max} is the maximum digit value available in the output color space, $D_{mid} = D_{Max} / 2$ and $S = D_{mid} / max(IM_{x,y})$.

The maximum value of IM is set to the maximum output value according to the chosen bit depth (e.g.255) and the zero (middle value) in IM is an estimate for the middle device digit. We use the maximum and zero to compute the slope S. In this way, the color space range of the final image is always centered around the middle output value and a maximum pixel is always present, implementing at the same time a histogram balance and a max normalization, separately in each chromatic channel.

Having scaled the output range using the max and mid values, it may not fill the color space. Tones around the very dark values could be lost. Alternatively, some pixels in O can result in negative values. In this case the values lower than zero are set to zero. This mechanism is visualized in Figure 33.4.

33.4   Retinex and ACE

The Retinex and ACE algorithms share a common approach and several computational characteristics, but differ in many non-trivial details. Since we are interested in comparing the fundamental principles of the two approaches, the following qualitative comparisons consider only the basic algorithms.

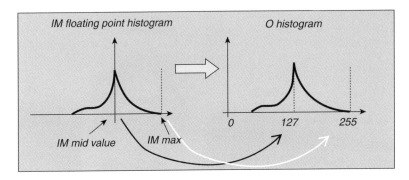

Figure 33.4   Color Space Range scaling phase.

An important common point is that both algorithms process the three chromatic channels separately, providing an automatic color correction.

Originally, ACE was devised to process uncalibrated images obtained from unknown sources such as the web and, to compute an improved image. It uses pixel subtraction with all the pixels in the image, distorted by the amplifying function r. This non-linearity is responsible for the final contrast; eliminating such distortion leads to very low contrast, as can be seen in Figure 33.3 (left).

In the Retinex model, the reset mechanism is devised to perform a normalization operation that mimics the human local-global normalization to the scene maximum and the local maxima. The locality of this effect depends on a number of parameters that respond to scene content (see Chapter 32).

A different non-linearity is introduced in ACE by the amplifying function r that emphasizes the contribution of each pixel difference. It amplifies the relative influence of brighter and dimmer zones around the pixel being computed, weighted by the distance d. This mechanism, repeated for every pixel, expands and compresses the color space range in different areas of the image, driven by the local image content.

The various Retinex implementations differ in the way they explore the image. ACE uses a distance-function-weighting contribution from a pixel compared with the other pixels in the image. From this point of view, it can resemble the Land (1983) Designator, but there are differences between the two approaches. ACE has a maximum referencing that Land's 1983 Designator has not, moreover in ACE the weighted averaging around the pixel to compute is amplified and distorted, in this way enhancing the role of pixel differences, thus enhancing final contrast.

The second and final stage prepares the computed image for visualization. Retinex produces output values for each chromatic channel in the range [0–1]; these values are scaled into the available range by taking into account the calibration of the output device (printer or display) (see Chapter 32.6.5).

The same mechanism can be applied on ACE as well, when it maps the output of its first phase, held in the intermediate matrix IM, into the final output image O, although this second phase scales in a different way the values computed in the first phase.

33.5   ACE Characteristics

One of the main effects of ACE is the scaling of the output image to the image's mean digit values. This *GrayWorld* behavior is a desirable feature for computer image enhancement.

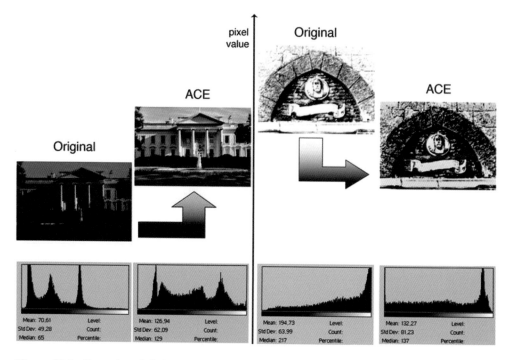

Figure 33.5 Examples of digit values equalization: (left) with under-exposed input, ACE increases the mean output digit value.value; (right) with over-exposed input, ACE reduces the mean output digit value.

33.5.1 ACE uses GrayWorld

ACE moves the mean value towards medium gray and stretches the histogram over the available output range. This allows an automatic regulation of the image mean lightness. As visible in Figure 33.5, with under-exposed input, ACE increases the mean output digit value; (left) with over-exposed input, ACE reduces the mean output digit value (right).

An important characteristic is the locality of the color correction and contrast adjustment. An example of the locality of ACE is presented in Figure 33.6, where the original and the computed images are shown together with the differences between the two in each RGB channel separately, visualized around the medium gray. Thus, in this visualization the medium gray values (128) indicate pixels untouched, the values smaller than 128 indicate pixels which value has been increased and values greater than 128 pixels which value has been decreased, in each chromatic channel.

33.5.2 ACE- De-quantization

The left column in Figure 33.7 shows a photograph that has been modified to have only 3 bits per color channel. We use only eight levels of R, G, and B, as the input image for both Retinex and ACE processing. Figure 33.7 shows the original and processed images, together with their channel histograms. For this test ACE uses a linear scaling in its second phase, in order not to add a lightness change, and the slope of r is set to 5.

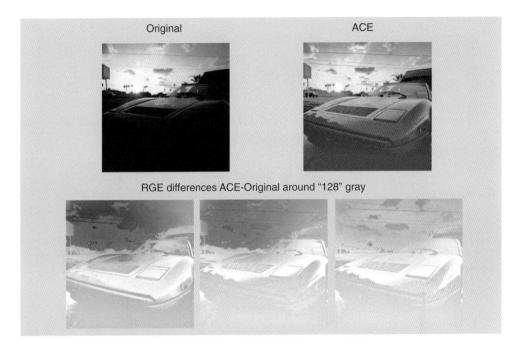

Figure 33.6 ACE's local effect: (upper left) original, (upper right) computed output images, (bottom) differences in each separate RGB channel. To enhance the visibility of the differences, they are visualized around digit 128.

Using Retinex or ACE to process an image with a limited color palette, results in a data driven color de-quantization. Namely, Retinex and ACE increase the number of gray levels in each channel. This interesting property derives from their local behavior, and depends on scene content.

We see in the histograms of Retinex that it distributes output values to the right of the input image histogram values. This is a result of reset to maxima. ACE distributes its output on both sides of the input, because of its mixture of max normalization and *GrayWorld*.

Using Retinex or ACE to process an image with a limited color palette, results in a data driven color de-quantization. Namely, Retinex and ACE increase the number of gray levels in each channel. This interesting property derives from their local behavior, and depends on scene content.

33.5.3 ACE on Simultaneous Contrast and Mach Bands

We have tested the algorithm on some interesting test targets in order to verify whether the ACE model behaves qualitatively like the human visual system.

Two visual configurations have been chosen: Simultaneous Contrast and Mach Bands. These configurations and the relative ACE outputs are shown in Figure 33.8. Notice that both the ACE computations follow qualitatively the expected visual appearance of the two configurations. Quantitative measurements of these shifts require input and output calibration. These results are in the direction of human visual processing and appear, in these prints, to far exceed the changes made by vision. Different ACE parameters could be used to get a better match to vision response.

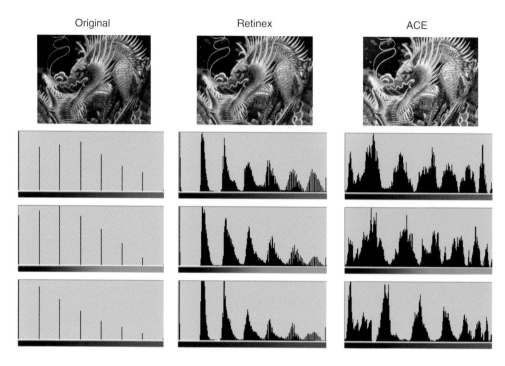

Figure 33.7   Retinex and ACE local color de-quantization property.

Figure 33.8   ACE on Simultaneous Contrast and Mach Bands.

33.5.4   Efficient ACE Processing

ACE compares every pixel with every other pixel in the image, one at a time. That is a time consuming computation load that makes ACE a $O(N^2)$ algorithm. (Rizzi et al., 2003)

Two multi-level approaches have been devised to lower ACE computation time (Gatta et al., 2006) (Artusi et al., 2006). The most interesting one is the Local Linear Look Up Table (LLL) speed-up technique (Gatta et al., 2006). The idea underlying the LLL method is to apply the ACE algorithm to a small sub-sampled version of the original image and to create three local LUT mapping functions (for R, G and B) for each pixel, between the two sub-sampled images (the ACE computed and the original sub-sampled one). These mapping functions are then used to quickly compute the original full-size image, generating each pixel in the final color corrected image, via a interpolation of the local LUT values, separately on each chromatic channel. A visual comparison of the basic and LLL-ACE algorithms does not reveal significant differences while the computational times are strongly reduced (Gatta et al., 2006).

33.6 RACE

RACE (Provenzi et al., 2008) is a combination of RSR, a MI-Retinex Algorithm (Provenzi et al., 2007) described in Chapter 32.7.5 and Automatic Color Equalization (ACE) (Rizzi et al., 2003) described above. The idea is to merge RSR and ACE using for both a series of sprays as a common spatial sampling method around the pixel of interest.

In both algorithms the mutual influence between pixels decreases as the inverse of their Euclidean distance. In ACE this local behavior is represented explicitly by the distance function d, while in RSR it is expressed implicitly through the spray's geometry. If spray density and weights have the same radial behavior (as happens for RSR and ACE), a sum performed in a localized spray is equivalent to a weighted sum in the original image.

The RSR basic algorithm is designed to be used with sprays, thus no modifications are made from the version described in Chapter 32.7.5. By contrast, ACE has to be slightly modified in order to be used with sprays.

33.6.1 ACE with Sprays

The ACE with sprays' first stage is essentially unchanged, except that differences are made between the pixel of interest (x,y) and all the pixels (x',y') in the sprays.

Having once established, the use of localized sprays, there is another change to implement: the color space range scaling. There is a simple way to remove the color space range scaling while retaining the local *GrayWorld* behavior: incorporating the term $1/2$ in r, i.e. considering this new slope function:

$$r_\alpha(I_{x,y} - I_{x',y'}) = \begin{cases} 0 & \text{if } -1 \le I_{x,y} - I_{x',y'} \le -\dfrac{1}{2\alpha} \\[2mm] \dfrac{1}{2} + \alpha(I_{x,y} - I_{x',y'}) & \text{if } -\dfrac{1}{2\alpha} < I_{x,y} - I_{x',y'} < \dfrac{1}{2\alpha} \\[2mm] 1 & \text{if } \dfrac{1}{2\alpha} \ge I_{x,y} - I_{x',y'} \le 1. \end{cases} \qquad (33.5)$$

Thus the resulting pixel is:

$$O_{x,y} = \frac{1}{N} \sum_{k=1}^{N} \frac{1}{n_k} \sum_{\substack{x',y' \in k^{th}-spray, \\ x',y' \ne x,y}} r_\alpha(I_{x,y} - I_{x',y'}) \qquad (33.6)$$

being n_k the number of spray pixels (center excluded) that lie inside the image.

33.6.2 RACE Formula

Now that the spray formulation of ACE, and the new slope function r_α are defined, we can combine RSR and ACE. The output value O representing the algebraic average between the local contributions of RSR and ACE, over the N spray contributions is:

Figure 33.9   (left) RSR, (middle) ACE, (right) RACE outputs.

$$O_{x,y} = \frac{1}{N}\sum_{k=1}^{N}\frac{1}{2}\left(\frac{R_{x,y}}{\max\left(R_{x',y'\in spray}\right)}\frac{1}{n_k}\sum_{x',y'\in spray}r_\alpha\left(I_{x,y}-I_{x',y'}\right)\right) \qquad (33.7)$$

Figure 33.9 shows an example of (left) RSR, (middle) ACE and (right) RACE (c) results. The averaging effect of RACE can be seen in this figure.

33.7   Other Vision-based Models

In addition, there are many image-processing algorithms that build their processes using neural spatial operators identified by neurophysiologists. Retinex mimicked vision, but did not synthesize the output image from models of horizontal, lateral geniculate, or cortical neurons. Marr (1982) and colleagues applied this approach to many different vision problems. Grossberg and colleagues used edge detectors to model Land's Color Mondrians and edge completion and object recognition. (Cohen & Grossberg, 1984; Grossberg & Mingolla, 1989; Grossberg et al., 1994). Blakeslee & McCourt (2001) used a multiscale model of cortical neurons to model assimilation experiments.

Vonikakis, et al. (2008) is an example of neural modeling for HDR imaging. It is a center-surround image-processing algorithm, which employs both local and global parameters. The local parameters, which affect the new value of a pixel significantly, are its intensity (center) and the intensity of its surround. The global parameters that affect the overall appearance of the image are extracted from image statistics. The surround is calculated using a diffusion filter, similar to the biological filling-in mechanism, which blurs uniform areas, preserves strong intensity transitions and permits partial diffusion in weaker edges. New pixel values combining local and global parameters, are inspired by the shunting characteristics of the ganglion cells of the human visual system. The algorithm is applied only to the Y component: http://sites.google.com/site/vonikakis/software

33.8   Summary

An extended family of successful algorithms has been fostered by the Retinex algorithm and other color spatial approaches that mimic the human vision system. Here, we presented a set of them, with comments about the similarities and the differences with Retinex. In some cases, they are based on original models, in some others they are a mixture of various approaches. This suggests a large series of algorithms that can be chosen, melded or designed according to specific needs. As we described in Chapter 1, and will discuss further in Chapter 35, the goals of the algorithm are essential in evaluating its success.

33.9   References

Artusi A, Gatta C, Marini D, Purgathofer W & Rizzi A (2006) Speed-up Technique for a Local Automatic Color Equalization Model, *Computer Graphics Forum*, **25**(1), 5–14.

Blakeslee B & McCourt ME (2001) A multiscale spatial filtering account of the Wertheimer-Benary effect and the corrugated Mondrian, *Vision Res*, **41**, 2487–2502.

Cohen MA & Grossberg S (1984) Neural Dynamics of Brightness and Perception: Features, Boundaries, Diffusion, and Resonance, *Perception & Psychophysics*, **36**, 428–56.

Gatta C, Rizzi A & Marini D (2006) Local Linear LUT Method for Spatial Color Correction Algorithm Speed-up, *IEE Proceedings Vision, Image & Signal Processing*, **153**(3), 357–63.

Gatta C, Rizzi A & Marini D (2007) Perceptually Inspired HDR Images Tone Mapping with Color Correction, *Journal of Imaging Systems and Technology*, **17**(5), 285–94.

Grossberg S & Mingolla E (1989) Neural Networks for Machine Vision, US Patent 4,803,736, Issued: Feb 7, 1989, Filed: Jul 23, 1987.

Grossberg S, Mingolla E & Ross WD (1994) A Neural Theory of Attentive Visual Search: Interactions of Boundary,Surface, Spatial, and Object Representations, *Psychological Review*, **101**(3), 470–89.

Hartline H, Wagner H & Ratcliff F (1956) Inhibition in the Eye of Limulus, *Journal of General Physiology*, **39**(5), 651–73.

Land EH (1983) Recent Advances in Retinex Theory and Some Implications for Cortical Computations: Color Vision and the Natural Image, *Proc. Natl Acad Sci USA*, **80**, 5163–9.

Marr D (1982) *Vision*, W. H. Freeman, San Francisco.

McCann JJ (1987) Local/Global Mechanisms for Color Constancy, *Die Farbe*, **34**, 275–83.

McCann JJ (1999) Lessons Learned from Mondrians Applied to Real Images and Color Gamuts, *IS&T/SID Color Imaging Conference*, **7**, 1–8.

Provenzi E, Fierro M, Rizzi A, De Carli L, Gadia D & Marini D (2007) Random Spray Retinex: A New Retinex Implementation to Investigate the Local Properties of the Model. *IEEE Transactions on Image Processing*, **16**(1), 162–71.

Provenzi E, Gatta C, Fierro M, & Rizzi A (2008) A Spatially Variant White Patch and Gray World Method for Color Image Enhancement Driven by Local Contrast, *IEEE Transactions on Pattern Analysis and Machine Intelligence*, Vol. **30**, No. 10.

Rizzi A & Chambah M (2010) Perceptual Color Film Restoration, *Society of Motion Picture and Television Engineers Journal*, **19**(8), 33–41.

Rizzi A, Gatta C & Marini D (2003) A New Algorithm for Unsupervised Global and Local Color Correction, *Pattern Recognition Letters*, **24**(11), 1663–77.

Rizzi A, Gatta C & Marini D (2004) From Retinex to Automatic Color Equalization: Issues in Developing a New Algorithm for Unsupervised Color Equalization, *Journal of Electronic Imaging*, **13**(1), 75–84.

von Kries J (1970) *Chromatic Adaptation, Sources of Color Science*, MacAdam ed. MIT Press, 109–19.

Vonikakis V, Andreadis I & Gasteratos A (2008) Fast Centre-Surround Contrast Modification, *IET Image Processing*, **2**(1) 19–34.

34

Analytical, Computational and Variational Algorithms

34.1 Topics

There are many ways to process images in addition to iterative algorithms. In this chapter we review different types of mathematical calculations using the Retinex Extended Family. These include analytical, computational and variational formalizations.

34.2 Introduction

In its original formulation, Retinex was a computational model, aimed at mimicking operations observed in our vision system. The idea was to build an appearance output record from scene content. Over the years, Retinex research has incorporated the advances in computer hardware, efficiencies in algorithm design, and taken advantage of special purpose image-processing hardware. These advances have reduced computation time by many orders of magnitude.

Transforming a model into something implemented on a machine or by other means is subject to simplifications, or small modifications, in order to make the implemented version more efficient, more simple, or sometimes more similar to how the scholar interpreted the original model. Retinex, from this point of view, has an interesting and complex story that is presented in Chapter 32. The present chapter reviews papers that move Retinex from an algorithmic process to more mathematically formalized expressions.

Moving to a mathematical formulation is a natural process, especially due to the large and active community of mathematicians that works intensively on imaging. However, other reasons are to have new tools to investigate and explore the models, or in some cases to attempt to have a series of rapidly assimilated methods to implement fast versions, or speed-up the existing ones, or also to interpret the model from different perspectives.

There is not a unique jump that takes Retinex from its algorithmic form to a shared and agreed mathematical formulation. This process can be seen as a continuum space in which the amount of formalization used has been functional to the goal and the background of the various researchers. Thus, different formalizations have been proposed. The following survey does not aim to be complete, it reports some interesting alternative formalizations in order to give to the reader a glance at the different perspectives.

The Art and Science of HDR Imaging, First Edition. John J. McCann, Alessandro Rizzi.
© 2012 John Wiley & Sons, Ltd. Published 2012 by John Wiley & Sons, Ltd.

In this chapter we describe analytical approaches that state the algorithm in a mathematical formulation which allow analysis of the algorithm's behavior. As well, we describe computational approaches, such as wavelet implementation of Retinex algorithms. Finally, we discuss variational formulations that describe various algorithms as optimization problems using differential equations.

34.3 Math in the Framework of the Human Visual System

As described in Chapter 32.6.3, there is a progression from Land's proposal of vision-based image processing to Stockham to Horn and Marr to Faugeras for models that mimic vision. In his widely cited paper, Faugeras (1979) points out the importance of considering human vision characteristics for better image processing. He says:

> "Since a human observer is likely to be the last element in the processing chain, it seems natural to become interested in the properties of the human visual system."

Extending Stockham's (1972) work, Faugeras proposes a model of human vision built as a sequence of blocks containing important psychophysiological data stressing a hypothetical similarity of the behavior of the human visual system with homomorphic filtering.

34.4 Analytical Retinex Formulas

The first analytical detailed equation for the MI-Retinex using Brownian Path and RSR (Provenzi et al., 2005) starts from the computational description of the random Brownian Path Retinex (Marini & Rizzi, 2000), and presents a complete formalization, that includes the mechanism of MI Retinex reset and analyzes the role of the threshold in the computation (Table 34.1).

The goal is to define an analytic equation, that describes all the operations that Marini and Rizzi's MI-Retinex performs on an image, despite the many implementation variants they used (Marini & Rizzi, 2000; Rizzi et al., 2002; Provenzi et al., 2007). They formulated Brownian Path Retinex with and without threshold, and used the formalization without threshold to analyze the role of parameters and multiple iterations on the same image.

34.5 Computational Retinex in Wavelets

The 2002 IS&T/SPIE Electronic Imaging conference, *"Retinex at 40"* was a joint session of Human Vision and Electronic Imaging VII; Color Imaging: Device-Independent Color, Color Hardcopy, and Applications VII; and Internet Imaging III. The 16 papers were published in a special issue of the

Table 34.1 The analytical and computational Retinex models.

Analytical	Idea	Author
Analytical	Description of Marini, Rizzi MI-Retinex	Provenzi, et al., 2005
Computational	Idea	Author
Wavelet	Implement of McCann (1999) in wavelet	Rising, 2002
Wavelet	Implement NASA Retinex in wavelet	Hu, 2004; Sun et al., 2010; Hao & Sun, 2010

Journal of Electronic Imaging (McCann, 2004). Among them, there is a formalization of Retinex using wavelet functions. Rising (2004) described the Frankle McCann Retinex using wavelet functions. It starts from the McCann multilevel version of Retinex, and through the use of wavelets implements its local and multilevel structure. The author presents Retinex as a redundant inverse wavelet decomposition, exploiting the similarities with the Haar transform in terms of neighborhood comparison and inner wavelet multilevel structure, with a non-linearity introduced by reset and threshold (Table 34.1).

Other wavelet formalizations were later developed based on the NASA Retinex by Hu (2004), Sun et al. (2010), and Hao and Sun (2010).

34.6 Retinex and the Variational Techniques

Horn (1974) implemented the threshold property of the Land and McCann (1971) Retinex using Laplacian differential equations. Andrew Blake (1985) proposed a formulation of Land's (1983) *Designator* as a Poisson partial differential equation. Terzopoulos (1986) proposed multi-resolution techniques for image analysis using partial differential equations. He processed the Black and White Mondrians and studied shape-from-shading and optical flow algorithms.

Years later, Guillermo Sapiro's book (2001) renewed interest in variational methods for digital imaging. Table 34.2 summarizes the variational formulations reviewed here.

Ron Kimmel and other HP colleagues (Elad et al., 2003) proposed variational methods, formulated constraints about illumination smoothness, and limited the range of the reflectance field. This resulted in a Quadratic Programming optimization, solved with an efficient multi-resolution algorithm.

Since Kimmel and colleagues chose to find the objects' reflectances in the scene, they are confronted with the ill-posed problem of separating reflectance and illumination. They called that ill-posed problem the "Retinex problem" (see Chapters 32.8.1 and 29). Nevertheless, a number of papers adopt this approach as the problem to solve.

In the context of what they called "Retinex type image enhancement", Shaked & Keshet, (2002, 2006) developed "robust envelope operators" and a method to speed up their algorithms, based on a non-linear illumination estimation followed by a manipulation module (Shaked, 2006).

Durand and Dorsey (2002) used bilateral filtering to attempt to separate reflectance from illumination. Later, Michael Elad (2005) refined the Durand and Dorsey work to solve the "Retinex problem" using

Table 34.2 Variational Retinex models.

Variational	Idea	Author
Laplacian Gradients Land's Designator Multi-resolution	Early Work	Horn (1974), Blake (1985), Terzopoulos (1986)
Poisson Equation	Separate Illumination and Reflectance	Kimmel et al. (2003) Elad et al., 2003
Bilateral Filter	Separate Illumination and Reflectance	Elad, 2005
ACE	Variational ACE	Beralmio et al., 2007
Vision	Color Constancy, Adaptation, Local Contrast	Palma-Amestoy et al., 2009
KBR	Kernel Base Retinex (KBR)	Bertalmio et al., 2009
Neural	Wison & Cowan Equation	Cowan and Bresslof, 2002 Bertalmio and Cowan, 2009
Brownian Path Extrema Retinex	Reset to Selected Pixels	Morel et al., 2010

two bilateral filters: one for reflectance and one for illuminance. They also proposed improvements in efficiency.

Marcelo Bertalmío, Vicent Casseles and colleagues have published a series of papers about the variational formalization of MI-Retinex and other related algorithms. They started from the variational formalization of ACE (Chapter 33; Bertalmío et al., 2007). They formalized a computational and variational version that gave very similar results. They addressed the wider problem of formulating in a variational form a more general method, identifying a class of functionals to model color constancy, local contrast enhancement, and light/dark adaptation. (Palma-Amestoy et al., 2009). They have also discussed the theoretical similarities and differences between these models and MI-Retinex and ACE, investigating the relationship between the two.

Later, they formulated a variational version they called the Kernel Based Retinex (KBR) (Bertalmío et al., 2009). It is based on the MI-Retinex and computes the conditional expectation of a random variable in a density field generated by a kernel function, hence the name. They developed a basic version and an anti-symmetrized one, discussing its similarities with ACE. They presented image processing results that improved renditions. This paper also presents an interesting review of the various Retinex implementations from the point of view of the dimensionality of the image sampling. The appendix contains an excellent analysis of the Land 1983 algorithm and the MI-Retinex. This paper discusses the role of the reset in MI-Retinex, and the importance of a postLUT to comply with the behavior of the visual system. It does not discuss the Land and McCann reset.

Cowan and colleagues derived variational models of the human cortex from physiological data (Wilson & Cowan, 1972; Cowan and Bresslof, 2002; Bertalmio & Cowan, 2009). The final paper in the series builds upon the Bertalmio formulation of the MI-Retinex. It shows that it has the same form as the Wilson & Cowan description of area V1 in the cortex. Thus they conclude that the location of the Retinex processing is in the cortex.

In a recent paper Morel et al. (2010) presents a variational formalization of several Retinexes. They state that to the best of their knowledge the Land & McCann reset described in Chapter 32.3.4 cannot be formalized in PDE. This is an important point because that reset is the most efficient implementation of *locality influence*. It modifies the output, in relation to the maxima, in every pixel comparison. As we saw throughout Sections C, D and E, the role of maxima is key to modeling appearance. The Land and McCann reset has been shown to have properties similar to measured *neural contrast*. It has been shown to act as a scenes-dependent antidote to glare (Chapter 32.5.2). When this efficient normalization process is combined with the Zoom Retinex (Chapter 32.6), it has O(N) computational efficiency. It is an extremely fast computational model, and is even more efficient when combined with both special purpose hardware, and Sobel's modifications.

Morel et al. continue to analyze the MI-Retinex and introduce a *Extrema Retinex* that is similar to a global normalization algorithm. They report results in Figures 4, 5 and 6 that are quite different from Retinex calculations for the same targets by McCann, using Land and McCann, Frankle and McCann and Sobel Retinexes, and Rizzi using MI-Retinex.

34.7 Summary

Despite the many attempts, in the history of Retinex, to simplify it, all these works show the richness, the complexity and relevancy of the Retinex model. Retinex is still a young algorithm with great potential for future research and implementations.

Here we have presented a partial list of the Retinex and Retinex-inspired implementations in different disciplines of digital imaging. They share the goal of investigating the Retinex Extended Family, and pay careful attention to the spatial aspects of color computation.

There remains an important challenge. The data presented in Sections C, D and E are in the form of observer response to test targets. The problem of integrating all these different measured appearances into a comprehensive model remains an important next step. If we could describe analytically the properties of *locality influence*, then we could design image-processing algorithms that fit the data. Such a model of vision that incorporated all the properties of normalization to maxima would give us a better target for characterizing the range of algorithms that aim at mimicking vision. Image enhancement, in itself, does not help to discriminate among all the algorithms. Everyone improves, in some way, the input image. Quantitative, and not only qualitative, predictions of the human vision system represent a more definite and challenging goal.

34.8   References

Blake A (1985) Boundary conditions of lightness computation in mondrian world, *Computer Vision Graphics & Image Processing*, **32**, 314–27.

Bertalmío M, Caselles V, Provenzi E & Rizzi A (2007) Perceptual Color Correction Through Variational Techniques, *IEEE Trans on Image Processing*, **16**(4), 1058–72.

Bertalmío M, Caselles V & Provenzi E (2009) Issues About Retinex Theory and Contrast Enhancement, *International Journal of Computer Vision*, **83**(1), 101–19.

Bertalmío M & Cowan J (2009) Implementing the Retinex algorithm with Wilson-Cowan equations, *J Physiol Paris*, **103**(1–2), 69–72.

Cowan J & Bresslof P (2002) Visual Cortex and the Retinex Algorithm, *Proc. SPIE Human Vision and Electronic Imaging VII*, **4662**, 278–85.

Durand F & Dorsey J (2002) Fast Bilateral Filtering for the Display of High-dynamic-range Images, *SIGGRAPH 2002*, 257–66.

Elad M, Kimmel R, Shaked D & Keshet R (2003) Reduced Complexity Retinex Algorithm via the Variational Approach, *Journal on Visual Communication and Image Representation*, **14**(4), 369–88.

Elad M (2005) Retinex by Two Bilateral Filters, in Scale Space and PDE Methods in Computer Vision, *Lecture Notes in Computer Science*, **3459/2005**, 217–29.

Faugeras O (1979) Digital Image Color Processing Within the Framework of a Human Visual System, *IEEE Trans. on Acoustic Speech and Signal Proc*, **27**, 380–93.

Hao M & Sun X (2010) A Modified Retinex Algorithm Based on Wavelet Transformation, Proc. Second Int Conf MultiMedia & Information Tech, Kaifeng, China.

Horn B (1974) Determining Lightness from an Image, *Computer Graphics and Image Processing*, **3**, 277–99.

Hu Q (2004) Image Quality Related Processing and Applications Based on Retinex Wavelet Theory, XXth ISPRS Congress, *Proc. Com IV, Geo-Imagery Bridging Continents*, Istanbul.

Kimmel R, Elad M, Shaked D, Keshet R & Sobel I (2003) A Variational Framework for Retinex, *Int. Journal of Computer Vision*, **52**(1), 7–23.

Land EH (1983) Recent Advances in Retinex Theory and Some Implications for Cortical Computations: Color Vision and the Natural Image, *Proc. Natl Acad Sci USA*, **80**, 5163–9.

Land E & McCann JJ (1971) Lightness and Retinex Theory, *J Opt Soc Am*, **61**, 1–11.

Marini D, Rizzi A (2000) A Computational Approach to Color Adaptation Effects, *Image and Vision Computing*, **18**(13), 1005–14.

McCann J (2004) Special Session on Retinex at 40, *J Electron Imaging*, **13**, 6–145.

Morel J, Petro A & Sbert C (2010) A PDE Formalization of Retinex Theory, *IEEE Trans on Image Process*, **19**(11), 2825–37.

Palma-Amestoy R, Provenzi E, Bertalmío M, Caselles V (2009) A Perceptually Inspired Variational Framework for Color Enhancement, *IEEE Trans on Pattern Analysis and Machine Intelligence*, **31**(3), 458–74.

Provenzi E, De Carli L, Rizzi A & Marini D (2005) Mathematical Definition and Analysis of the Retinex Algorithm, *J Opt Soc Am A*, **22**, 2613–21.

Provenzi E, Fierro M, Rizzi A, De Carli L, Gadia D & Marini D (2007) Random Spray Retinex: a New Retinex Implementation to Investigate the Local Properties of the Model, *IEEE Trans on Image Processing*, **16**(1), 162–71.

Rising III H (2004) Analysis and Generalization of Retinex by Recasting the Algorithm in Wavelets, *J Electron Imaging*, Special Section on Retinex at 40, **13**, 93.

Rizzi A, Marini D & De Carli D (2002) LUT and Multilevel Brownian Retinex Colour Correction, *Machine Graphics and Vision*, **11**(2/3) 153–68.

Sapiro G (2001) *Geometric Partial Differential Equations and Image Analysis*, Cambridge University Press, Cambridge 0-521-79075-1.

Shaked D & Keshet R (2002) Robust Recursive Envelope Operators for Fast Retinex, Hewlett-Packard Research Laboratories Technical Report, HPL-2002-74R1.

Shaked D & Keshet R (2006) Robust Recursive Envelope Operators for Fast Retinex-type Processing of Images, European Patent EP1 668 593B1.

Shaked D (2006) Interpolation of Non-Linear Retinex Type Algorithms, HP Laboratories Israel, Technical Report, HPL-2006-179.

Stockham T Jr (1972) Image Processing in the Context of a Visual Model, *Proc. IEEE*, **60**, 828–42.

Sun B, Li J, He J & Zhu X (2010) Application of Retinex Wavelet Moment Features for Complex Illumination, *Proc. SPIE*, **7668** OY.

Terzopoulos D (1986) Image Analysis Using Multigrid Relaxation Methods, *IEEE Trans on PAMI*, **8**, 129–39.

Wilson H & Cowan J (1972) Excitatory and Inhibitory Interactions in Localized Populations of Model Neurons. *Biophys J*, **12**, 1–24.

35

Evaluation of HDR Algorithms

35.1 Topics

Spatial image processing, such as Retinex, ACE, and spatial-frequency filters use the entire image in rendering scenes. These algorithms process use captured scene radiances as input. Then they use the spatial information to synthesize a new image for rendition by a display or print. Spatial algorithms can convert identical input values into different output values. We discuss techniques most appropriate for measuring the success of spatial algorithms.

35.2 Introduction

In the previous chapters we reviewed and categorized many algorithms. We have only scratched the surface. We have limited the discussion to models of human visual sensations and ideas described as derivatives of Retinex. Unfortunately we were not able to include a number of recent articles. If you search the data bases for the Patent Offices of the United States, the European Union and Japan, you find more than 300 hits for the word Retinex. That list includes many passing references that are not very relevant, but also shows that many companies are working on topics discussed here. As well, there are many papers and patents on closely related topics that do not use the word Retinex.

How do we evaluate all these old and new algorithms?

We would like a simple calculation that provides a *figure of merit* quantifying the performance of our favorite HDR algorithm, relative to its alternatives. However, we found that goal impractical for two reasons:

> First, HDR algorithms are in the middle of the imaging chain, and their success is affected by pre- and post-processing. As we saw in Section B, camera digits are not proportional to scene radiance.

> Second, there are a variety of distinct goals for different spatial algorithms:
> - to reproduce the scene exactly
> - to find the objects' reflectance, and/or illumination

The Art and Science of HDR Imaging, First Edition. John J. McCann, Alessandro Rizzi.
© 2012 John Wiley & Sons, Ltd. Published 2012 by John Wiley & Sons, Ltd.

- to make the best HDR picture
- to mimic human vision for HDR rendering.

With different goals, there are different *ground truth* data for each type of algorithm. If the goal is accurate reproduction of all pixels, then radiometry will prove its accuracy. Color media, using printer dyes, phosphors, LCD filters and LED emitters, have markedly different spectra and limited range, so truly accurate reproduction of every pixel is not practical.

If the algorithm's goal involves reflectance and illumination then quantitative analysis of its success is easy to measure using light meters. The calculated output must equal radiometric measurements of the scene. If the algorithm predicts physical properties then one has to study the digits not the pictures. Sometimes we see articles that use processed pictures of scenes used as evidence of an algorithm's performance. Looking at pictures adds the viewer's visual image processing to that of the algorithm, so appearances do not correlate with digit value.

If the terms reflectance and illumination are only used metaphorically, without commitment to the actual definitions in physics, then the goal is ambiguous. It joins the simple, but hard to define, goal of making desirable renditions. The majority of pixel-based HDR algorithms attempt to make better pictures, without a defined objective numerical *ground truth* for every image. We can use the subjective techniques for finding most preferred individual image, but in such experiments the conclusions are image dependent (Kuang, et al., 2007). Subjective comparisons can identify success for individual scenes, but cannot be used to draw general conclusions beyond the algorithm's performance for that specific scene.

The Retinex family of algorithms has the explicit goal of mimicking vision. This goal provides a more specific *ground truth* framework, because we can compare appearance measurements of each target with algorithm output. The challenge is to mimic vision for all scenes: HDR, LDR, all sizes of grays' test areas in white, gray, and black surrounds. In this chapter we will discuss four different approaches to measure quantitatively the success of an algorithm.

35.3   Quantitative Approaches to Algorithm Evaluation

Over the years there have been a number of different quantitative test targets used to evaluate HDR algorithms. These examples all use a number of different targets. They all have measured appearance data. The evaluation is a simple comparison of algorithm output vs. observed appearance. They follow along the same experimental design as the McCann, Land, and Tatnall (1970) and the McCann, McKee, and Taylor (1976) measurements of Color Mondrians.

- **Lightness Test Targets:** In the 1960s we made a series of over a dozen black-and-white test targets to guide our work on lightness models (McCann et al., 1970). We carefully measured all the scene luminances and observers matched the lightness of all the areas in all targets. We described the result of the *Simultaneous Contrast* and the *Gradients and Edges* targets in Chapter 32.
- **Ratio Metric:** In colorimetry, we describe color differences between the original and its reproduction as the distances between their positions in the same color space. As a part of evaluating Retinex Gamut Mapping we described a technique that replaces pixel comparisons with edge-ratio comparisons.
- **Quantitative Evaluation of 3-D Mondrians:** Carinna Parraman's watercolor painting of the LDR & HDR pair of 3-D Mondrians provides us with quantitative measurements of observed appearances (Chapter 29). This data provides us with *ground truth* for these very complex scenes. A successful HDR algorithm must be able to predict the appearances of both the LDR and HDR halves of the painting.
- **Locality Test Targets:** The measured lightness data found in Section D provides the *ground truth* for analyzing an algorithm's ability to mimic locality influence. Future studies can use this data to

evaluate models that attempt to predict both *Simultaneous Contrast* and *Assimilation* test targets using the same algorithm and parameters.

35.4   Lightness Test Targets

As we described in 32.5, we devised experiments in the 1960s to measure the appearances of lightnesses with a wide variety of scene content. This set of targets included variations in reflectances, illuminations both uniform and gradients, and visual phenomena. In particular, we made test targets that showed appearances that departed from reflectances. Figure 35.1 shows a series of over a dozen black-and-white test targets to guide our work on lightness models. (McCann et al., 1970). The targets included variations in scene average luminance, gradients in illumination, variations of Simultaneous Contrast, extremes in background, and combination of edges and gradients with the same luminance changes.

We carefully measured all the scene luminances and observers matched the lightness of all the areas in all targets. We described the result of the *Simultaneous Contrast* and *Gradients and Edges* targets in Chapter 32. Funt et al. (2002) used the targets to find parameters of a model.

There are two different ways to use this data. In the 1960s, we studied the effects of model parameters on each area in each display as individual events. Although time consuming, it gives a good sense of parameter properties for all parts of the test target. We saw such analysis in 32.5. An alternative approach is to accumulate the average performance statistics of all areas from all targets. This technique assumes the best algorithm is found by looking at the global average of all data.

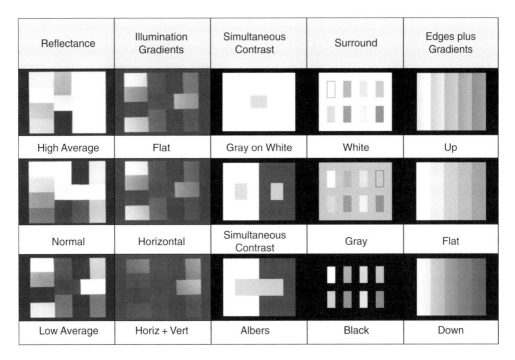

Figure 35.1   Lightness test targets used to study spatial comparison models, and their locality influence properties.

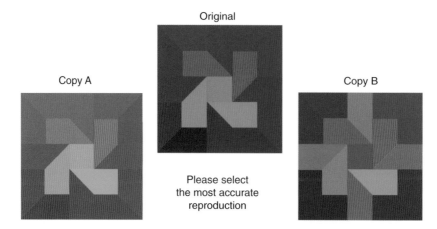

Figure 35.2 Original and two copies. Although Copy A looks like the original, Copy B has exactly the same colorimetric error for every pixel.

35.5 Ratio Metric

McCann (1998) made an Original, and two copies (Figure 35.2). The colors in the *Copy A* were selected to be significantly different from those in the Original. Each area in *Copy A* is 10 units lighter, 10 units less red, and 10 units more yellow in CIEL*a*b* space. The combined distance is $\Delta E = 17$. For each area the color difference was $\Delta E = 17$ between the *Original* and *Copy A*. *Copy B* was also made with each area $\Delta E = 17$ compared to the Original, but it was designed have the color changes go in many different directions. *Copy B* significantly changed the local relationships, while *Copy A* preserved them.

Copy A is a fairly good reproduction, considering that it has a $\Delta E = 17$ for each area. *Copy B* was made so that the $\Delta E = 17$ changes were in many different directions in color space. In this case the ΔE components were chosen to change the appearance of the display. The outer corner patches moved closer in color to each other. The other, mid-side patches moved closer in color to the inner areas. The net effect is that *Copy B* does not look like the Original. It looks like a different display. Nevertheless, if judged by the ΔE L*a*b* Color Metric, *Copy B* is exactly as good a reproduction as *Copy A* (McCann, 2002a).

If average ΔE could be used as Color Metric for quality of reproduction, half the observers would select *Copy A* as the best reproduction and half the observers would pick *Copy B*. If spatial parameters, namely the relationship between different areas within the test target are important, then observers will select *Copy A*, preserving the spatial relationships. All non-colorblind observers reported *Copy A* is better than *Copy B*.

35.5.1 Spatial Color Metric

We have seen in this text that there is a fundamental difference between pixel-based digital imaging systems and human vision. Appearance is the relative response of all the receptors across the field of view. The metric for color appearance needs to build on colorimetry and accumulate relationships all across the image.

When printer and display technologies introduce different color gamuts, the problem becomes more interesting. In fact, most displays have much larger gamuts than prints in high lightness at full saturation. Most printers have larger gamuts than displays in low lightnesses at full saturation. Typically, one finds that the common volume of a three-dimensional color-space is half the combined display and

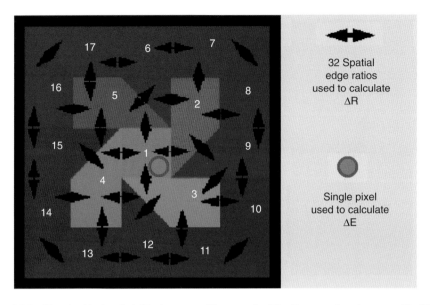

Figure 35.3 The double-headed, black arrows illustrate the 32 adjacent edges between 17 different color patches; red ring identifies a single pixel used in colorimetry.

printer volume. A test image that represents all parts of display plus printer color space will reproduce accurately only half the pixels on a printer, or on a display (Marcu 1998; Morovic & Luo, 1999).

A spatial approach to solving the problem is to use the information learned from the comparison of *Copy A* with *Copy B*. The human eye cares more about the relationships of the parts of the image than it does about the absolute value of the match. Figure 35.3 shows the Original image with identification numbers. Area 1 is the gray square in the center; the numbers are assigned in a clockwise spiral up to Area 17 on the top left. The round black circle in the bottom right corner of Area 1 represents the input information used in colorimetry to calculate the tristimulus values of Area 1. Relative colorimetry compares the information from a single pixel to the ratio of media white to illuminant; that is information that cannot be derived from the image itself. The media white and the illuminant have to be measured independently (Hunt, 2004).

The double-headed black arrows in Figure 35.3 show the 32 comparisons possible between adjacent areas in the image. This is the information the human eye uses to calculate color appearance. Our measurements are in CIE XYZ color space. We can propose a Color Metric more like human vision by comparing X from one area with X' of an adjacent area in the same image. We can calculate a corresponding spatial comparison for the Original image. We can compare the two with a ratio of the pair of edge ratios.

$$\left[\left(\frac{X_{Copy}}{X'_{Copy}}\right)\Big/\left(\frac{X_{Orig}}{X'_{Orig}}\right)\right] \tag{35.1}$$

Figure 35.4 is a diagram of the standard colorimetric comparison to measure ΔE L*a*b*, and ΔR using spatial comparisons. The left half of the figure shows the copy above the Original for standard *ΔE* L*a*b* evaluation.

The procedure uses the measurements of corresponding pixels in the copy and the Original. The value of *ΔE is:*

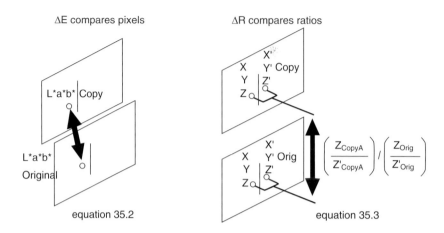

Figure 35.4   ΔE uses corresponding pixels; ΔR compares corresponding ratios.

$$\Delta E = \sqrt{\left(L^*{}_{Copy} - L^*{}_{Orig}\right)^2 + \left(a^*{}_{Copy} - a^*{}_{Orig}\right)^2 + \left(b^*{}_{Copy} - b^*{}_{Orig}\right)^2} \tag{35.2}$$

The right half of Figure 26.3 shows the calculation of *ΔR* (equation 35.3). As described in Chapter 32, the spatial comparison (Z_{Copy}/ Z'_{Copy}) replace the tristimulus value Z_{Copy}. The comparison between *Copy* and *Original* is the ratio of ratios in equation 35.1. The quality metric is shown in equation 35.3.

$$\Delta R = \sqrt{\left(1 - \frac{X_{Copy} - X'{}_{Copy}}{X_{Orig} - X'{}_{Orig}}\right)^2 + \left(1 - \frac{Y_{Copy} - Y'{}_{Copy}}{Y_{Orig} - Y'{}_{Orig}}\right)^2 + \left(1 - \frac{Z_{Copy} - Z'{}_{Copy}}{Z_{Orig} - Z'{}_{Orig}}\right)^2} \tag{35.3}$$

We use these ratio-of-ratios to formulate a spatial metric ΔR analogous to ΔE for pixels. We can calculate ΔR for each ratio in the image.

The edge ratio results for the 32 edges are plotted in Figure 35.5. *Copy A*, with the same shifting for all the colors (top right) has 32 nearly constant edge ratios. The ratio of ratios is very close to 1.0 for all 32 edges in X, Y, Z. In fact, the departures from ratios of 1.0 are experimental errors introduced by color media limitations. *Copy B*, with different shifts for each area, shows significant changes in the 32 spatial comparisons (bottom right).

We can average ΔR to accumulate a net quantity that describes the entire image. Average ΔR for *Copy A* is 0.17 ± 0.11 and 1.05 ± 0.83 for *Copy B*. The spatial metric ΔR predicts that *Copy A* is six times better than *Copy B*. The displays were made to have equal ΔE's. The spatial metric analysis of *Copy A* and *Copy B* corresponds with observer reports, namely that *Copy A* is significantly better than *Copy B* (McCann, 2002a).

This particular spatial metric is very primitive. X, Y, Z are proportional to cone response, but are not isotropic in color appearance. The point here is that spatial comparisons can discriminate between *Copy A* and *Copy B* while pixel comparisons cannot.

The comparison of *Copy A* and *Original* shows an additional important point. Although *Copy A* has the same edge ratios as the Original, it does not match the *Original*. Edge ratios are an important tool for a successful metric, but not a complete metric by themselves. Human vision appearance depends on both spatial and absolute quanta catch information (Chapter 21).

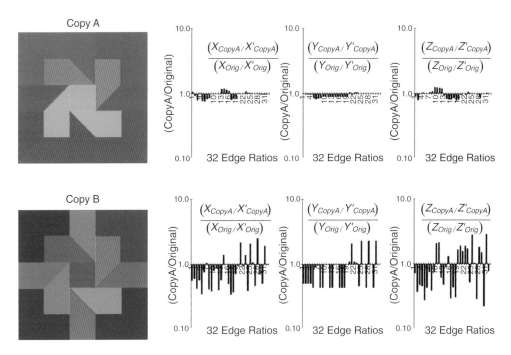

Figure 35.5 The 32 edge ratios. The spatial metric ΔR can discriminate between Copy A and Copy B, while colorimetry ΔE cannot.

35.5.2 Experiment in Munsell Space

It order to remove any color space distortions in the demonstration shown in Figure 35.2, we can make new displays using the Munsell Notation, instead of L*a*b*. Figure 35.6 shows a set of nine different Umbrella displays. We call the one in the middle [E], the Original. All the others are reproductions. Which of the eight different reproductions are acceptable and which are not?

For all eight reproductions each individual triangular patch differs from the original by one chip in the Munsell Book of Color. That means that each individual chip is a constant distance from the original in the Munsell Uniform Color Space. Thus, changes in lightness, chroma and hue are equal. Along the left-to-right axis *[D-E-F]* the *Original* and reproductions vary only in lightness (*D* patches are one Munsell chip lighter; *F* patches darker). Top-to-bottom *[B-E-H]* reproductions vary only in hue (*B* patches are shifted counterclockwise; *H* one Munsell chip clockwise). Along the upper-left to bottom-right axis *[A-E-I]* the *Original* and reproductions vary only in chroma (*A* is one Munsell chip less saturated; *I* is more saturated). Reproductions *[ABDFHI]* are still reasonable with the one-chip color shifts. Reproductions *C* and *G* are examples of individual one-chip color shifts in hue, lightness and chroma. Unlike systematic shifts, individual shifts create unacceptable distortions of the original. An error of 1 Munsell chip is acceptable if it is global, but not if it is local (McCann, 2002b).

The analogy to color constancy is very compelling. Changing the colors independent of the neighbors disrupts the spatial ratios (Figure 35.2). Changing the local ratios in the context of a color constancy experiment is the same as changing the reflectances of the areas. Changing reflectances of individual papers (local shifts) cause big changes in appearances, while changes in illumination (global shifts) cause small changes.

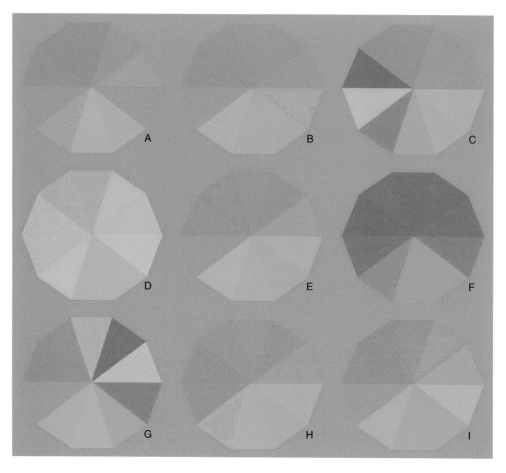

Figure 35.6 Eight reproductions of the Original [E]. Each individual triangular patch differs from the Original by one chip in the Munsell Book of Color. DEF has uniform shifts in lightness; BEH has uniform shifts in hue; AEI has uniform shifts in chroma. C and G show *individual one-chip color shifts in hue, lightness and saturation.*

Evans (1974) describes a "consistency principle" for color reproductions. He says:

"As long as the chromatic relations within a given scene are consistently like those of the original scene, the colors will actually seem to be the same in both, unless of course, direct point-for-point comparisons are made. The actual discrepancies can be enormous, providing there is no inconsistency."

Evans attributes this observation to poor color memory. By introducing spatial comparisons we can add computational properties to Evans' principle. The suggestion from these experiments is the use of spatial comparisons will significantly improve both color-evaluation and color-gamut calculations. Colorimetrically they will have larger color-difference errors, but they will look better. The same color constancy mechanism, that reduces large physical shifts in illumination to small appearance changes, can be employed to make gamut-limited reproductions look better (see Chapter 32.7.3). As well, spatial comparisons in Munsell Color Space are a meaningful measure of color accuracy.

35.6   Quantitative Evaluation of 3-D Mondrians

McCann et al. (2010) describe test images, measurements of scene characteristics, and examples of a set of flexible tools for quantitative evaluations of spatial color algorithms. Quantitative measurements of spatial algorithms evaluate the true performance of the spatial process. This paper works in parallel with http://sites.google.com/site/3dmondrians/ that provides appendices for detailed data.

35.6.1   Rendition Quality Metric

Can we find an objective analysis of image rendering using error metric analysis? First, we would need to measure the error for each pixel in a complex image. The error is a distance between the spatial processing output and the ideal *ground truth*. We also need to compare these errors in a uniform color space. In such a spacc, apparent changes in hue, lightness and chroma must all be equal to distances in the 3-D space. In uniform spaces, such as Munsell, the distance in the space represents the size of the change in appearance, while in XYZ, RGB, and sRGB spaces distance does not equal change in appearance. In addition, camera digits follow sRGB guidelines, but do not always follow the standard in regions near the limits of their color space along the color gamut. Built-in color enhancement firmware distorts these near-gamut regions of color space. In order to convert camera digit to colorimetric XYZ accurately, one needs detailed proprietary information of the signal processing in each camera. It is impractical to assume that we can transform the rendered camera response back into scene XYZ values and then convert them into accurate, uniform color space coordinates. We might get reasonable results near middle gray, but human vision and XYZ normalize to maxima. Cameras' RGB values are highly modified by camera firmware, particularly in chroma around the gamut limits.

The second problem is that we need a goal image; we need an array of perfectly rendered pixels. How does one find the ideal rendered image of the scene?

35.6.2   3-D Mondrians

The 3-D Mondrian work was based on a series of experiments from the CREATE Project (2010). (Chapter 29, Parraman, et al., 2010; McCann, et al., 2009) The set of experiments used a scene with a Low-Dynamic-Range portion next to a High-Dynamic-Range portion in the same room at the same time. Both LDR and HDR parts were made of wooden blocks painted with 11 different paints. The LDR blocks were placed inside an illumination cube so as to be as uniform as possible; the HDR blocks had two highly directional lights (see http://sites.google.com/site/3dmondrians/).

This LDR and HDR 3-D test target have been characterized many different ways. We measured the following properties of the display:

- Measurements of object's spectral reflectances;
- Light coming from the facets (XYZ);
- Multiple exposure photographs using a number of different cameras (digits);
- Magnitude estimates of appearance of block facets;
- Reflectances of Watercolor painting for measurements of appearances.

We used a number of these scene measurements to discuss possible evaluation techniques of spatial image processing (McCann et al., 2010).

35.6.3   Spatial Color Examples

Figure 35.7 shows examples of images from the 3-D Mondrian experiments: normal digital images, spatial processed images, and Carinna Parraman's watercolor paintings. It shows different renditions

Figure 35.7 (top row) LDR; (bottom row) HDR images. The columns show (left) normal digital photographs; (center) Vonikakis spatial image processing; (right) watercolor rendition of appearance. Reproduced by permission of Vassilios Vonikakis.

of LDR and HDR CREATE scenes (rows). The left column shows control photographs taken with a Panasonic DMC FZ5 digital camera. The middle column shows the LDR and HDR outputs of a spatial algorithm (VV). The right column shows the Carinna Parraman watercolor painting of the scenes (rendition of scene appearance) (Vonikakis et al., (2008)).

35.6.4 Evaluations of LDR and HDR 3-D Mondrians

McCann et al. (2010) and its accompanying website provide the details of six different evaluations of the experimental data.

- **Scene vs. Camera sRGB values:** A portion of the scene is a flat piece of paper in bright uniform illumination (LDR color circle. We measured the XYZ values of the light coming from the papers. We averaged the digital values in a camera image which in principle is in sRGB space. We converted Scene XYZ to relative sRGB and scaled to the range of 0–255. The sRGB values for the gray scale for scene and camera record are in close agreement. The six color segments show considerable camera distortion of scene information (see Figure 35.8).
- **Effect of illumination on appearance:** The radiances from the same reflectance paints change in different illumination. This data measures the physical change in illumination.
- **Unwanted range compression of LDR scene:** There are many algorithms that compress dynamic range. This technique tests whether the algorithm compresses the LDR scene in an inappropriate manner.
- **Wanted Range Compression HDR scene:** By comparing the rendition of the HDR and using the LDR as a control, we can measure the algorithm's dynamic range compression.

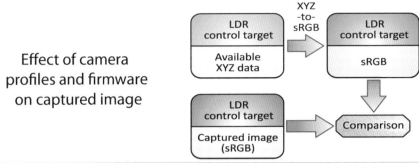

Effect of camera profiles and firmware on captured image

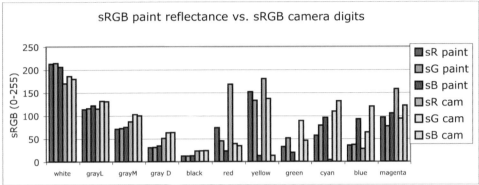

Figure 35.8 (top) Diagram of comparison; (bottom) Plot of sRGB values for the 11 paints.

- **Appearances: LDR vs HDR:** Using the Parraman watercolor, we compare changes in appearance caused by going from LDR to HDR illumination.
- **Appearance vs. algorithm rendering:** Using the Parraman HDR watercolor as ground truth, we can compare the dynamic range compression by a human with that of the algorithm.

The full report of the 3-D Mondrian experiments in the CREATE program are found on the web. (see http://sites.google.com/site/3dmondrians/).

Using these techniques McCann et al. (2010) analyzed two cameras with different algorithms. These examples show two significant results:

> Both sets of images were adversely affected by the cameras pre-LUT and profiles. The cameras performed as they were designed, but the digital values in the images were not accurate s-RGB records of the scene. Further work with RAW and film images are needed to get undistorted input records of the scene. If the camera's digital values cannot be converted to accurate scene radiances, then it limits and distorts subsequent image processing and evaluation.

> The algorithms tested did not compress the LDR image, while significantly compressing the HDR image. This is an important success in HDR spatial imaging. Many HDR algorithms, when applied to LDR scenes, make the LDR scene rendition worse.

35.7 Locality Test Targets

We have seen that cameras can mimic vision in Chapter 32.7.7 and 33.5.3. We have seen examples of Simultaneous Contrast, Adelson's B&W Mondrian Tower experiment, Sinha's double gradients and others. Such results are neither trivial coincidences, nor essential requirements for the design of a camera. If we step back a bit, we have discussed the importance of using vision data to guide us in finding spatial image processing algorithms that work for all types of scene content. For example, it is a bad idea to make cameras with the spectral sensitivity of the cone pigments with their considerable crosstalk (McCann, 2005). Cameras should not be a copy of vision (Fedorovskaya et al., 1997). Cameras should be inspired by vision to have some of its desirable properties. The more we understand vision, the more tools we have available for cameras.

There is a significant challenge in vision that, so far, has not been solved. There is no bottom-up model of both Simultaneous Contrast and Assimilation. Given the array of scene radiances, can we calculate the appearances of all the observation data in Section D? Contrast works well with large gray patches, while Assimilation is best with smaller test areas near the human resolution limit. The measured lightness data found in Section D provides the *ground truth* for analyzing an algorithm's ability to mimic locality influence. Future studies can use this data to evaluate models that attempt to predict both *Simultaneous Contrast* and *Assimilation* test targets using the same comprehensive algorithm and parameters. Anything we learn about *locality influence* is readily applied to spatial image processing in cameras.

(See Wiley web site http://www.wiley.com/go/mccannhdr for data.)

35.8   Summary

Figure 35.9 is a block diagram of options for evaluating spatial algorithms. The process begins with a scene that is captured by a camera and processed with an algorithm.

We have four independent branches for evaluating four distinct goals:

- The *faithful reproduction* option is easiest to evaluate. All one has to do is use a tele-radiometer to measure the X, Y, Z values of each area in the scene. Accurate instrumentation is available. The problem is that reproduction technology cannot reproduce all areas over the dynamic range of scenes. This problem leads to discussion of the best compromises (Hunt, 2004).
- The beauty contest approach is widely used to measure observer preferences. It works well for identifying the preferred rendition for one scene, but cannot be expanded to all scenes without considerable experimentation.
- The third block in that column is time of calculation. Many factors, such as computer components, special image processors, still vs. video, combine to determine the hardware limits. As well, specific

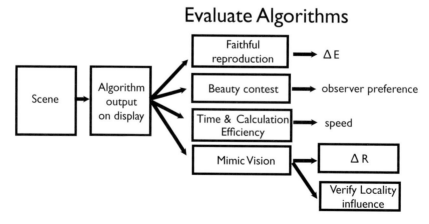

Figure 35.9   Four types of evaluation.

operations, used in the algorithms, introduce an additional set of limits. Time to compute the image is a critical design feature, particularly for video.

- The bottom block is the additional goal for Retinex processing. How well does the algorithm mimic vision? At first that may seem to be an unnecessary burden for people interested in computer graphics, rather than human vision. Is there a compelling need for this additional requirement? The other side of that coin is that mimicking vision can be your compass in finding the best algorithm. A single algorithm that does well with hundreds of scenes, is much better than hundreds of algorithms that do well with one scene.

35.9   Lessons from Quantitative Studies of HDR in Cameras

There are a number of things we would do differently if we were designing digital HDR photography today using the clean sheet of paper approach:

- We would make sensors with fewer, larger pixels. Larger pixels improve the dynamic range of the sensor. The only reason for so many megapixels in cameras is marketing. Most high-end cameras today use pixels much smaller than the optical resolution limits of their lenses (Kriss, 2011). Commercial displays, computer and video, have limited number of pixels with severe information losses from compression. There is no way to see a complete undistorted 12 megapixel digital image except by making a 30 by 40 inch print. Most of the data generated in high-pixel-count images is thrown away in compression, or transmission losses before reaching the observer.
- We would do away with multiple exposures. All the usable information above the veiling glare limit can be captured in single exposures, as done in color negative films. Multiple exposures create the need for a lot of messy work fusing the different images. They have to have both spatial correction introduced by camera motion and tone scale adjustments to combine the different tone scale quantizations.
- We would avoid the color space spatial distortion of scene information introduced by approximate sRGB compromises. We saw above in 35.6.4 that cameras introduce significant distortion of spatial information. The sRGB values of the scene are not recorded accurately in camera digits. These errors may seem unimportant when you study them using pixel-based logic. These designed errors are often described as compromises. From the spatial-comparison perspective they are major distortions of the information needed to process HDR scene successfully. Of particular importance is that these compromises misrepresent the critical scene data near white. For spatial processing, these compromises alter the information used by models of neural contrast. In order to process HDR as well as vision does, the input data must represent accurately the spatial information in scene content.
- We would use a single spatial algorithm designed to process low-contrast foggy scenes, normal LDR scenes, and HDR scenes using the same locality-influence parameters. The spatial algorithm needs to respond to image content.

35.10   References

http://sites.google.com/site/3dmondrians/

CREATE Project (2010) http://www. create.uwe.ac.uk

Evans R (1974) *The Perception of Color*, John Wiley & Sons, Ltd, Chichester, 205.

Fedorovskaya E, de Ridder H & Blommaert F (1997) Chroma Variations and Perceived Quality of Color Images of Natural Scenes, *Color Research & Application*, **22**, 96–110.

Funt B, Ciurea F & McCann J (2002) Tuning Retinex parameters, *Proc. SPIE*, **4662-43**, 358–66.

Hunt R (2004).*The Reproduction of Color*, 6th edn, John Wiley & Sons, Ltd, Chichester.

Kriss MA (2011) How many pixels does it take to make a good 4"x6" print? Pixel count wars revisited, *Proc SPIE*, **7876**, 14.

Kuang J, Yamaguchi H, Liu C, Johnson G & Fairchild M (2007) Evaluating HDR Rendering Algorithms, *ACM Transactions on Applied Perception (TAPI)*, **4**(2), 9.

Marcu G (1998) Gamut Mapping in Munsell Constant Hue Sections, *Proc. IS&T/SID Color Imaging Conference*, Scottsdale, Arizona, **6**, 159.

McCann JJ (1998) Color Theory and Color Imaging Systems: Past, Present and Future, *J Imaging Sci & Technol*, **42**, 70.

McCann JJ (2002a) *A Spatial Color-Gamut Calculation to Optimize Color Appearance, in Colour Image Science: Exploiting Digital Media*, ed. MacDonald LW &. Luo MR, John Wiley & Sons, Ltd, Chichester, 213–33.

McCann JJ (2002b) Color Appearance: A Spatial Computation, *Proc ICIS02*, Tokyo, 14–25.

McCann JJ (2005) The history of spectral sensitivity functions for humans and imagers: 1861 to 2004, *Proc. SPIE*, **5667**, 1–9,

McCann JJ, Land EH & Tatnall S (1970) A Technique for Comparing Human Visual Responses with a Mathematical Model for Lightness, *Am J Optom*, **47**, 845–55.

McCann JJ, McKee S & Taylor T (1976) Quantitative Studies in Retinex Theory: A Comparison Between Theoretical Predictions and Observer Responses to "Color Mondrian" Experiments, *Vision Res.*, **16**, 445–58.

McCann JJ, Parraman C & Rizzi A (2009) Reflectance, Illumination and Edges, *Proc. CIC*, Albuquerque, **17**, 2–7.

McCann JJ, Vonikakis V, Parraman C & Rizzi A (2010) Analysis of Spatial Image Rendering, *Proc. IS&T/SID Color Imaging Conference*, **18**, 223–28.

Morovic J & Luo R (1999) *Developing Algorithms for Universal Colour Gamut Mapping, in Colour Imaging Vision and Technology*, Ed. L. W. MacDonald and M. R. Luo, John Wiley & Sons, Chichester, 253–82.

Parraman C, McCann JJ & Rizzi A (2010) Artist's Colour Rendering of HDR Scenes in 3-D Mondrian Colour-constancy Experiments, *Proc. IS&T/SPIE Electronic Imaging*, San Jose, **7528**, 1.

Vonikakis V, Andreadis I & Gasteratos A (2008) Fast Centre Surround Contrast Modification, *IET Image Processing*, **2**(1), 19–34.

36

The HDR Story

36.1 Topics

This chapter brings together the themes of the text. Our HDR world is the result of nonuniform illumination. We benefit from rendering the HDR world in LDR media. There are many experiments measuring different aspects of dynamic range, and many demonstrations of improved scene rendition. All of this would be very confusing, if we did not have two observations that help resolve the many paradoxes. First, optical veiling glare, that depends on the content of the scene, limits the range of light on cameras' sensors, and on retinas. Second, neural contrast, the spatial image processing in human vision, counteracts glare with variable scene-dependent responses.

The optical and neural processes found in human vision shape the goals of HDR imaging. All demonstrations of HDR imaging increase the apparent contrast of details lost in the shadows and highlights of conventional images. They change the spatial relationships by altering the local contrast of edges and gradients. The goal of HDR imaging is displaying calculated appearance, rather than accurate light reproduction. By using this strategy we can develop universal algorithms that process all images, LDR and HDR, by mimicking human vision.

36.2 Straightforward Technology Stories

Texts on digital engineering are usually stories that start with a specific invention and proceed step by step in a straight-forward progression to the state-of-the-art today. The story of the solid state transistor began in 1954. Gordon Moore (1965) made the striking prediction that the number of transistors per area in integrated circuits would double every two years. His article showed four data points for years 1962 to 1965. For the next 46 years the transistor size data has fit Moore's Law. Many different advances have contributed to this progress. This is an example of a straight-forward technology story.

36.3 The HDR Story is Defined by Limits

What makes HDR imaging so interesting is that it is not just a technology story. For many colleagues the multiple exposure story started in 1997 with the new premise of capturing scene radiance. With known exposure times we could convert camera digits to scene radiance and recreate the original scene using HDR displays.

But, as it is true of any scientific theory and technological embodiment, there are limits. These limits have to be measured in order to verify the hypothesis and the reduction to practice.

- The dynamic range in scene acquisition is limited by glare. As Jones and Condit (1994) pointed out the dynamic range of cameras is determined by the sum of the unwanted glare and the wanted image of the scene. Glare sets the range limit of cameras. That limit varies with the content of the scene: beach scenes have a dramatically smaller range on a camera's image plane than starry night scenes. The effect of glare cannot be removed accurately by calculation (ISO, 1994). Our straight-forward technology story has been limited by camera glare.
- Displays also have limits. We could bypass acquisition limits using synthetic HDR images calculated by computer graphic techniques. Even with the dramatic advances in HDR displays, technology limits the accuracy of perfect scene reproduction. For example, if we made a display that emitted a perfect HDR image, we would have to view it in complete darkness to eliminate front surface reflected light from the room.
- Our vision system is also limited by glare. The light on the retina is the sum of the image of the scene plus unwanted scene-dependent glare. However, we have a second scene-dependent *neural contrast* mechanism that acts to counteract glare. The scenes that generate the most veiling glare have the lowest retinal contrast, and the greatest apparent contrast. Both veiling glare and neural contrast are image-dependent. Both are substantial effects. Their large magnitude is concealed because they effectively cancel each other. Veiling glare is seldom apparent. Neural contrast is so powerful that the effect we see, called Simultaneous Contrast, is the remainder after cancelation.
- Accurate reproduction of light is not necessary; reproduction of appearance is. Renaissance painters knew that a certain spatial distribution of tones would render the appearance of a dynamic range higher than the one reproducible on the canvas. Since the early days of photography the goal has been to render the appearance, and not the radiance from the scene. Equivalent appearance of scenes can be generated without reproducing the original's quanta catch on the retina.
- Color and lightness appearances are controlled by scene dependent spatial mechanisms. The limits imposed by the human visual system cannot be analyzed by pixel-based techniques. These limits are controlled by the two opposing spatial mechanisms: glare and neural contrast.
- The physical dichotomy of reflectance and illumination initiates the light on the retina that leads to vision. However, it is the dichotomy of edges and gradient, in that retinal image, that generates HDR appearances and color constancy.

Summarizing, the Goal of the digital HDR reproduction should be displaying calculated appearance, rather than the accurate physical light distribution.

36.4   HDR Works Well

Regardless of these limits, HDR images made with digital image processing have shown considerable improvements in the rendition of highlight and shadow areas. Digital HDR works, and works very well. How?

We see how HDR works every time we modify a photograph with Photoshop®. We alter the output of the digits in computer memory to change the image on the display. If we cannot see a detail in the shadow well, we modify the local contrast to enhance the edges. We avoid artifacts by adjusting the gradients. We apply the dodging and burning tools described so clearly by Ansel Adams. We modify the spatial content of the image until it looks right.

In fact, that digital process is what painters did with mixing paints in the Renaissance. They selected the pigment mixture that looked right on the painting. When we turn to more complex spatial algorithms we often find the same process. We may be using very complex electronic imaging techniques, but we evaluate the results with the same test. Does the image look right? We may have a theory that the algorithm can find reflectance, but we seldom prove its ability to calculate the actual physical reflectance. Too often, we just call the result reflectance because it looks right.

As we now see, we have traveled in a circle that began historically in the Renaissance with Chiaroscuro painting, and continued with multiple exposure photography in the 1850s. Mees and Jones added subjective criteria in 1920 that moved the goals away from quanta catch reproductions. Early analog and digital algorithms in the 1960s introduced the idea of calculating sensations using spatial comparisons, and writing calculated appearance on film. These models were influenced by the measurement of human vision's response to scene content. This strategy integrates art, science, and technology to find the general HDR solution for all scenes.

Digital imaging technology has provided remarkable improvements in image processing power. We described a small fraction of algorithms that render scenes in this text. We believe that this power needs to be used to search for the general solution of HDR imaging. Namely, we need to work on algorithms that process all scenes succesfully. We need to work on the objective measures of the success of our algorithms. The key to objective evaluation is better techniques to compare the algorithm's output to human visual response. Only with better and more efficient means of testing the algorithm's *locality* influence, will we be able to advance our ability to render all LDR and HDR scenes with a constant algorithm. We need to replace the Renaissance painter's "looks right" criteria with the scientific analysis of spatial processing.

36.5   References

ISO Standard (1994) Optics and Optical Instruments, Veiling Glare of Image Forming Systems: Definitions and Methods of Measurement, ISO 9358: 1994.

Jones LA & Condit H (1941) The Brightness Scale of Exterior Scenes and the Computation of Correct Photographic Exposure, *J Opt Soc Am*, **31**, 651–78.

Moore GE (1965) Cramming more components onto integrated circuits, *Electronics*, **38**(8), 114–17.

Glossary

Black & White Mondrian experiment: Edwin Land's neural contrast experiment using achromatic papers and HDR illumination. The light was placed near a black paper, and far from a white; the position of the light was adjusted until the white paper luminance equalled the black paper luminance. Observers reported white and black appearances from identical stimuli. Tone Scale mapping cannot render equal inputs both lighter and darker, as required by this scene.

brightness: magnitude estimates of the appearance of light, such as Hipparchus's estimates of stellar magnitude; an attribute of a visual sensation in which a stimulus appears to emit more, or less, light (CIE).

brightness (video): adjustment that controls the black level of video signal.

camera response functions: measured camera-digit response to glare-free scene radiance measurements.

chiaroscuro: the painting of scenes having HDR non-uniform illumination; rendition of light and shadows by the use of gradations and edges of color; the term in Italian means "light-dark".

chroma: a dimension of Munsell Color Space, perpendicular to lightness and hue, that goes from achromatic to the most colorful samples.

CIE colorimetry standards: models of color based on cone sensitivity functions measured using spots of light. They are psychophysical color systems that can predict if spectral spots of light will match. They cannot predict appearances of complex HDR scenes in non-uniform illumination, such as 3-D Mondrians.

color matching functions (CMF): psychophysical measurements of the start of human color response in the cone receptors.

color space: a three-dimensional space with amounts of long-wave, middle-wave, and short-wave light as axes.

> **color space (1 – Munsell):** isotropic appearance space; distances in hue, chroma and lightness are equally spaced appearances; average distances in this space have meaning.

The Art and Science of HDR Imaging, First Edition. John J. McCann, Alessandro Rizzi.
© 2012 John Wiley & Sons, Ltd. Published 2012 by John Wiley & Sons, Ltd.

color space (2 – LMS): axes are the quanta catch of the long-, middle-, and short-wave cone pigments with intraocular absorptions, distance in this space is non-uniform in appearance.

color space (3 – XYZ): axes are the normalized quanta catch of the color matching functions derived from color matches using spots of light on a black background (aperture mode); distance is non-uniform in appearance.

color space (4 – L*a*b*): axes are transforms (cube roots and differences of cube-roots of X, Y, Z); the form of the axes are similar to those in Munsell space; distance is uniform in lightness; distance is non-uniform in chroma and hue; distance is more uniform than X, Y, Z.

color space (5 – MLab): axes are linear transforms of position in Munsell space, so as to have quantitative values in the L*a*b* format; distances in hue, chroma and lightness are equally spaced in appearance.

color space (6 – RGB): axes are the quanta catch of narrow-band camera sensors designed to reduce crosstalk; non-linear transforms of X, Y, Z; digit color space used in cameras, computers, displays and printers; distance is non-uniform in appearance.

color space (7 – sRGB): axes are a digital device standard for color; non-linear transforms of X, Y, Z; distance is nearly uniform in lightness; distance is non-uniform in chroma and hue.

computer graphics (CG): visual images synthesized by computer programs rather than by scene capture.

contrast: a description of image content with many different definitions.

contrast (assimilation): from psychology, the name of a class of neural contrast experiments, namely when a gray test area looks lighter with adjacent white areas.

contrast (Michelson): from physics, ratio $(max - min)/(max + min)$; description of the range of sine-wave gratings, edges, or images.

contrast (neural): from physiology, human post-retinal image processing; the name of the mechanism, or the set of mechanisms, that transforms retinal contrast into appearances.

contrast (photographic): slope of film's response function to light (optical density vs. log luminance); higher slope means a higher contrast image.

contrast (retinal): from psychophysics, the calculated image on the retina after intraocular scatter made by convolving all scene radiances with the human eye's glare spread function (GSF).

contrast (simultaneous): from psychology, the name of a class of neural contrast experiments, namely when a gray test area looks darker with adjacent white areas; (see also assimilation).

contrast (video): from circuit designs, the gain that adjusts the luminance of the white.

crosstalk: loss of spectral information from scene radiance separation caused by overlapping spectral sensitivities. Color cameras and films minimize spectral overlap compared to color matching functions, and cone sensitivity curves. Vision compensates for crosstalk by opponent color processing. Crosstalk causes the departures from perfect color constancy.

Designator: Edwin Land's term for lightness predictions from center-surround spatial operators.

discounting the illumination: a 19th century hypothesis for explaining color constancy.

double density: most reliable way to increase the dynamic range of test targets by superimposing film images. Two films double the optical densities, and square the % transmissions.

Duplicity theory: 19th century hypothesis that rods generate achromatic vision at low light levels, and cones generate color vision at high light levels.

dynamic range: the range of useful light described as a ratio of maximum/minimum responses to light. It has to be measured with a light meter; it cannot be calculated from the number of quantization levels used in a digitizer.

high dynamic range (HDR): non-uniform illumination expands the range of light beyond that controlled by surface reflections; range of light from a scene with non-uniform illumination: e.g., a complex room illuminated by a single candle, sun and shade, indoor and outdoor scene, multiple light sources with different spectra.

low dynamic range (LDR): the range of light controlled by surface reflection: e.g. paints, pigments and dyes, in uniform illumination; limited range of light caused by uniform illumination, or restricted range of reflectances, or both.

reflectance dynamic range: the range of light reflected by objects in a scene in perfectly uniform illumination; range of reflection prints.

headroom: describes additional bits per pixel to allow for future digital alterations without artifact problems in H&D scene captures.

glare: unwanted stray light that adds to the wanted image of scene radiance at the pixel of interest; the sum of unwanted light from all other pixels in the scene, including light from outside the image's field of view.

glare spread function (GSF): the point spread function for the human eye; psychophysics, the response of the eye to a point source of light as spread by the eye's optics, intraocular scatter, and retinal reflections; veiling glare on the retina as a function of angular separation from a point source of light.

GrayWorld: wide variety of image processing assumptions that relate average scene data to achromatic outputs; holds true for only a few special-case scenes.

ground truth: scientific vernacular referring to objective reality of a statement. Measured scene radiance is the ground truth for algorithms that calculate radiance from camera's scene capture data.

H&D curve (Hurter and Driffield's film response function): plots of film's density vs. log exposure.

Hipparchus Lines: plots of appearance caused by overall changes in illumination; using constant test targets the measurements of appearance changes from adjustments of spatially uniform illumination; psychophysical brightness.

hue: a dimension of Munsell Color Space, perpendicular to lightness and chroma, that forms a color circle.

illumination: light; electromagnetic radiation falling on objects in the scene; light from illuminants, and reflections from other objects.

irradiance: a measure of the energy from the number of photons continuously falling on an area; measured [energy/area]; (see radiance).

lightness: a dimension of Munsell Color Space that goes from white to black; matches made to an equally spaced achromatic standard, such as Munsell Values; an attribute of a visual sensation in which a stimulus appears to emit more, or less, light in proportion to that emitted by a similarly illuminated area perceived as a "white" stimulus (CIE); the triplet of L, M, S lightnesses predicts color appearances (Land).

locality influence: describes the property of human neural contrast that is neither local, nor global; spatial comparisons of image segments depends on the segments' relative luminances, sizes, separation, and enclosure; algorithm's design locality influence to mimic vision; (see surround (locality) influence).

lookup tables (LUTs): memory tables used to convert digital input values to output values (e.g. 256 memory locations for 8-bit data); efficient, inexpensive digital transform of data; controls the internal values between maximum and minimum.

> **Three 1D-LUTS:** efficient, inexpensive digital transform of color data; using three 8-bit LUTS (3*256 memory locations for 24-bit data); controls the internal values within the color space.

> **One 3D-LUT:** very powerful digital transform of color data; (e.g. 256^3 memory locations for 24-bit data); controls the internal values within the color space with independent control of each segment of the color space; used with less memory and interpolation in electronic imaging devices to control color profiles.

luminance: photometric unit of light coming from the scene to the eye, or meter.

magnitude estimation: psychophysical technique that asks observers to estimate appearances by assigning them numerical values. First used by Hipparchus to estimate stellar brightnesses in the 2nd century BC.

metamers: stimuli that are spectrally different, but visually identical.

modes of appearance: from 19th century psychology, an explanation for variable appearances from identical retinal quanta catches; various manners in which colors can be perceived depending on the spatial light distribution that mediates perception.

> **mode of appearance: (aperture):** non-located mode, by which color is perceived when divorced as completely as possible from all spatial attributes, as through an aperture.

> **mode of appearance: (object):** located mode, by which tangible objects are perceived.

> **perceptual frameworks:** assumed mechanisms using cognitive processes to influence an object's appearance; assumptions of human top-down image processing for lightness and color.

Mondrians: Edwin Land's complex test targets inspired by a Piet Mondrian painting in the Tate Gallery, London. The targets used about 100 uniform reflectance papers in controlled illumination to create HDR scenes and color constancy experiments. A requirement of all Mondrians is that all areas have different sizes, shapes and spatial patterns, so as to minimize afterimages.

Multiple Exposure to Scene Luminance (ME2SL): Debevec and Malik introduced to the Computer Graphics community a calculation using multiple-exposure data to measure high-dynamic-range scene luminances. The long history of multiple exposures in traditional photography never made this claim because of glare. For photographs of stars at night in a black sky, cameras can capture more than 4 log units of accurate information from this low-glare camera image. For beach scenes, glare limits accurate luminance measurements in cameras to less than 1.5 log units.

neural contrast: vision's post-receptor spatial image processing (Section C); scene-dependent property of spatial comparisons, or neural image processing. It varies the apparent contrast of edges depending upon scene content; (see also glare). The counteracting spatial mechanisms, *glare* and *neural contrast* play important roles in human vision.

optical density (OD): = log [1/transmittance], or log [1/reflectance].

perception: the mode of mental functioning that includes the combination of different sensations and the utilization of past experience in recognizing the objects and facts from which the present stimulation arises (see also: sensation). The distinction between sensation and perception centers on the roles of cognition and recognition in the perception of objects in complex images.

photometry: a field of psychophysics which measures light using standard human wavelength sensitivity functions.

pixel (contraction of picture element): digital imaging devices segment a scene's image on the sensor into millions of pixels.

pixel-based imaging: (film photography model) all pixels have the same response to light over the expanse of the sheet of film.

photopic luminosity standard V_λ: the nominal standard sensitivity of the eye to different wavelengths at high light levels.

quanta catch: the physical sensor response that integrates the light's spectral radiance values and sensor's spectral sensitivity.

radiometry: a field of physics which measures electromagnetic radiation.

radiance: the measurement of radiation coming continuously from a particular object to a particular point in space; [energy/ (angle*area)].

  camera's image-plane radiance: light falling on the camera's sensor after optical and intra-camera reflections; scene radiance plus veiling glare.

  retinal radiance: light falling on the retina after intraocular scatter.

  scene radiance: meter measurement of light from the scene using special precautions to shade the meter's optics.

ratio-product algorithms: the family of algorithms that calculate lightness appearances from input images, e.g., scene radiances, camera scene captures, or retinal images. The components of such a model are:

  1 – ratio: means of measuring spatial comparisons; initial spatial comparison of two image segments used to measure their ratio of radiances; image segment size varies from a single pixel to the average of one-half of the input image; for computational efficiency, the difference of log radiances is often used.

  2 – threshold: means of removing invisible gradients of light from images; sub- threshold ratios are set to unity.

  3 – product: means of propagating spatial comparisons across an image; the product of a path of ratios equals the ratio of the first to last image segments.

  4 – reset: used with setting initial values, e.g. output array set to maximum output value; if spatial comparison process exceeds maximum, then that value is reset to maximum value.

  5 – average: combining the output of the previous iteration with the current one.

  6 – ratio modification operator (RMO): a threshold to limit the maximum size of ratios.

RAW image: a partially processed digital camera image that includes some, but not all of the camera manufacturer's firmware processing.

Red & White Projections: Edwin Land's color experiments showing dramatic examples of color neural contrast.

reflectances (physics): The ratio of the flux reflected from a surface to that incident on it.

rendering intent: goal for image modification.

replica: a copy of a painting, by the original's artist, using the same media, or color materials. Replicas have the same physical properties for controlling light, and share the identical color space of the original.

reproduction: copies of original art in different media, such as a computer-screen copy of a painting. Reproductions have different physical properties for controlling light, and have a different color space than the original.

Retinex (contraction of retina and cortex): Edwin Land's coined word describing a theoretical biological structure that makes color separation lightnesses using spatial comparisons; L, M, S lightnesses determine color appearance.

Retinex (computational model of lightness): Algorithms that calculate apparent lightness of all scene areas using the entire scene for input, i.e., spatial comparisons.

Retinex (Extended Family): A wide range of algorithms that augment, modify, and extend the original Retinex Algorithm.

  Land and McCann (1971): original algorithm using ratio, threshold, product, reset and average components in discrete paths. Resets at each ratio iteration.

  Frankle and McCann (1983): introduced multi-resolution techniques for computational efficiency using ratio, threshold, product, reset and average components; described fast hardware and pyramid techniques; replaced Purcell's paths with global iterations.

  Sobol (2002): first Retinex algorithm used in commercial amateur cameras; used ratio limits (RMO).

  Rising (2002): McCann [99] in wavelets.

  Designator (Land, 1986): algorithm to calculate lightness using spatial comparisons without reset.

  NASA Retinex (1996): implemented Land's Designator, and later a multi-scale version.

  Gamut Retinex (1999): maps large-gamut original into small-gamut reproduction using spatial comparisons, rather than pixel values.

  MI-Retinex: alternative mechanism; for each path or spray it calculates the final pixel value equal to the ratio of that pixel's input value to the maximun in the path or spray.

  Brownian Paths (1993): first MI-Retinex algorithm searches each path for local maxima; the mid-point-displacement technique approximates Brownian paths.

  Random Spray Retinex (RSR) (2007): uses random set of points cast around the pixel of interest to identify the highest value used as local maximum to compute output.

  Variational Implementations: formulation with partial differential equations of MI-Retinex.

  Retinex-inspired algorithms: a set of algorithms based on different approaches.

  ACE: (Automatic Color Equalization) (2003): algorithm uses distance-weighted corrections and distorted difference among pixels values.

  RACE: (Random Spray Retinex + Automatic Color Equalization) (2008): combines ACE and RSR computations.

  Retinex extensions: wide variety of algorithms applied to the HDR Black & White Mondrian, and Color Mondrians constancy experiments.

reset: powerful spatial image-processing component for introducing locality influence.

Land and McCann: resets ratio-product value to the maximum at each ratio iteration.

MI-Retinex: resets each path, or spray to maximum.

scaled integrated reflectance: measurements made by spatial comparisons of the light from a sample to the light from the maximum using cone pigment sensitivity functions. These ratios are scaled by a cube-root function to correct for scatter, and give equal lightness spacing.

sensation: the mode of mental functioning that is directly associated with stimulation of the organism (see also: perception).

silver halide: light sensing crystals used in photographic films.

spatial comparisons: alternative paradigm to colorimetry; it builds images from many spatial comparisons, rather than individual pixel values; spatial synthesis.

spatial comparisons (abrupt luminance changes): generate large changes in appearance, in color and lightness.

spatial comparisons (gradual luminance changes): generate small changes in appearance, in color and lightness; usually subtle, but detectable gradients in appearance; used by chiaroscuro painters and photographers to synthesize HDR appearances using LDR media.

surround (effect of): neural contrast modifies the appearance of the pixel of interest in response to the rest of the scene content.

surround (global algorithms): uses interactions between all pixels in the field of view.

surround (local algorithms): uses interactions between adjacent, or nearly adjacent pixels.

surround (locality influence): describes the property of human neural contrast that is neither local, nor global; spatial comparisons of image segments depends on the segments' luminance values, sizes, separation, and enclosure; algorithms design locality influence to mimic vision.

Tatami: name used to describe simplified array of colored papers; simpler complex image than a Mondrian; Ed Purcell used this name because they resembled Japanese floor mats.

tone scale curve: Mees used "tones" to explain in simple terms the progression from white to black and how the shape of the photographic tone-scale response curve affects the photographic print; popular language substitute for Hurter and Driffield's film response function.

tone scale curve (digital): transform to control the image values between maximum and minimum; frequently a LUT.

uniform color space (UCS): colors samples as they appear at the end of the human color processing mechanism in an isotropic color space.

veiling glare: unwanted stray light in the image on an image sensor caused by optics, camera body (or intraocular media), and the surface of the sensor.

Zone System: Ansel Adams' process to measure scene luminances and adjust film response function, so as to capture in a single exposure all the usable scene information.

Author Index

The Art and Science of HDR Imaging, First Edition. John J. McCann, Alessandro Rizzi.
© 2012 John Wiley & Sons, Ltd. Published 2012 by John Wiley & Sons, Ltd.

Subject Index

The Art and Science of HDR Imaging, First Edition. John J. McCann, Alessandro Rizzi.
© 2012 John Wiley & Sons, Ltd. Published 2012 by John Wiley & Sons, Ltd.